Applied Mathematical Sciences | Volume 18

U. Grenander

Pattern Synthesis

Lectures in Pattern Theory Volume 1

Springer-Verlag New York · Heidelberg · Berlin
1976

U. Grenander
Division of Applied Mathematics
Brown University
Providence, Rhode Island

AMS Classifications: 05B30, 05B40, 05B99, 28A75, 60G35, 60G99, 68A45

Library of Congress Cataloging in Publication Data

Grenander, Ulf.
Pattern synthesis.

(His Lectures in pattern theory; v. 1)
(Applied mathematical sciences; v. 18)
Bibliography: p.
Includes index.
1. Pattern perception. 2. Pattern recognition
systems. I. Title. II. Series: Applied
mathematical sciences; v. 18)
Q327.G7 vol. 18 510'.8s [001.53'4 76-209

Printed in the United States of America

ISBN 0-387-90174-4 Springer-Verlag New York · Heidelberg · Berlin
ISBN 3-540-90174-4 Springer-Verlag Berlin · Heidelberg · New York

PREFACE

This book consists of updated and considerably expanded lecture notes. I am grateful to F. John, J. P. LaSalle, L. Sirovich and G. Whitham for accepting this preliminary manuscript for the series Applied Mathematical Sciences published by Springer-Verlag.

Many colleagues and students have helped with their comments and suggestions. Among them are D. E. McClure, I. Frolow, J. Silverstein, R. A. Vitale, and I would like to thank especially W. Freiberger for all his encouragement and help in many different ways.

The work has been supported by the Division of Mathematical and Computer Sciences of the National Science Foundation, and I appreciate the positive interest shown by John Pasta, Kent Curtis, Bruce Barnes and other program directors, over the years.

I am indebted to E. Fonseca for her untiring and careful preparation of the manuscript, to E. Addison for her skillful help with the many diagrams, and to K. H. Avery, D. P. Libutti, K. MacDougall and S. V. Spinacci for the final typing. I gratefully acknowledge permission to reproduce figures, as mentioned in the text, to: American Elsevier Publishing Company, Inc., A. J. Larkin, and M. Gardner.

I dedicate this work to my wife, Emma-Stina.

June, 1975 Ulf Grenander, Providence

TABLE OF CONTENTS

INTRODUCTION

In this volume we shall present a mathematical study of regular structures.

In 1966 this author suggested, at a scientific meeting in Loutraki, Greece, that it would be possible to create a general theory of patterns starting from some basic ideas that were sketched and illustrated by some simple examples. The suggestion was presented in the form of a research program which had barely been started at the time.

Now, nine years later, a substantial part of the program has been carried out by the author and his colleagues in the Division of Applied Mathematics, Brown University. It will be reported in a series of books of which this is the first volume.

The second volume is planned to appear soon after the first one. A good deal of the work on which it is based has been completed but only a part of it has been written up. In the not too distant future a third, more comprehensive volume, will give a systematic account of the theory. A preliminary list of content of it is given in the Appendix. The list contains twenty chapters, organized into four parts. The present volume corresponds to Part II, Chapters 3-6.

To avoid misunderstandings it should be pointed out from the beginning that this is not primarily a text on pattern recognition. Our main concern is not how patterns are recognized in nature or how they should be recognized by machines and algorithms. Instead, the object of our study will be the patterns themselves in terms of a precise formalism which will serve as the conceptual basis for synthesizing and analyzing patterns, helping us to understand how they are formed and processed. In the next volume this tool will be exploited systematically for the description, approximation, restoration, and eventually also recognition of patterns.

As outlined in Grenander (1967) we shall start from the following four principles. The theory will be _atomistic_: patterns are formed from simple building blocks - the signs or generators. They are our indivisibles, or atoms, and will be chosen in each particular application as simply as possible. These primitives may be abstract symbols, sets, relations, or functions, and the reader will meet a large variety of generators in the following chapters.

Given the generators, we shall insist on definite rules limiting the way they may be combined together. These laws will lead to the typical regularities of the patterns and represent their _combinatory_ structure.

The resulting regular configurations are abstract constructs, not necessarily observable in all their details. It depends on the _observational_ set-up to what extent the regular configurations can be identified by the observer. The resulting observable corresponding to a set of regular configurations is called the image and the main object of our study is the set of images together with the relations that hold between them: the image algebra.

The image represents what can be observed under ideal conditions, neglecting errors of measurements, limitations of the instruments or apparatus involved, and imperfections in the model. A theory of patterns that would not take into account how actual patterns behave would be severly limited in its applications. We must, therefore, assure that the theory is <u>realistic</u> so that it can deal with real patterns. In other words, we must consider how the pure patterns are transformed into actual ones by some deformation mechanism.

In this way we shall study patterns in general - the pattern of patterns - but the discussion will be made as concrete as possible based on particular cases. To illustrate them we will use configuration diagrams that occur throughout the book. There is nothing intrinsically geometric about the approach; the patterns can just as well be logical, or consist of strings, or describe behavior. In Bohm (1969), the reader can find a thought provoking discussion of the need for a pattern formalism, but from a different perspective.

Part I of the more definitive volume that will appear later (see Appendix) will describe in more detail how the study of patterns in science and technology, in the arts and humanities, has led us to insist on these four guiding principles characterizing our approach to patterns as atomistic, combinatory, observational, and realistic. Some information on this can be found in Grenander (1969), (1970).

In the present volume (Part II of Appendix) we shall study how patterns are formed from the generators by combination, identification, deformation: pattern synthesis. Much of this is conceptual analysis and examines the logical structure of the pattern formalism that we employ.

The next volume will be devoted to the inverse problem: pattern analysis. There we start from the other end, with the observed actual patterns, and try to describe and analyze them. Here, the discussion changes character and becomes both more mathematical and more computational, rather than conceptual, and with more emphasis on empirical results.

The reader will see that throughout this work we have tried to utilize computational experiments in parallel with mathematical analysis. Often we start by a numerical study to gain some intuitive insight. When this leads to some conjecture we go ahead and try to prove it analytically.

As our work progressed, this way of employing the computer for experimental mathematics became successively more powerful. It would have been hard, if not to say impossible, to do without it. Much of the computing was done interactively, in the programming language APL, and with the aid of various computer graphics devices. The role of efficient interactive computer facilities cannot be overemphasized in this context.

In the proofs we have occasionally omitted measure theoretic and other qualifications. This has been done where progress results in no essential loss of rigor. However, there are some cases in the text where the omissions are serious. When this is so, it has been mentioned explicitly, for example in Sections 2.10 and 3.6. We hope that future work will fill these gaps in our discussion.

As is obvious from the title of this book, we will be studying the <u>theoretical</u> aspects of patterns, and applications play a subordinate role in this first volume. It will be seen, however, that the main motivation behind this theoretical investigation comes from subject matter areas: geomorphology, anatomy, neurophysiology, spectrography, anthropology, and many others. This influence has dominated the development of the theory.

The first two volumes should be considered as only a preliminary presentation. They contain material presented by the author in lecture form at Brown University. This has been done twice in a one-semester course, and once as a two-semester course. Some of the work done by our group at Brown University has been written up in the form of thirty-five research reports (see Bibliography) but only some small fragments have appeared in print. The theory is still only at its beginning, but has been growing at a rapid rate during the last few years; this rapid growth will no doubt continue as the theory is applied to new subject matter areas.

The presentation is not quite as polished and careful as might be wished, and therefore it is likely that this volume contains numerous obscurities, mistakes, and inconsistencies. After all, pattern theory is a fairly new branch of applied mathematics and it would be unreasonable to expect anything else at this stage. The author would appreciate comments, advice, and corrections.

CHAPTER 1

THE GENERATORS

1. Generators and Their Properties.

The primitive objects used to build configurations and images are called the _generators_. The set of generators will be denoted by G and the generic symbol for one of them will be g, $g \in G$.

The generators are the units carrying information and since they signify certain primitive statements they will sometimes be spoken of as the _signs_.

The whole set G is made up of disjoint _generator classes_ G^α, $G^\alpha \subset G$, where α is a general index, the _generator index_,

$$G = \bigcup_\alpha G^\alpha, \; G^\alpha \text{ disjoint.} \tag{1.1}$$

The interpretation of this partition is that generators that are qualitatively alike will be put into the same class (see the discussion in Section 3.2).

Generators are the simplest entities, the building

blocks. They may have properties, and if so, the properties are of two types.

The first one consists of <u>attributes</u>. To a generator is associated an attribute $a = a(g)$, where the values of $\underline{a}$ may be integers, real numbers, vectors, and so on. One part of the attribute is the generator index α, but there will usually be others carrying more specific information.

The second one consists of <u>bonds</u>. To a given generator g corresponds the <u>arity</u> $\omega(g)$ which is a non-negative integer or infinity. The arity tells us what is the maximum number of connections between the generator and other generators. How these connections should be made is the subject of Chapter 4.

The arity is the sum of the <u>in-arity</u> $\omega_{in}(g)$, and the <u>out-arity</u> $\omega_{out}(g)$,

$$\omega(g) = \omega_{in}(g) + \omega_{out}(g). \tag{1.2}$$

They express the maximum number of incoming and outcoming connections to and from the generator respectively.

To each such (potential) connection corresponds a <u>bond value</u>, usually denoted by β with an appropriate subscript. As will be seen in the next sections the nature of the β's varies a good deal depending upon what sort of application we have in mind.

The set of bonds of any g, suitably enumerated, forms the <u>bond structure</u> of the generator. The bond structure does not specify the values associated with the bonds.

In addition to the properties of a generator we also need an <u>identifier</u> or a name to keep the different generators that we use apart from each other.

A word of caution concerning equality of generators.

We may want a particular generator to appear more than once in a configuration. We then take identical copies of it and keep them separate by using identifying marks as parts of the attribute. It should be clear from context when this is intended.

We shall use a <u>graphical formalism</u> as in Figure 1.1 to give an intuitive idea of the properties of the generator. In the figure the in-arity is 2 with the bond values β_1 and β_2, the out-arity is 3 with the bond values β_3, β_4, and β_5, so that the total arity is 5.

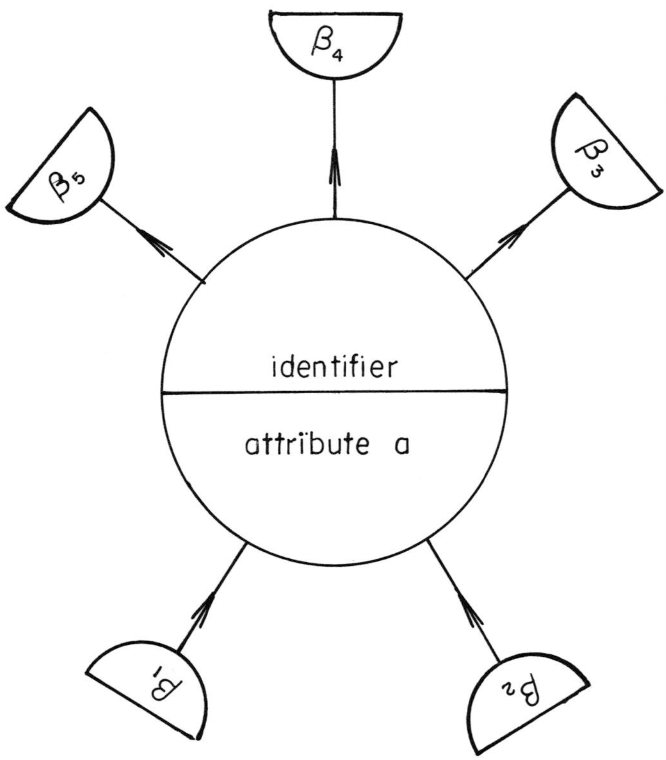

Figure 1.1

This graphical representation should not be understood as showing a generator surrounded by its bonds: the bonds form part of the generator itself.

In most applications we shall have certain mappings from G to itself which do not seriously affect the information carried by the generator. They are the similarity transformations and should satisfy

Definition 1.1. The set S of mappings $s:G \to G$ is said to form the similarity transformations if

(i) S is a semi-group with identity under composition of transformations,

(ii) s carries any G^{α} into itself for any $s \in S$ and any generator index α,

(iii) S does not affect the bonds (but possibly the bond values).

Note that the bond values of sg may differ from those of g.

The generators are our atoms, the indivisibles, but just as physical atoms have internal structure and can be split into elementary particles, in the same way the generators may sometimes be further divided. Indeed, we shall see later (Chapter 3) that it will occasionally be natural to have the images in one pattern formalism make up the generators in a higher-level pattern formalism. Whether we deal with such a hierarchical organization or not, at any level, the generator will be treated as a primitive.

To define a generator we can operate in two different modes. The simplest one, and least interesting, is when the generator is given abstractly without paying any attention to any medium in which it (possibly) is embedded. It would then just be denoted by an abstract symbol. The opposite case, which we shall encounter frequently, occurs when the generator is defined relative to a medium, an information carrier. The generator then has a concrete interpretation. In one and the same pattern formalism nothing prevents us from having abstract and concrete generators present at the same time, see for example Case 3.4.3, although this will rarely be the case.

Occasionally we shall have a neutral generator denoted by ε and called the null generator with special properties. In that case ε will be allowed to belong to all the generator classes so that these are no longer strictly disjoint.

We are going to encounter a multitude of generators of the most different appearances. To give the reader an idea of some possibilities we give a few examples in the next two sections, without however specifying the bonds since this is more naturally done in Chapter 2 dealing with connections of generators resulting in configurations. Hence the descriptions of generators is not complete in Section 1.2 - 1.3.

Given two sets of generators, G and G', associated with the same similarity transformation semi-group S, let us consider a mapping $h: G \to G'$. We shall say that h is bond invariant if for any $g \in G'$ we have the same bonds and bond values for g and $g' = h(g)$ and $sg \to sg'$ for all s and g.

With such a mapping it will be possible to replace generators from G by those from G' without harming the combinatory structure.

2. Abstract Generators.

The examples in this section give generators abstractly,

without reference to any medium or background in which they may have an interpretation.

Case 2.1 (properties). The generators consist of a list of symbols, say capital letters

$$G = \{A,B,C,D,...\} \tag{2.1}$$

where $g = A$ means the statement of a certain property A being applicable to the situation at hand, and so on, say A = "red", B = "blue", C = "large", D = "not blue", etc. The similarity transformations are built from identity and negation applied to any generator. The generator classes would group color properties into one class, size properties into another, and so on.

Another set of generators would be the set $G = \{♂, ♀\}$ which would appear in the patterns of ancestry, say in population genetics. In the latter case we would have attributes expressing the genetic properties carried by the generators.

Case 2.2 (Morse code). Here $G = G^1 \cup G^2$ with

$$\left.\begin{aligned} G^1 &= \{\text{dot, dash}\} \\ G^2 &= \{\text{short space, letter space, word space}\}. \end{aligned}\right\} \tag{2.2}$$

The attribute $\underline{a}$ is given by

$$\left.\begin{aligned} a(\text{dot}) &= 1 \\ a(\text{dash}) &= 3 \\ a(\text{short space}) &= 1 \\ a(\text{letter space}) &= 3 \\ a(\text{word space}) &= 5. \end{aligned}\right\} \tag{2.3}$$

This is one of several Morse conventions. The meaning of this attribute is of course the duration of the generator (in some unit of time) but at present we look at the generators as just symbols supported by no medium. There are two generator classes corresponding to a signal and space. Similarity transformations will be introduced first when we treat the generators concretely in this case.

<u>Case 2.3</u> (elements in chemistry). Here the set of generators could be

$$G = \{H, He, Li, Be, B, \ldots\} \tag{2.4}$$

and one of the attribute functions, the atomic weight, would take the values

$$\left.\begin{aligned} a(H) &= 1.00 \\ a(He) &= 4.00 \\ a(Li) &= 6.94 \\ a(Be) &= 9.01 \\ a(B) &= 10.8 \end{aligned}\right\} \tag{2.5}$$
$$\ldots$$

<u>Case 2.4</u> (formal language). A simple case is when

$$G = G^1 \cup G^2 \tag{2.6}$$

where G^1 is a finite list of symbols (representing Boolean variables) and

$$G^2 = \{\wedge, \vee, \sim,), (\} \quad . \tag{2.7}$$

This example will illuminate the importance of incorporating bonds in the generators. If we just define the generators as in (2.6), (2.7) it would be impossible to "attach" bonds to them and get meaningful configurations. Instead we shall have to subdivide G finer when the bonds are introduced; see Section 2.6.

Case 2.5 (natural language). The set G would be given by a dictionary, the generator classes G^{α} would be the word classes, and attributes could be number, gender, case, and so on, with the corresponding ranges. This could be criticized as the last case was and by the same argument.

3. Concrete Generators.

We shall not encounter abstract generators very often. The usual situation is that the generator is defined relative to some medium X. The background space, X, can be practically anything: the real line, multidimensional Euclidean space, Hilbert space, or subsets of any of these spaces.

In the background space we may have certain transformations $X \to X$ that will be used to characterize invariance properties of patterns. They should form a semi-group with identity and induce mappings in G which will be taken as the similarity transformations. It will be convenient to use the same symbols s and S both for the mappings $X \to X$ and for $G \to G$.

Definition 3.1. If the generators are elements of the background space X they are called point generators.

Case 3.1 (numerical attributes). Let $X = R$ and let a generator consist of an identifier and a value in R. The s-transformations consist of scale changes $x \to kx$, $k > 0$.

The identifiers may be length, width, frequency, etc. It should be pointed out that an s may affect only one generator class and leave the other generators unchanged and k-values may be different for different classes.

Similarly, it may be necessary to define S as the

direct product of different S^α for the different generator indices α. For numerical attributes that have no preferred origin, for example location measurements, S^α may be chosen as the translation group. For others with a preferred origin, for example size measurements, the scale changes mentioned may form the natural S^α.

Very similar is

Case 3.2 Logical attributes where the background space is $B = \{0,1\}$, so that a generator has an identifier and the value 0 or 1. The similarity transformations consist of identity and $x \to 1-x$.

Definition 3.2. *If the generators are subsets of* X *they are called set generators.*

If this is the case the similarity transformations are extended in the obvious manner from X to G by pointwise application.

Case 3.3 (convex sets). Let $X = R^\ell$ and G = family of all convex sets in R^ℓ with $S = EG(\ell)$ = Euclidean group in ℓ-space.

As we will see later, it is natural to look at these generators as belonging to the second level of a hierarchy where the first level generators are half-planes; see Section 3.5.

Case 3.4 (spheres). Let $X = R^3$ and G^α consist of the spheres with arbitrary center (x_1,x_2,x_3) and radius r^α, any $r^\alpha > 0$. Similarity transformations consist of $EG(3)$ applied to (x_1,x_2,x_3).

Case 3.5 (plane geometry). Here $X = R^2$ and

$$\left.\begin{aligned} G^{(1)} &= \text{all points} \\ G^{(2)} &= \text{all lines, rays, line segments} \\ G^{(3)} &= \text{all circles, circular arcs} \end{aligned}\right\} \quad (3.1)$$

and $S = EG(2)$ and uniform scale changes.

Case 3.6 (features). Consider a background space X and a family F = false, T = true. A subset of the form

$$g^f = \{x \mid f(x) = T\} \quad (3.2)$$

is called a feature generator. We let S consist just of identity and complementation. The nature of $\mathcal{F}$ will of course depend upon the application and will vary from case to case.

Case 3.7 (mosaics). Let $X = R^2$ and consider a partition of it into sets g such that all the g's are congruent modulo a given discrete translation group. This group plays the role of S.

An example of this is shown in Figure 3.1 where the translation group is made up of the translations $(x_1,x_2) \to (x_1+k_1+2k_2, x_2+2k_2)$; k_1 and k_2 integers.

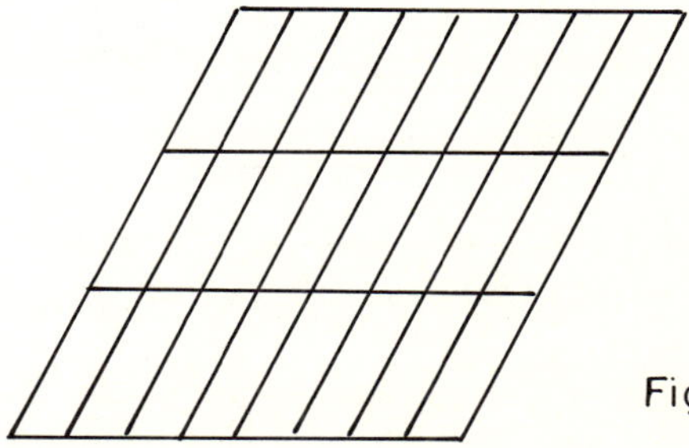

Figure 3.1

The generators are the parallellograms in the picture.

<u>Case 3.8</u> (objects with environment). In R^3 consider a set of objects g. These objects are described by their shape, location and orientation. Some shapes make up a generator class. Other attributes may be color, smell, texture. The natural similarity transformations are again those in EG(3).

<u>Case 3.9</u> (arcs). In $X = R^2$ let G be a collection of analytic arcs in X and $S = EG(2)$. Each arc is described by its intrinsic equation $\rho = \rho(\ell)$, where ρ is curvature and ℓ is arc length. A set $\{\rho^\alpha\}$ of analytic functions is given and an arc $g \in G^\alpha$ if its intrinsic equation is $\rho = \rho^\alpha(\ell)$ over some ℓ-interval. Total arc length is another attribute of such a generator.

<u>Case 3.10</u> (arrows). In some topological space X let g be an ordered pair (x_1,x_2) thought of as an arrow from x_1 to x_2. Let S consist of topological mappings from X onto itself and with the extended definition $s:(x_1,x_2) \to (s(x_1), s(x_2))$.

It is clear that point generators are special cases of set generators if we identify a point in X with a subset containing just that point.

We now come to the type of generators that occurs most often.

<u>Definition 3.3</u>. <u>Let the generators consist of mappings from the background space</u> X <u>into a contrast space</u> Y. <u>We then speak of contrast or function generators</u>.

In particular we shall speak of linear (often temporal), plane, or spatial generators, when X is the real line, the plane, or three-dimensional Euclidean space respectively.

If g is the mapping $y = g(x)$ and S is a group of

similarity transformations we shall define sg through the mapping $y = g(s^{-1}x)$.

If a contrast generator $g = g(x)$ is defined only on a subset $E \subset X$ we shall define sg to mean the function $g(s^{-1}x)$ restricted to the set sE.

Case 3.11 (gray level). If Y is the positive real line, or a subset of it, the corresponding contrast generators can be considered as gray level pictures.

Case 3.12 (color pictures). If Y is the positive octant of R^3, the corresponding contrast generators can be considered as three-color pictures where each coordinate means the strength of red, green, blue respectively.

Case 3.13 (orthogonal generators). Let $Y = R^1$ and μ be a measure on X. We speak of orthogonal generators if $G = \{g_1, g_2, \ldots\}$ where the g_ν are orthogonal functions with respect to μ in the $L_2(\mu)$ geometry.

Case 3.14 (waveforms). Let X and Y be the real line and let G be a given set of real-valued functions on X satisfying $L^\alpha g = 0$, where L^α is a given linear differential operator not involving x for each value of the generator index α. S = translation group on X.

A variation of the last case introduces a new element that will be seen to be crucial in the synthesis of configurations.

Case 3.15 (restricted waveforms). Same as in the last case except that we consider only the restriction of a solution to $L^\alpha g = 0$ to some interval $[a,b]$. This interval is not fixed but can vary with the generator so that the values a and b

are attributes of g.

A more general version of Case 3.3.15 in several dimensions is

Case 3.16 (restricted contrast generators). In the background space $X = R^k$ let the g's be continuous real valued functions defined on closed subsets belonging to a family of sets closed under translations of X. These translations also form the similarity group on G.

We now look at a more special case which will also be expanded and generalized.

Case 3.17 (arithmetic operators). Let $X = Y$ be the real line and any g a function defined on some subset $X_g \subseteq X$ and taking values in some subset $Y_g \subseteq Y$.

As before we insist on looking at X_g and Y_g as parts of the generator.

More generally in a multi-dimensional analog we introduce

Case 3.18 (general operators). Any generator is an operator with ν (variable) inputs $x_1, x_2, \ldots, x_\nu$ and μ (variable) outputs $y_1, y_2, \ldots y_\mu$. The domain of any x_i is some space X_i, the range of any y_j is some space Y_j. In particular there are assignment operators with no inputs (but usually with attributes). The similarity transformations operate only on the assignment operators and leave all other generators unchanged. Its operation on an assignment operator will usually change its attributes but we insist that α should remain as it was and the X's and Y's should not increase.

This case will return again and again due to its high degree of flexibility. We single out one specialization of it as follows.

<u>Case 3.19</u> (operators on stochastic variables). This is the same as in Case 3.18 but with the X_i and Y_j defined as sets of stochastic variables.

Another special case that will be encountered later on is

<u>Case 3.20</u> (projections). Let $X = Y =$ separable, real Hilbert space H and let generators be multiples of projection operators P, $g = kP$, $k \in R$, where the P's occurring in two generators are either equal or orthogonal. The semi-group of similarity transformations are formed from multiplication by (real) scalars. Generator classes are defined in terms of proportional P's.

Set generators are special cases of function generators if we define subsets in terms of their indicator functions.

These three types of generator do not exhaust all the possibilities we shall study. Let us mention just one additional example that will be of interest.

<u>Case 3.21</u> (measures as generators). In a topological vector space X we consider a certain σ-algebra $\mathscr{A}$ of Borel subsets of X and a collection G of measures defined on $\mathscr{A}$. S is the operation on a measure g induced by the translations of X.

Say that G consists of contrast generators so that g is a function defined on some subset X^g of a background space X, and let S be a group for the moment. The family $\mathscr{F}(G)$ of domains X^g should be closed under S of course.

Let B be an S-invariant subset of X and say that <u>the bonds are determined by</u> X^g <u>and that the bond values of</u>

g <u>can be uniquely calculated knowing the values of</u> $g = g(x)$ <u>for</u> $x \in X^g \cap B$. In other words, if two generators have the same domain their bond values are the same if the two functions coincide on those points of B where they are defined.

Now we define a new set G' of generators by introducing an S-invariant subset $X_0 \subset X$, $B \subset X_0$, and shall let the new generators g' be defined as functions on subsets of X_0. More precisely for any $g \in G$ we define a new generator g' as the restriction of g to $X^g \cap X_0$ and with the same bond structure.

If both g_1 and g_2 are mapped into $g' \in G'$, then we must have $X^{g_1} \cap X_0 = X^{g_2} \cap X_0$ which implies $X^{g_1} \cap B = X^{g_2} \cap B$. Hence g_1 and g_2 have not only the same bonds but also the same bond values. Forming $(sg_1)'$ and $(sg_2)'$ it follows that they have the same bond structure but as functions on X_0 they are equal so that they represent the same element in G'. Hence sg' is uniquely determined and s leaves bonds unchanged in $G : S$ as defined in G' is uniquely defined and a legitimate group of similarity transformations.

It then follows that the mapping $G \to G'$ preserves bond structure and $(sg)' = sg'$, it is bond invariant.

Special instances of this construction will appear in various forms in later chapters.

4. <u>Sources of Generators</u>.

Consider a device generating generators from G: $g_1, g_2, g_3, \ldots$. Such a source can be deterministic or stochastic.

A <u>deterministic source</u> produces the sequence of generators according to a rule with uniquely determined result. The simplest case (for finite or denumerable G) is complete enumeration where the order is determined arbitrarily or in terms of a certain attribute. The result would just be G itself.

A stochastic source is a discrete-time stochastic process taking values in G. The simplest case is

Case 4.1 (simple stochastic source). The different generators in the sequence are stochastically independent with one and the same probability distribution q over G.

If the *a priori* measure q is known it will make it possible to talk about average properties of generators. Let a be an attribute on G taking values in R^k. The average value of it is then

$$E(a) = \int_G a(g)q(dg) \tag{4.1}$$

if this integral converges.

Such averages describe, in probabilistic terms, the occurrence of certain properties of the generators. It is also useful to describe properties of q itself, in particular how concentrated or dispersed it is.

Several quantities will be used for this. We shall consider only the discrete case at present, so that the measure q attributes weights q_k to a finite or denumerable number of generators.

Shannon's measure of information is one:

$$H = -\sum_k q_k \log_2 q_k. \tag{4.2}$$

H is non-negative and vanishes only if all the probability mass is concentrated on a single generator. Large values of

H mean that q is spread out and the information content given by selecting one generator at random is high.

Two other measures are defined by

$$c_p = \sum_k q_k^p \tag{4.3}$$

where p is a positive number. If $p = 2$ this is <u>Yule's criterion</u> with the obvious interpretation that c_2 is the probability that two generators independently generated by the simple stochastic source are equal. The smaller c_2 is, the more dispersed q is. It is clear that $c_2 \le 1$ with equality if all the mass is concentrated in one generator. The case $p = 1/2$ is less intuitive but it will be seen that it tells us how difficult it is to break up G into subsets, clusters, in such a way that the clusters are small on the average (Chapter 16, see Appendix). The smallest value of $c_{1/2}$ is one, again meaning that we have a one-point distribution. Schwarz's inequality tells us that the upper bound for the criterion is $\sqrt{n}$, where n is the cardinality of G. For finite n this bound is attained for $q_k = 1/n$, uniform distribution.

A somewhat more complex way of producing generators is the following one.

<u>Case 4.2</u> (Markov stochastic source). The sequence $g_1, g_2, \ldots$ forms a stationary Markov chain with given transition probabilities.

In this case too one can use information measures, although in a slightly more complicated manner compared to the simple stochastic source.

CHAPTER 2

THE CONFIGURATIONS

1. Rules and Restrictions.

In the combinatory theory of patterns the building blocks - the generators - are combined in a structured manner into *configurations*.

A configuration is specified by its *content* and its *structure.* Two configurations are considered to be identical only if both their contents and their structures coincide.

To define which configurations should be considered as *regular* or *admissible*, we could proceed in two ways. We could start from the set of all configurations and single out those that satisfied certain given constraints. This would be a *restrictive definition*. Alternatively, we could start with the empty set and add new configurations successively by applying a generative rule. This would be a *generative definition*. Whichever way we should choose is a matter of convenience and which is the more natural one in a given case.

Anyway, we shall denote by $\mathcal{R}$ the system of rules, or restrictions, or both, that determines what configurations should be considered regular. The set of regular configurations obtained from $\mathcal{R}$ will be denoted by $\mathcal{C}(\mathcal{R})$, or by $\mathcal{C}_n(\mathcal{R})$ if the number of generators is specified to be n. $\mathcal{C}(\mathcal{R})$ expresses the regularity of the patterns.

We have not insisted on G or $\mathcal{R}$ or $\mathcal{C}(\mathcal{R})$ to be finite in any sense. In particular for infinite configurations some additional constraints of non-combinatory character will be needed in order to make it meaningful to the observer. The condition will usually be a convergence criterion, but since this question really concerns images, the discussion of it will be postponed to Chapter 3.

In Chapter 5 we shall also introduce complexity measures for images, but already here we shall introduce

Definition 1.1. For a given set of generators and two systems $\mathcal{R}_1$ and $\mathcal{R}_2$, we shall say that the structural complexity of the regular configurations under $\mathcal{R}_1$ is greater than that of the regular configurations under $\mathcal{R}_2$ if $\mathcal{C}(\mathcal{R}_1) \supset \mathcal{C}(\mathcal{R}_2)$.

In contrast to the structural complexity, we shall use the term numerical complexity of a configuration c belonging to a given $\mathcal{C}(\mathcal{R})$ to mean simply the number of generators included in c.

2. The Bond Structure.

For a finite configuration c we shall write for the content of it

$$\text{content}(c) = \{g_1, g_2, \ldots, g_n\} \tag{2.1}$$

where the right hand side is just a set, completely unstructured.

The structure of a configuration is a set σ of connections between all or some of the bonds of the generators in its content. If the bonds have been enumerated β_{ij}; $i = 1,2,\ldots,n$; $j = 1,2,\ldots\omega(g_i)$, the set σ can be given as a list with entries of the form $(\beta,\beta') = ((i,j),(i',j'))$ corresponding to a connection between the bonds β and β'. Alternatively, we can give σ as a square incidence matrix of order $\Sigma\omega(g_i)$ with ones and zeroes according to whether two bonds are connected or not.

Usually we shall not consider all possible σ's but only a certain class, for example vector structure, tree structure, etc. The set of all σ's allowed will be denoted by Σ and is called the <u>connection type</u> of the configurations in $\mathcal{C}(\mathcal{R})$ in question.

To formalize this we introduce the concept of connection type.

<u>Definition 2.1</u>. <u>A connection type</u> Σ <u>is the union of sets</u> Σ_n <u>where each</u> Σ_n <u>is a set of graphs over</u> n <u>nodes</u>.

For $\sigma_1 \in \Sigma_{n_1}$, $\sigma_2 \in \Sigma_{n_2}$ we consider all graphs in $\Sigma_{n_1+n_2}$ that can be obtained as the union of σ_1 and σ_2 by adding a set σ of connections between the two original graphs.

The set of all those σ, for given σ_1 and σ_2, will be denoted by $\Sigma(\sigma_1,\sigma_2)$ and we shall use σ as the generic symbol combining two such graphs. Each time an expression is formed by repeated use of σ-operations, it will be assumed that each sub-expression belongs to Σ.

In most cases it will be clear from context what $\Sigma(\sigma_1,\sigma_2)$ is and in special cases we will use other, more informative symbols instead of σ.

If Σ allows for all the degenerate combinations consisting of an isolated arbitrary generator, then Σ is called <u>monatomic</u>.

As examples of connection type we mention some of the more important ones.

Σ = FREE meaning that each $\sigma \in \Sigma$ is empty. Then no connections are set up and the configurations are unstructured, just sets of generators.

If $\omega_{in}(g) \equiv \omega_{out}(g) \equiv 1$ we may choose Σ = LINEAR ORDER, meaning that the n generators should be connected by an out-bond from g_i to an in-bond of g_{i+1}, $i = 1,2,\ldots,n-1$.

If $\omega_{in}(g) = 1$ and $\omega_{out}(g)$ is arbitrary, Σ = TREE means that the generators should be combined into trees with arrows entering in-bonds and exiting from out-bonds.

Σ = PARTIAL ORDER means that the connections should form a partial order when account is taken of the direction of the arrows. Occasionally we shall meet connection types for which $\sigma \in \Sigma$ implies that the connectivity belonging to any subset of the configuration is also in Σ. We shall then speak of a <u>monotonic</u> Σ. The use and handling of configuration types will be made clearer when we discuss concrete cases, see Section 2.3 - 2.9. A reader may prefer to skip the somewhat abstract discussion in the present section and go directly to 2.3 and only return to the general discussion in 2.2 after seeing these cases. Given a Σ we can always extend it to become monotonic by including every sub-combination of the original combination. In this way we get the <u>monotonic extension</u> $\mathcal{M}(\Sigma)$ of the given connection type.

For example if Σ = LINEAR ORDER then $\mathcal{M}(\Sigma)$ consists of disjoint sets, each one of which forms a linearly ordered set.

More precisely, if Σ allows a certain combination of generators with some or all the bonds connected, then it should also allow sub-combinations resulting from opening some of the bonds and deleting some of the configurations. When a generator is deleted, then, of course, its bonds are automatically deleted too.

Once content(c) and

$$\text{structure}(c) = \sigma \tag{2.2}$$

are given for a configuration, its regularity will depend upon how the connected bonds fit each other. This will be determined by an agreement relation or bond relation, ρ, which depends upon the two bonds in question and is written $\beta\rho\beta'$.

The bond relation shall be S-invariant so that if $\beta\rho\beta'$ then $\beta_1\rho\beta_1'$ where β_1 and β_1' are the transformed bond values resulting from applying s to respective generators.

This relation can be something very simple, for example just equality. In general it need not be symmetric, say when ρ is an inclusion relation, or transitive, say when ρ is given in terms of several inequalities with different direction.

Definition 2.2. A configuration c with structure $\sigma \in \Sigma$ is regular if and only if $\beta\rho\beta'$ for any connection $(\beta,\beta') \in \sigma$.

For a configuration $c \in \mathscr{C}(\mathscr{R})$ some of its bonds are connected by σ - the internal bonds. Its remaining bonds are the external bonds. For the set of external bonds and bond values we shall write ext(c).

Regular configurations will be represented graphically by a configuration diagram in which generators are shown as large circles (usually with identifiers and attributes) and small half circles for the bonds (often with the bond values). If two bonds are connected, it will be indicated by a small circle with a diameter across it. Many such diagrams will occur in the following.

We can now extend the notions of arity and similarity transformations from G to $\mathscr{L}(\mathscr{R})$. The arity of c is the number of its external bonds, and just as for generators, it is made up of the in-arity and the out-arity.

The domain of any $s \in S$ is extended to $\mathscr{L}(\mathscr{R})$ by defining, if content(c) = $\{g_1, g_2, \ldots, g_n\}$,

$$\text{content}(sc) = \{sg_1, sg_2, \ldots, sg_n\} \tag{2.3}$$

and

$$\text{structure}(sc) = \text{structure}(c) \tag{2.4}$$

The statement (2.4) is consistent with (2.3) since similarity transformations do not affect bond structure, see Definition 1.1.

Since each connection involves two bonds, we have

$$\omega(c) = \text{arity}(c) = \sum_i \omega(g_i) - 2\#(\sigma) \tag{2.5}$$

and similar relations for the in- and out-bonds.

In addition to the incidence matrix of σ, order $\sum\omega(g_i)$, we also need an "incidence" matrix between the generators. It is of order n and the entry (i,i') is equal to the number of out-bonds from g_i to in-bonds of $g_{i'}$. It can be calculated from the matrix of σ, but not vice versa.

We also extend the combinatory structure to configurations. Consider two configurations $c_1, c_2 \in \mathscr{C}(\mathscr{R})$ and the sets $B(c_1)$ and $B(c_2)$ consisting of the external bonds of c_1 and external bonds of c_2. Let σ_{12} be a list of connections from $B(c_1)$ to $B(c_2)$ such that no multiple connections are established, only pairwise ones. Then we write $c_1\sigma_{12}c_2$ for the combined configuration with

$$\text{content}(c_1\sigma_{12}c_2) = \text{content}(c_1) \cup \text{content}(c_2) \tag{2.6}$$

and

$$\text{structure}(c_1\sigma_{12}c_2) = \text{structure}(c_1) \cup \text{structure}(c_2) \cup \sigma_{12}. \tag{2.7}$$

Then it follows that $c_1\sigma_{12}c_2 \in \mathscr{C}(\mathscr{R})$ if and only if

$$\left.\begin{array}{ll} \text{(i)} & \text{structure}(c_1\sigma_{12}c_2) \in \Sigma \\ \text{(ii)} & \beta\rho\beta' \text{ for all new bonds connected by } \sigma_{12}. \end{array}\right\} \tag{2.8}$$

Instead of the list σ_{12} we may use the rectangular incidence matrix for the connections in σ_{12}. Often, it will be clear from the context what σ_{12} is and we may then use some symbol like "o", ",", "+" instead of σ_{12} or its incidence matrix.

It should be noted that in this way we may establish binary operators on configurations but that they are in general not defined throughout $\mathscr{C}(\mathscr{R}) \times \mathscr{C}(\mathscr{R})$ but only on a subset determined by the connection type and agreement relations.

We can now draw some simple conclusions about the properties of $\mathscr{C}(\mathscr{R})$, our <u>configuration space</u>.

<u>Theorem 2.1.</u>

(i) <u>The similarity transformations take the regular</u>

<u>configurations into regular configurations,</u> $s: \mathscr{C}(\mathscr{R}) \to \mathscr{C}(\mathscr{R})$

(ii) <u>if</u> $c_1, c_2 \in \mathscr{C}(\mathscr{R})$ <u>and</u> $c_1\sigma_{12}c_2 \in \mathscr{C}(\mathscr{R})$ <u>then</u>

$$s(c_1\sigma_{12}c_2) = (sc_1)\sigma_{12}(sc_2) \in \mathscr{C}(\mathscr{R})$$

(iii) <u>consider connections</u> $\sigma, \sigma', \sigma'', \sigma'''$ <u>such that</u> $\gamma = (c_1\sigma c_2)\sigma' c_3$ <u>and</u> $\gamma' = c_1\sigma''(c_2\sigma''' c_3)$ <u>make sense in terms of correct connection type. Then</u> γ <u>and</u> γ' <u>are both regular or both non-regular and coincide, if</u> $\sigma \cup \sigma' = \sigma'' \cup \sigma'''$.

<u>Proof</u>: To verify (i) it is enough to observe that structure (sc) = structure(c) $\in \Sigma$ and since ρ is S-invariant all bond agreements $\beta\rho\beta'$ for c remain true after s has been applied to all the generators in content(c).

For (ii) we note that s does not affect the connections. Because of (2.3) we have

$$\begin{aligned} \text{content}\,[s(c_1\sigma_{12}c_2)] &= \text{content}(sc_1) \cup \text{content}(sc_2) \\ &= \text{content}\,[(sc_1)\sigma_{12}(sc_2)]\,. \end{aligned} \tag{2.9}$$

Hence the equality in (ii) holds.

The relation (iii) is proved just as (ii) since both structure and content are the same for both configurations and it only remains to verify whether all the bond connections satisfy the agreement relation ρ or not. The answer will be the same in both cases so that either the configurations are both regular or none of them is. Hence associativity holds in this restricted sense.

If, for two regular configurations c_1 and c_2 we have

$$\left.\begin{aligned} \text{content}(c_1) &\subseteq \text{content}(c_2) \\ \text{structure}(c_1) &\subseteq \text{structure}(c_2) \end{aligned}\right\} \tag{2.10}$$

(Recall that structure(c) is a <u>set</u>), we shall write

$c_1 \subseteq c_2$ and say that c_1 is a <u>subconfiguration</u> of c_2. This defines a partial ordering in $\mathscr{C}(\mathscr{R})$. If $c_2 = c_1\sigma_{12}c'$ with $c_1, c_2 \in \mathscr{C}(\mathscr{R})$, then $c_1 \subseteq c_2$, so that <u>composition of configuration is a monotonic operation</u> with respect to the given partial order; it always adds information or, more precisely, never destroys any.

Consider two configurations c_1 and c_2 in $\mathscr{C}$ with Σ-structure and with the same external bond structure but not necessarily in $\mathscr{C}(\mathscr{R})$. The latter condition is often, but not always, a consequence of the first. The bond values need not be the same.

Now combine c_1 and c_2 in the same way with another configuration c such that the resulting connection type belongs to Σ. If it then follows that the resulting configurations are either both in $\mathscr{C}(\mathscr{R})$ or none of them is, and this is so for any c, we shall say that c_1 is <u>congruent</u> with c_2. This establishes an equivalence relation so that $\mathscr{C}$ is partitioned into disjoint <u>congruence classes</u>. If we use the empty configuration as c it follows that c_1 and c_2 being congruent implies that they are either both regular or none of them is.

To find out whether c_1 and c_2 are congruent or not, it is clear, first, that if one is in $\mathscr{C}(\mathscr{R})$ and the other not, then they are not congruent. Also, if they are both non-regular so that one or more connections violate ρ in each, then the combined configurations must also be non-regular automatically. Hence, we need only consider the case when both are regular.

The set of regular configurations will be written as a

quadruple

$$\mathscr{C}(\mathscr{R}) = (G,S,\Sigma,\rho) \tag{2.11}$$

or if we combine the Σ-structure and the bond relation ρ into the <u>rules</u>

$$\mathscr{R} = (\Sigma,\rho) \tag{2.12}$$

we get the triple

$$\mathscr{C}(\mathscr{R}) = (G,S,\mathscr{R}). \tag{2.13}$$

When we only consider regular configurations of given cardinality n we shall write for the configuration space

$$\mathscr{C}_n(\mathscr{R}) \subset \mathscr{C}(\mathscr{R}). \tag{2.14}$$

Sometimes certain regular configurations occur often as sub-configurations. It will then be convenient to consider them as indivisibles - as generators - with given fixed internal bonds. They will be spoken of as macro-generators.

3. Free Configurations.

The simplest but also least interesting connection type is when Σ = FREE, no connections are established, so that any configuration is regular and has no internal structure, it is just a set. This will be spoken of as <u>free configurations.</u>

For free configurations any combination of generators is regular which implies that all configurations are congruent and there is only one congruence class, $\mathscr{C}(\mathscr{R})$ itself.

As an example let us mention the following:

<u>Case 3.1</u> (finite point sets). In a given background space X let the generators be arbitrary finite sets of points in X. With Σ = FREE any combination of generators is regular.

The reason why we mention this highly unstructured case

at all is that it may appear even in structured situations as the result of deformations. Typically deformations loosen up the structure and information is destroyed, so that even when we have a rigid combinatory set up, the case mentioned may be the natural one to describe deformed images, see Chapter 4.

Another unstructured situation involves function spaces.

<u>Case 3.2</u> (linear subspace). On a background space X with a measure μ consider the L_2-space $L_2(X;\mu)$. Let the generators consist of elements of a fixed subspace of $L_2(X;\mu)$ with Σ = FREE as the connection type.

This will also come up as the result of simple deformation mechanisms.

Another unstructured case is:

<u>Case 3.3</u> (Euclidean patterns). With generators as in Case 1.5 and with Σ = FREE we get the typical configurations of plane geometry.

Note however, that when we study connected line segments, circles passing through a given point and other constraints, we need bonds for the incidence relations and another connection type will be required.

Since free configurations are just sets, it is natural to use the symbol "$\cup$" (union) for the connection symbol σ, with the proviso that two copies of a generator may be treated as distinct by the use of additional markers in certain contexts.

Combination of this type is of course both associative and commutative.

4. Linear Connection Type.

Now let Σ consist of linear orderings so that the regular configuration with n generators looks like Figure 4.1.

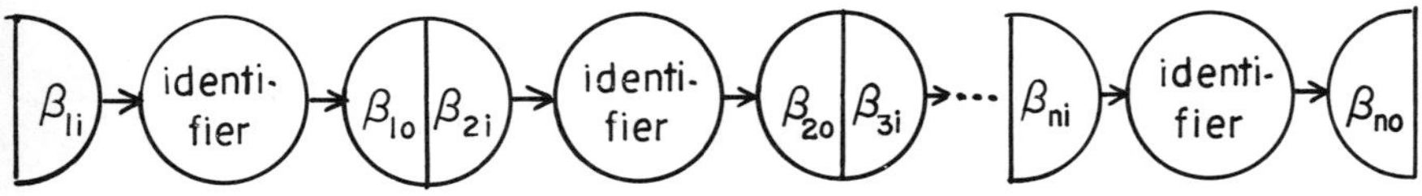

Figure 4.1

Each generator as well as each configuration has in-arity and out-arity equal to one, and each of the $2(n-1)$ internal bonds must satisfy the relation ρ in order that the configuration be admissible.

For linear connection type we shall use ',' (comma) for the connection symbol σ corresponding to concatenation of ordered strings. It is associative but not cummutative.

Case 4.1 (restricted waveforms). Let the generators be the restricted waveforms of Case 3.3.15 and the in-bonds with the values

$$\left.\begin{aligned} \beta_{ki} &= (a_k, g(a_k)) \\ \beta_{ko} &= (b_k, g(b_k)) \end{aligned}\right\} \tag{4.1}$$

with ρ meaning equality. This means that g_k can be connected to g_{k+1} if $b_k = a_{k+1}$, contiguity, and $g(b_k) = g(a_{k+1})$, a continuity condition.

A variation of this $\mathcal{L}(\mathcal{R})$ is obtained by strengthening the constraints by making the bonds include values of derivatives at the end-points up to some order p.

As an example let each L^α be a differential operator of order p with constant coefficients. The coefficients may vary between different generator classes G^{α}. It then follows that a configuration c in $\mathscr{C}(\mathscr{R})$ involving only generators from the same G can describe a function that is itself some generator in G but in a different domain. Note, however, that c and g are not identical, their internal structures differ. Later on we shall come back to this question when studying the identification rules that make up images.

Case 4.2 (arithmetic sequences). Let us consider arithmetic operations as in Case 3.17 of the following types. $G^{(1)}$ consists of assignment operators with no in-bond and one out-bond, its attribute is the real number assigned by it and its out-bond a real number x. $G^{(2)}$ consists of many arithmetic operators with one in-bond and one out-bond, both of them subsets of the set of integers, representing the domain and range of the operator, respectively. The relation ρ should be the inclusion relation.

This set up gives two sorts of configurations, either a string of arithmetic operators, or one which starts with an assignment and is followed by arithmetic operators. In the first case $\omega_{in}(c) = 1$, in the second case it is 0. In both cases $\omega_{out}(c) = 1$.

We can think of the second one as describing a particular arithmetic sequence and the first one more abstractly as a general arithmetic sequence with unspecified first term.

Strings generated by finite state automata also belong to this connection type and we refer the reader to Section 2.6 for a brief description of the concepts and notation used below. Generators will belong to the set G of objects depicted in Figure 4.2 so that they have

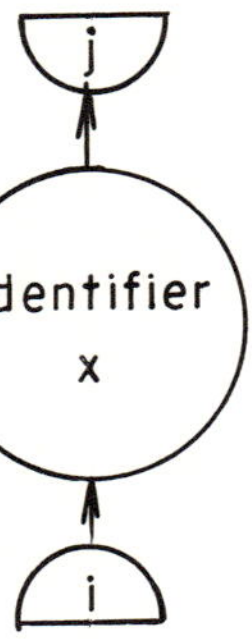

Figure 4.2

$\omega_{in}(g) = \omega_{out}(g) = 1$ with bond values i and j denoting states. An attribute of g is x, a terminal symbol. The interpretation of g is the transition from state i to state j while writing the symbol x. We can also think of it as the corresponding rewriting rule. Let s be a permutation of the states $i = 1,2,3,\ldots$ so that the rewriting rule $i \to xj$ is transformed into $i' \to xj'$ with the new numbering of states. We extend the definition of s to G by defining sg for g as in Figure 4.2 to be the generator corresponding to the transformed rewriting rule.

Case 4.3 (substrings of finite state languages). The generators are as above, ρ is equality, and Σ is of linear order type. The similarity group S is given through the permutation group and its extension to G.

With this definition $\mathcal{L}(\mathcal{R})$ becomes the set of substrings, not necessarily grammatical but with correct state transitions from a finite state language. Admissible configurations will have in- and out-arity equal to one and may look as in Figure 4.3 which represents the terminal string aaba with in-bond value i and out-bond value m.

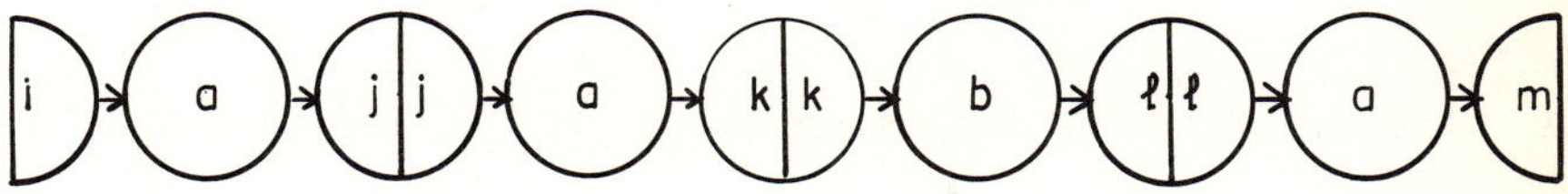

$\omega(c)=2$
$n=4$

Figure 4.3

Let us now add two special generators to G. The first is the <u>assignment</u> operator g_0 with $\omega_{in}(g_0) = 0$, $\omega_{out}(g_0) = 1$ with the out-bond 1. The second one the <u>acceptance</u> generator g_v with $\omega_{in}(g_v) = 1$, $\omega_{out}(g_v) = 0$ and in-bond = the terminal state. If we do this the set of regular configurations of arity zero will be the grammatical strings of the language.

5. Bounded Arity Greater Than Two.

In the cases in the last two sections we have encountered configurations with arity equal to 0, 1, or 2. Let us now consider situations with higher but bounded arity.

<u>Case 5.1</u> (arithmetic sequences of higher order). Let the generators be arithmetic operations with two inputs and two outputs, both of them real and let the connections type be as in Figure 5.1. The generator g_k has in-arity 2 with the two bonds β_{k1}, β_{k2} and out-arity 2 with the two bonds β_{k3}, β_{k4}. In-bonds are domains and out-bonds ranges as in Case 3.18. If content(c) = $\{g_1, g_2, \ldots, g_n\}$ then for $1 \leq k \leq n-1$ a $\sigma \in \Sigma$ connects β_{k3} to $\beta_{k+1\ 1}$ and for $1 \leq k \leq n-2$ β_{k4} to $\beta_{k+2\ 2}$ so that we have for any $c \in \mathscr{L}(\mathscr{R})$

$$\left.\begin{aligned} \omega_{in}(c) &= 3 \\ \omega_{out}(c) &= 3 \\ \omega(c) &= 6 \end{aligned}\right\} . \tag{5.1}$$

The relation ρ means inclusion: range of one included in the domain of the next.

What are the congruence classes here? Consider two configurations c_1 and c_2 in $\mathscr{L}(\mathscr{R})$ and combine them via σ, all the time with Σ-structure. Because of the special connection we have here this means that either c_1 and c_2 are joined to the right of c or to the left via σ; let us say to the right.

Let us say that the out-bonds of c_i have values $\beta_{i1},\beta_{i2},\beta_{i3}$. We use the convention that β_{i1} and β_{i2} go to the first generator of c and β_{i3} to the second one if there is one. Similarly β_1,β_2,β_3 for the corresponding in-bonds of c. We must insist that for any c with $\text{card}(c) \geq 2$ the Boolean relation holds

$$\bigwedge_{k=1,2,3} (\beta_{1k}\rho\beta_k) = \bigwedge_{k=1,2,3} (\beta_{2k}\rho\beta_k) \tag{5.2}$$

or if $\text{card}(c) = 1$

$$\bigwedge_{k=1,2} (\beta_{1k}\rho\beta_k) = \bigwedge_{k=1,2} (\beta_{2k}\rho\beta_k) \quad . \tag{5.3}$$

It now depends upon what are the possible combinations of external bond values for regular configurations. The following situation is of interest.

Remembering that bond values in the present case represent domains and ranges, let us say that they are certain subsets $B_1,B_2,..$ of a given space B, and that all combinations are possible as external bonds of regular configura-

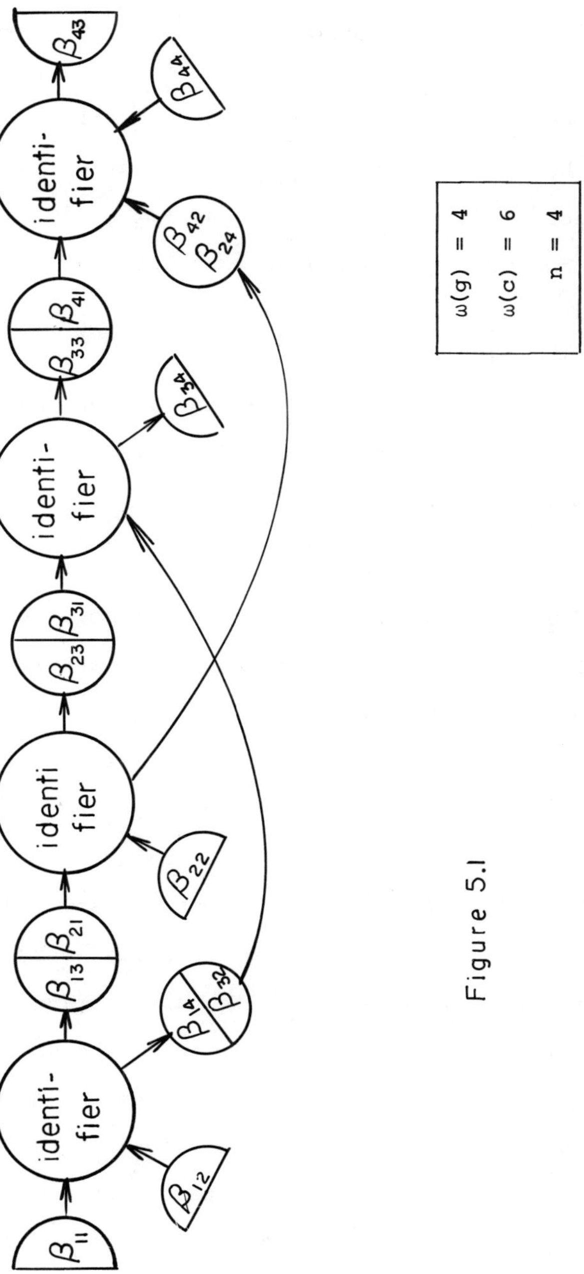

identi-
fier
β_{11}
β_{12}
β_{13}
β_{21}
β_{14}
β_{32}
β_{22}
β_{23}
β_{31}
β_{34}
β_{33}
β_{41}
β_{42}
β_{24}
β_{44}
β_{43}
$\omega(g) = 4$
$\omega(c) = 6$
$n = 4$

Figure 5.1

tions.

Now we can claim that congruence here means that all external bond values at corresponding bonds must coincide. To prove this say that they do coincide: $\beta_{1k} = \beta_{2k}$ for all k. Then obviously (5.2) and (5.3) are identically true. On the other hand if for some $\beta_{1k} \neq \beta_{2k}$, say $\beta_{1k} \not\supset \beta_{2k}$, then we can choose $\beta_k = \beta_{1k}$ for all k. If we do, then the left hand conjunction in (5.2) is true but since $\beta_k = \beta_{1k} \not\supset \beta_{2k}$ so that $\beta_{2k}\rho\beta_k$ = FALSE and c_1 and c_2 cannot be congruent, which completes the proof.

If the B_k can only appear in certain combinations for regular configurations, exact fit $\beta_{1k} = \beta_{2k}$ may not be needed to insure congruence.

In our next example we meet

<u>Case 5.2</u> (growth patterns in space-time). Let the generators be points in the plane (with numerical attributes such as density and identifiers) with S as a group of topological mappings of R^2 onto itself. In this case it will be convenient to let some of the connections be bi-directional. For $n = M \times T$ the connection type is illustrated in Figure 5.2 with $M = 7$ and $T = 4$. The arity of any generator is 4 and there are two types of bonds indicated by solid lines, β_1, and dotted lines, β_2. We have not indicated the values themselves in this picture - it should be obvious that $\beta_1\rho\beta_1$, $\beta_2\rho\beta_2$ in this case. An individual generator with its bonds is shown in (b). The bond relation is equality here.

A generator z_{mt} is connected bi-directionally by β_1 bonds to $z_{m+1,t}$ and $z_{m-1,t}$ where addition in the first subscript is understood modulo M. It is connected to $z_{m,t-1}$

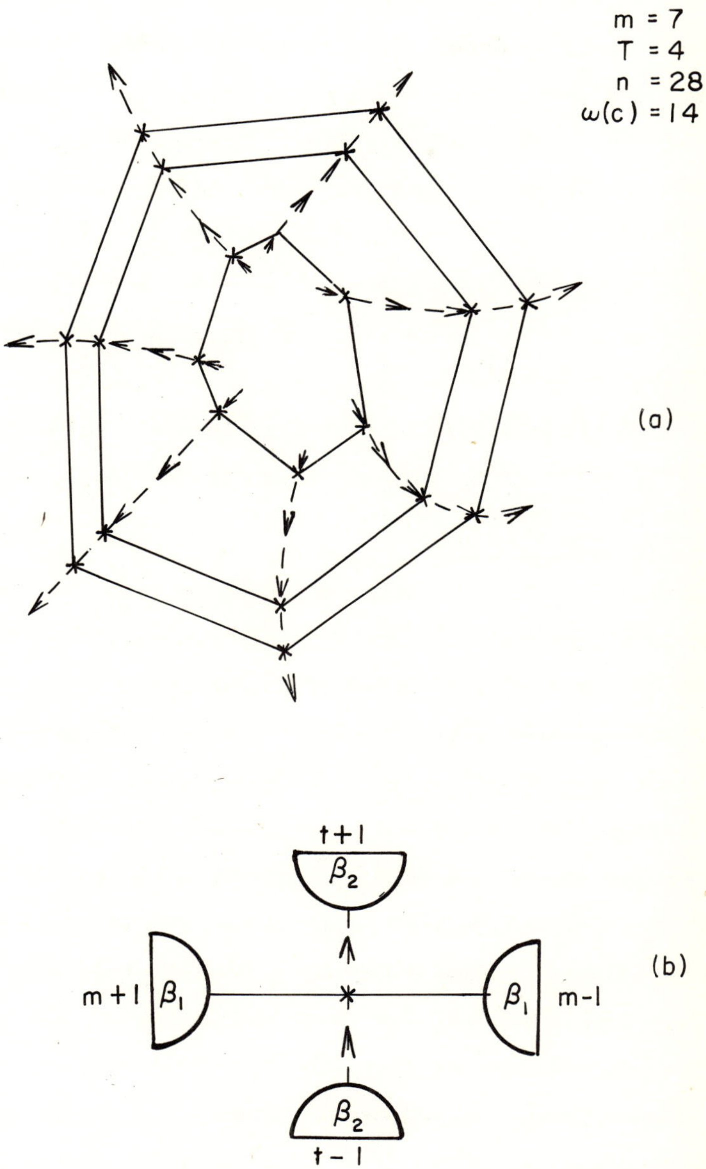

Figure 5.2

and $z_{m,t+1}$ uni-directionally as indicated by the arrows. It follows that the arity of c is $2m$, half of it in-bonds and half of it out-bonds.

The interpretation of the Σ-type is that growth (or decay, or a combination of both) occurs when time t increases. Hence the generators with a given attribute t give the contour of the organism at this time. The points with common attribute m are the homologous points and represent growth lines.

Alternatively, we could have used the macro-generator corresponding to constant m-values as generators, the growth curves. They would then be combined by a different connection type relating the homologous points, to lead to a growth pattern. We shall return to this topic in Section 3.8.

6. Tree Type Connections.

Turning to Σ consisting of trees we encounter cases where the arity of the configurations is not bounded. The out-arity of any generator is equal to one but the in-arity is an arbitrary non-negative integer. For a configuration c with content$(c) = \{g_1, g_2, g_3, g_4\}$ the internal structure could be as in Figure 6.1.

The generators are arranged in a tree topology with the out-bond (possibly) connected by a unidirectional arrow to one of the in-bonds of another generator. This implies that $\omega_{out}(c)$ is always one for any $c \in \mathscr{C}(\mathscr{R})$ while $\omega_{in}(c)$ can be arbitrary and is not bounded.

Given two configurations c_1 and c_2 a combination $c_1 \sigma c_2$ can be specified simply by $\sigma = (i,j)$ where i identifies the out-bond of one of c_1 and c_2 and j identifies one of the in-bonds of the second configuration.

Congruence classes can be discussed here just as they

were in Section 5.

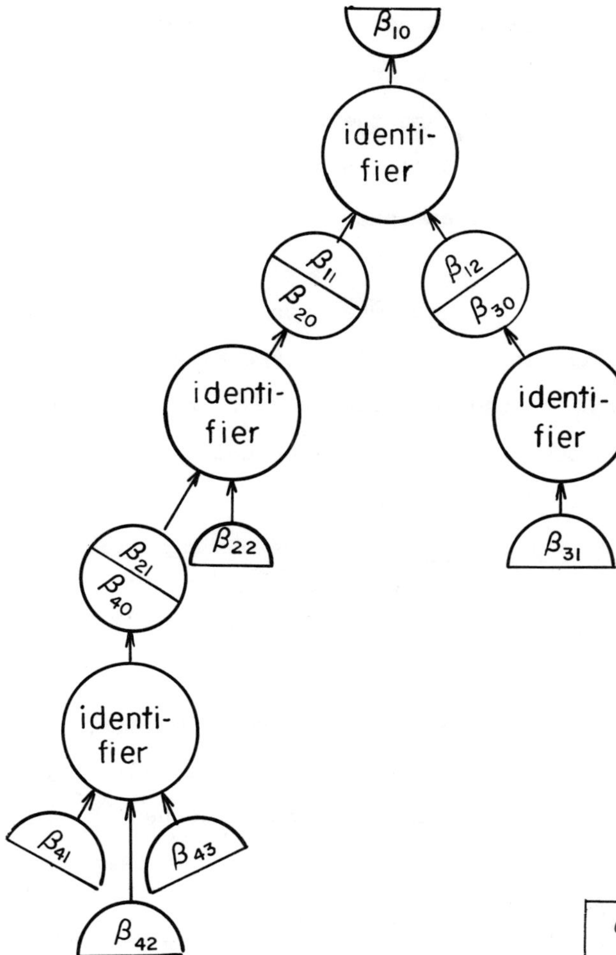

$\omega_{out}(c) = 1$
$\omega_{in}(c) = 5$
$n = 4$

Figure 6.1

<u>Case 6.1</u> (river systems). In the plane consider directed segments as generators where we write $g(x_1,y_1;x_2,y_2)$ for the coordinates of its two end points. The out-arity is of course one, but we also stipulate that the in-arity be two. Look at

the sub-configuration in Figure 6.2 with 1 out-bond and 4 in-bonds as an example. The bond relation says that two in-bonds should always be equal. Here S would be a group of topological mappings of the plane onto itself.

For actual river systems the segments will be replaced by other arcs and with the additional constraint that the arcs do not intersect except at end points and the in-arity may be greater than two.

Before turning to a case when context-free languages are looked at as configurations with tree type connections we shall briefly recall some of the concepts needed. For more detailed information about context-free languages and related topics, the reader is referred to Ginsburg (1966).

In a <u>phrase structure grammar</u> the basic notions are

the set V_T of terminal symbols, the vocabulary or lexicon
the set V_N of syntactic variables or non-terminals, including in particular the initial symbol σ
the set $\mathscr{R}$ of rewriting rules where each rule has the form $V_N^* \to V^*$. (6.1)

The notation A^*, for any set A, represents the collection of all finite strings formed from elements in A and we write $V = V_T \cup V_N$.

In linguistics the sets V and $\mathscr{R}$ are always assumed to be finite, to make infinite use of finite means, to quote a famous motto. Finiteness will play a rather subordinate role for us, but it must be assumed in the present case.

This is how the rules are applied. For two strings, a

and b in V^*, we shall write $a \to b$ if there are strings α, β, x, y such that $a = x\alpha y$, $b = x\beta y$ and the relation $\alpha \to \beta$ belongs to the rules in $\mathscr{R}$. Extending the relation " $\to$ " even further, one says that $a \to b$ if $a = b$ or if there exist an n and strings $z_0, z_1, z_2, \ldots, z_n$ such that $z_0 = a$, $z_n = b$, and $z_i \to z_{i+1}$ for $i = 0,1,2,\ldots,n-1$. The sequence $z_0, z_1, \ldots, z_n$ is called a <u>derivation</u> of b.

<u>The grammatical strings are the strings in</u> V_T^* <u>that can be derived from the initial symbol</u>. An important class of phrase-structure grammars is the set of <u>context-free grammars</u>. This means that the rules in $\mathscr{R}$ are all of the form $a \to b$ where $a \in V_N$.

<u>Example.</u> Well-formed expressions in propositional calculus can be obtained as follows.

Let the terminals be

$$V_T = \{x_1, x_2, \ldots, x_n; \vee, \wedge, \tilde{},), (\} \tag{6.2}$$

with the syntactic variable σ and the rules

$$\mathscr{R} = \{\sigma\to x_1, \ldots, \sigma\to x_n, \sigma\to(\sigma \vee \sigma), \sigma\to(\sigma \wedge \sigma), \sigma\to(\tilde{}\sigma)\} \text{ .} \tag{6.3}$$

The grammatical strings will then be all well-formed expressions in propositional calculus. This example should be compared to Case 1.2.4.

A celebrated theorem (see e.g. Ginsburg (1966), Chapter 2) states that context-free languages are identical to sets of strings accepted by a <u>pushdown automaton</u>. Such an automaton is given in terms of

$$\left.\begin{array}{l}\text{a set } K \text{ of states with an initial state } k_0 \\ \text{and a subset } F \text{ of final states} \\ \text{a set } V_T \text{ of inputs} \\ \text{a set } \Gamma \text{ of pushdown symbols with an initial} \\ \text{pushdown symbol } \gamma_0 \\ \text{a mapping } \delta\colon K\times(V_T \cup \{\varepsilon\})\times \Gamma \rightarrow \text{ finite subsets} \\ \text{of } K \times \Gamma^* .\end{array}\right\} \quad (6.4)$$

The sets K, V_T, Γ are supposed to be finite and not empty and ε stands for the empty vector. The automaton operates as follows. Consider the triple (p,w,α) where $p \in K$ is the state of the automaton, $w \in V_T^*$ is the input string, and $\alpha \in \Gamma^*$ is on the pushdown tape. By a move $(p,xw,\alpha z) \rightarrow (q,w,\alpha\gamma)$, where $x \in V_T \cup \{\varepsilon\}$, $q \in K, z,\gamma \in \Gamma^*$ one means the operation of going to state q, replace z by γ, and expend the input x. This is allowed if $\delta(p,xw,z)$ contains (q,γ). An input string is <u>accepted</u> by the automaton if, starting in state k_0 and with γ_0 on the pushdown tape, it can go in a finite number of moves to a state in F.

A subset of context-free grammars is the <u>right linear grammars</u>, which means that all the rules in $\mathscr{R}$ are of the form $x \rightarrow a$ or $x \rightarrow ay$ with $a \in V_T$ and $y \in V_N$.

The right linear grammars are equivalent to <u>finite-state automata</u>, and one also speaks of finite-state grammars and languages. Such an automaton is given by

$$\left.\begin{array}{l}\text{a set } K \text{ of states with a set } K_0 \text{ of start states} \\ \text{a set } V_T \text{ of inputs with a set } F \text{ of final states} \\ \text{a mapping } \delta \text{ from } K \times V_T \text{ into subsets of } K.\end{array}\right\} \quad (6.5)$$

The automaton works as follows. An input string w is accepted if it is the empty word or if there is a sequence $k_0,k_1,\ldots,k_n$ in K, and $w = x_1x_2\ldots x_n$, $x_k \in V_T$, such that

$$\left.\begin{array}{l} k_0 \varepsilon K_0 \\ k_i \varepsilon \delta(k_{i-1}, x_i) \\ k_n \varepsilon F \ . \end{array}\right\} \qquad (6.6)$$

Note also the following well-known relation (see Ginsburg (1966), Chapter 2). Consider the congruence relation in the special case of a context-free language, so that two strings x and y are congruent if, for any strings u, v, one has either uxv grammatical and uyv or none of them. Then the language is of finite-state type if and only if the number of the corresponding equivalence classes is finite.

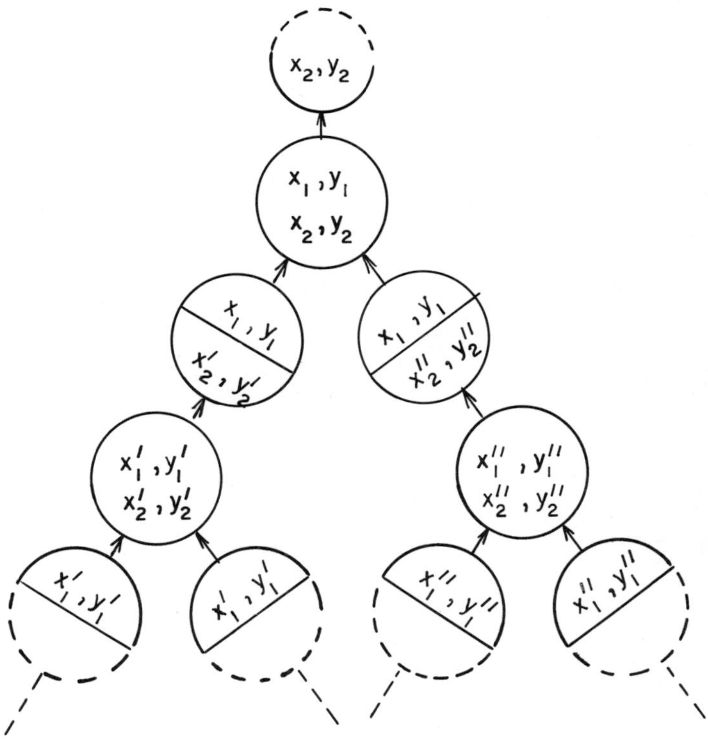

Figure 6.2

ω(c) = 5
n = 3

Case 6.2 (subderivations of context-free languages). The generators correspond to the rewriting rules and have out-arity one with the bond value equal to the syntactic variable being rewritten. The in-arity is the number of symbols in the rewritten string and the bond values the corresponding symbols in the appropriate order. Σ is tree structure and ρ equality.

Then $\mathscr{C}(\mathscr{R})$ will be grammatical subderivations. Any regular configuration will have out-arity one but arbitrary in-arity.

If we also add an acceptance generator with out-arity zero, in-arity zero and the bond value equal to the initial symbol, and assignment operators of in-arity one, out-arity zero, and the bond equal to one of the terminal symbols, then the regular configurations of arity zero are the derivations of grammatical strings in the language.

Let us return for a moment to Case 1.2.2, one of the Morse code systems. It may appear as if it had essentially linear structure but it is more appropriate to view it as having Σ = tree structure. Indeed it can be viewed as context-free with the syntactic variables $i, w, \ell, u, 1, 3, 5$ with the interpretation

$$\left.\begin{array}{l} i = \text{initial symbol} \\ w = \text{word} \\ \ell = \text{letter} \\ u = \text{signal} \end{array}\right\} \qquad (6.7)$$

and the terminal symbols

$$\left.\begin{array}{l} \text{dot} \\ \text{dash} \\ 1 = \text{short space} \\ 3 = \text{letter space} \\ 5 = \text{word space} \end{array}\right\} \qquad (6.8)$$

and the rewriting rules

$$\left.\begin{array}{l} i \to w \\ i \to w5i \\ w \to 1 \\ \text{..} \to \ell 3w \\ \ell \to u \mid u\ell u \mid u\ell u\ell u \mid u\ell u\ell u\ell u \mid u\ell u\ell u\ell u\ell u \\ u \to \text{dot} \\ u \to \text{dash} \; . \end{array}\right\} \qquad (6.9)$$

<u>Case 6.3</u> (Linnaean patterns). On a background space X let $\mathscr{F}$ be a collection of functions on X. For such a function f we denote its range by $R(f)$. On $\mathscr{F}$ a simple order relation is given and denoted by " $\prec$ ".

A generator is given through a function f and one in-bond β and an arbitrary finite number of out-bonds $\beta_1, \beta_2, \ldots, \beta_n$. Each bond is a truth function and

$$\beta_i = \beta \wedge (f=\phi_i) \qquad (6.10)$$

with some $\phi_i \in R(f)$ and $\phi_i \neq \phi_j$ for $i \neq j$.

The connection type Σ is TREE so that the configurations have tree structure and this structure is assumed to be consistent with the " $\prec$ " relation along up-going branches.

The bond relation from an out-bond β' to an in-bond β'' should say that β'' implies β' : ρ = IMPLY.

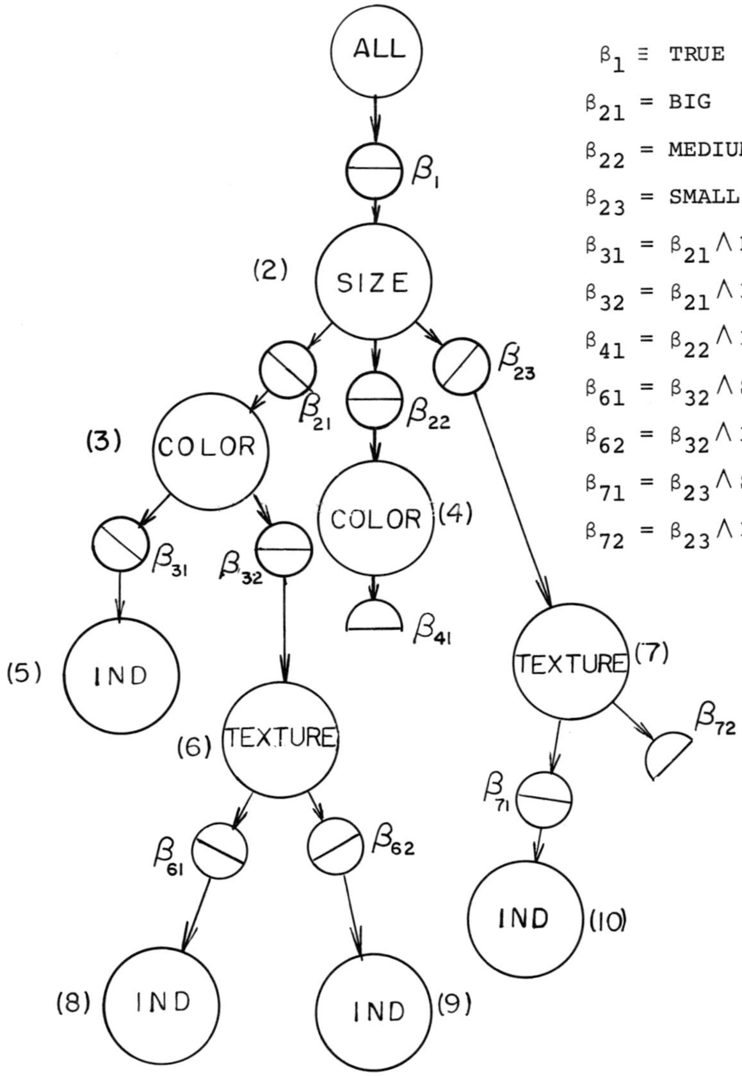

$\beta_1 \equiv$ TRUE

β_{21} = BIG

β_{22} = MEDIUM

β_{23} = SMALL

$\beta_{31} = \beta_{21} \wedge$ BLUE

$\beta_{32} = \beta_{21} \wedge$ RED

$\beta_{41} = \beta_{22} \wedge$ RED

$\beta_{61} = \beta_{32} \wedge$ SMOOTH

$\beta_{62} = \beta_{32} \wedge$ ROUGH

$\beta_{71} = \beta_{23} \wedge$ SMOOTH

$\beta_{72} = \beta_{23} \wedge$ ROUGH

Figure 6.3

Among the generators there is one, ALL, with in-arity zero and one out-bond identically true. It is the greatest one according to the order relation. There is another one, IND for individual, which is the smallest one. It has no out-bond and one in-bond. The latter one is the truth function of a particular element $x \in X$.

As an example see Figure 6.3 for which G = {ALL, SIZE, COLOR, TEXTURE, IND} with this ordering.

7. Partial Order Connection Type.

We now choose the Σ-type such that the directed arrows in a σ define a partial order with the usual axioms. There is a particularly important case defined as follows.

Case 7.1 (operator configurations). Let the generators be as in Case 1.3.18 (general operators) and with in-arity ν if there are ν inputs for a generator. The corresponding bond values are the domains $X_1, X_2, \ldots, X_\mu$ in the appropriate order. Similarly the out-arity is μ if there are μ-outputs and the bond values are $Y_1, Y_2, \ldots Y_\mu$. A connection σ is allowed if the directed arrows in it form a partial order. The bond relation ρ is taken to be inclusion so that $\beta\rho\beta'$ if $\beta \subseteq \beta'$.

Let us look as an example at Figure 7.1 where we have used as generators addition(+), multiplication(x), squaring(sq), assignment to the value 1, natural logarithm(ln), in each case with bond values. The bond values appearing in the figure are R = real line, R_+ = non-negative reals, $R_+ +1$ = reals not smaller than one and the one point set {1}. The σ happens to be a tree in this figure, but parallel branches e.g. can very well occur as long as the partial order connection type is not violated.

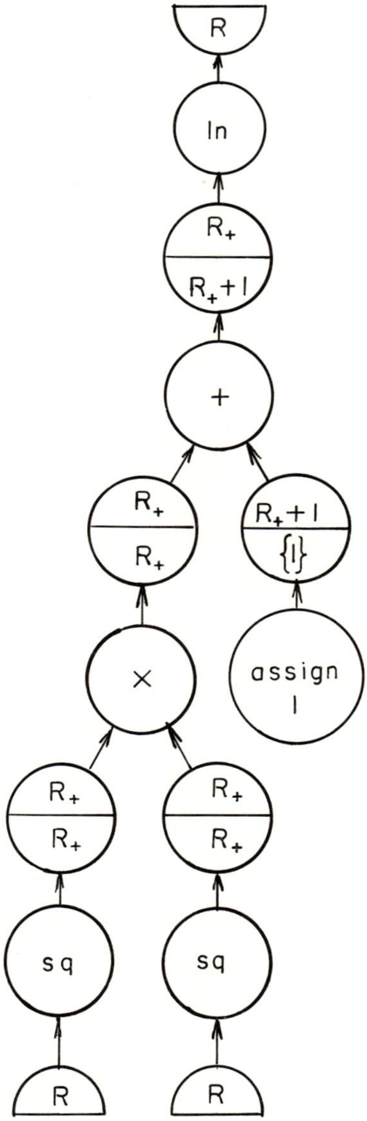

$\omega_{in}(c) = 2$

$\omega_{out}(c) = 1$

$n = 6$

Figure 7.1

The interpretation of such regular configurations is that they represent possible computing schemes (not allowing looping): this is apparent since a domain of one computing module g includes the range of a preceding module.

Remember that for this sort of generator the similarity transformations affect only assignment operators, so that in the figure the only effect would be to add positive numbers to the value 1 in the right-most assignment generator and change its bond values correspondingly.

8. Other Connection Types.

We shall present a few cases with configurations with other Σ-structure here; more will appear later. First let us look at family "trees". If both paternal and maternal descent are being traced, family "trees" are not trees in the usual sense if the generator is the marriage. The marriages will be assumed to be monogamous and we shall only consider cases where a subject is married at most once during his or her lifetime. The out-arity is two, the first bond value equal to ♂, the second to ♀. The in-arity is arbitrary with bond values ♂ and ♀.

Case 8.1 (family patterns). Generators as above, and for any σ we shall insist that it has no loops, but the further restrictions in Σ will be left unspecified till later when different mating rules are discussed. The bond relation ρ should mean equality.

A typical regular configuration is shown in Figure 8.1. It should be mentioned that alternatively one could let the generators be the subjects rather than the marriages between them; this will result in another structure.

Now let the generators be features g from a family

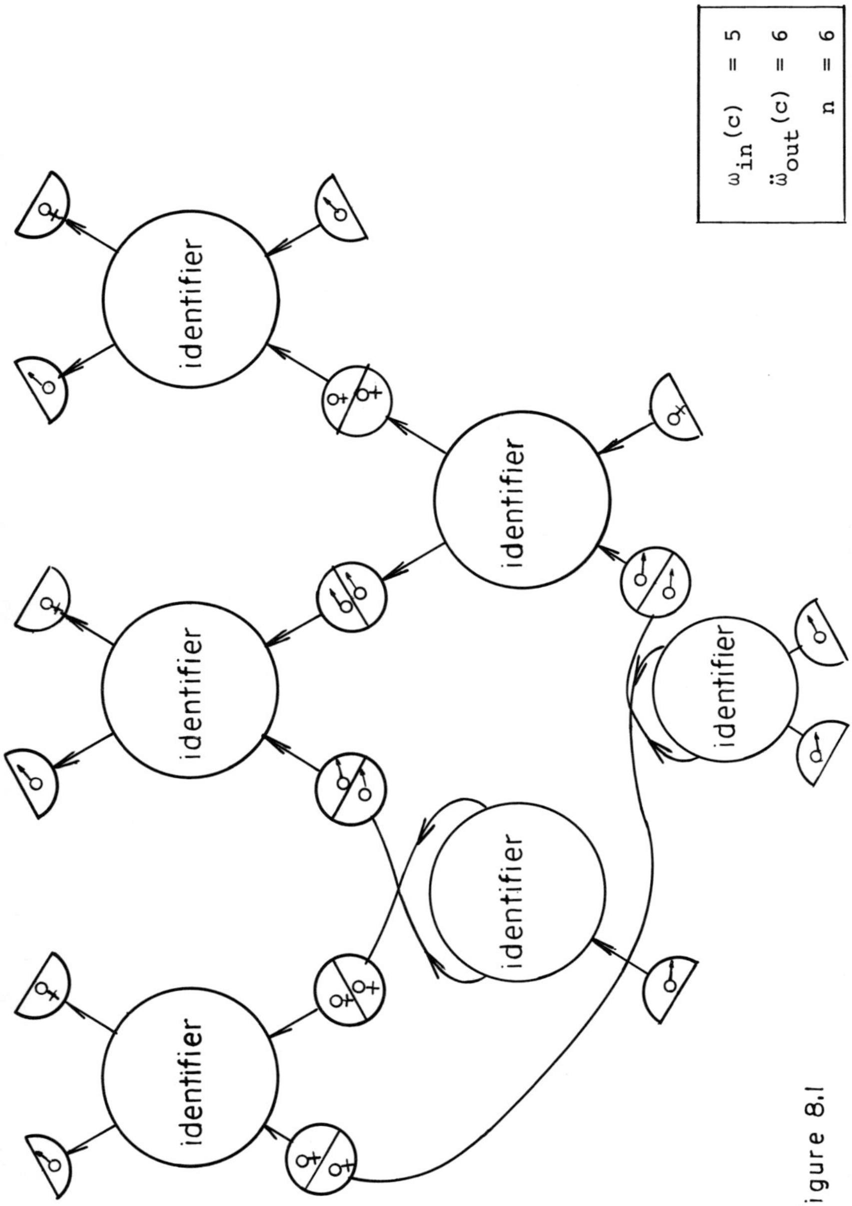
identifier
identifier
identifier
identifier
identifier
identifier
$\omega_{in}(c) = 5$
$\ddot{\omega}_{out}(c) = 6$
$n = 6$

Figure 8.1

$G = \mathscr{F}$ (see Case 1.3.6) and with generator classes G^{α}.

Case 8.2 (feature configurations of order p). Let any σ be of the following form. It allows only for $n = kp$ generators in a configuration, where k is a natural number, and the connections are all possible ones in each of k subconfigurations with p generators in each. The arity of any g is $2(p-1)$, with $p-1$ in-bonds and $p-1$ out-bonds, the value of each is the generator index α of g. The bond relation ρ means "not equal to". A typical admissible configuration would look as Figure 8.2 where $p=4$, $k=3$, and $\alpha_1,\alpha_2,\alpha_3,\alpha_5$ are dif- generator indices and similarly $\alpha_1,\alpha_3,\alpha_5,\alpha_6$ and $\alpha_1,\alpha_2,\alpha_4,\alpha_5$.

In this case the arity for configurations in $\mathscr{C}(\mathscr{R})$ is always zero. We have not indicated the bonds in the figure and the connections are all bi-directional. It follows that $c_1 \sigma c_2$ is always in $\mathscr{C}(\mathscr{R})$ if both c_1 and c_2 are.

It will be seen in Section 3.9 that the interpretation of the bond relation ρ is that inconsistencies and/or superfluous specifications be avoided.

The next case starts with the masaics of Case 1.3.7 but here we shall let G consist instead of continuous functions on the parallelograms in Figure 1.3.1. To any such generator we associate 8 bonds, both in- and out-bonds at the same time, so that strictly speaking the arity is 16 rather than 8, but since we shall have a symmetric ρ relation and symmetric σ's we prefer to work with bi-directional arrows and the arity will thus be taken to be 8 (double arrows). Consider the shaded parallelogram in Figure 8.3 with the four sides numered 1,3,5,7 and the four corners numbered 2,4,6,8. The bonds $\beta_1,\beta_2,\ldots,\beta_8$ are given as values of the function associated with g and

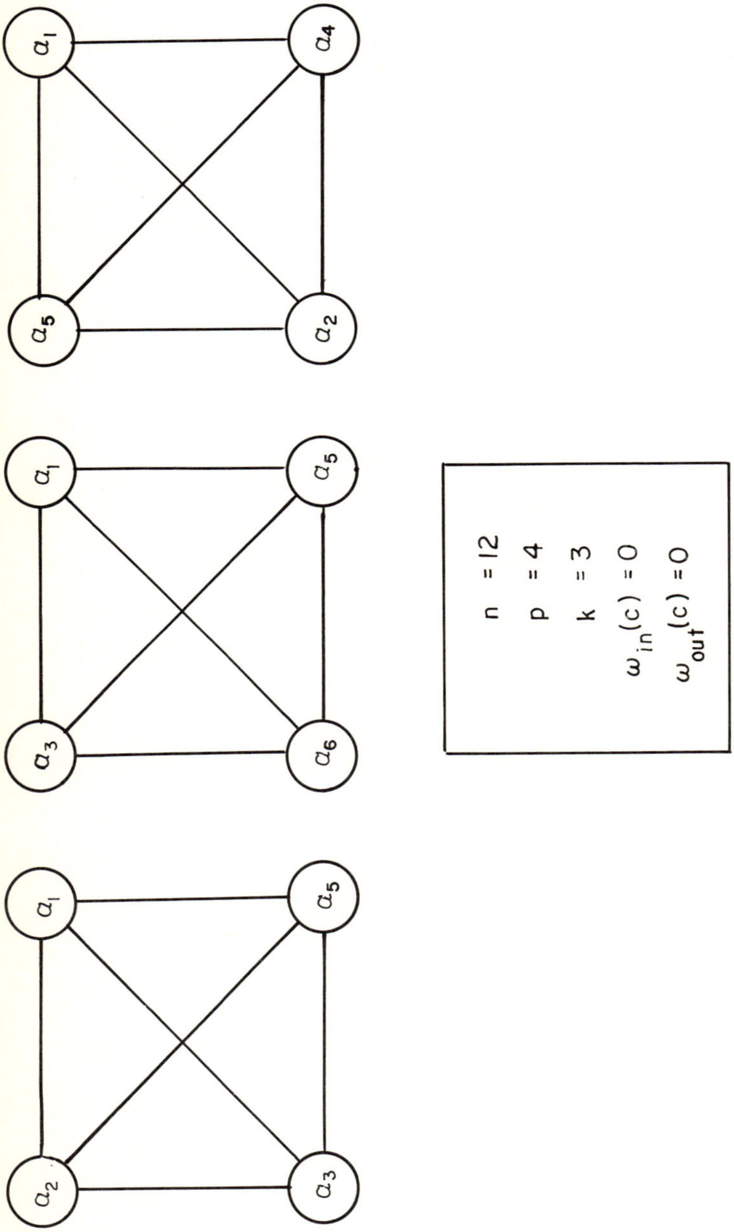
a_1
a_4
a_5
a_2
a_1
a_5
a_3
a_6
a_1
a_5
a_2
a_3
n = 12
p = 4
k = 3
$\omega_{in}(c) = 0$
$\omega_{out}(c) = 0$

Figure 8.2

restricted to the corner or side in question. The bond relation should mean equality.

Two generators with bonds $\beta_1, \beta_2, \ldots, \beta_8$ and $\beta_1, \beta_2, \ldots, \beta_8$ respectively may be joined by one or several of the connections

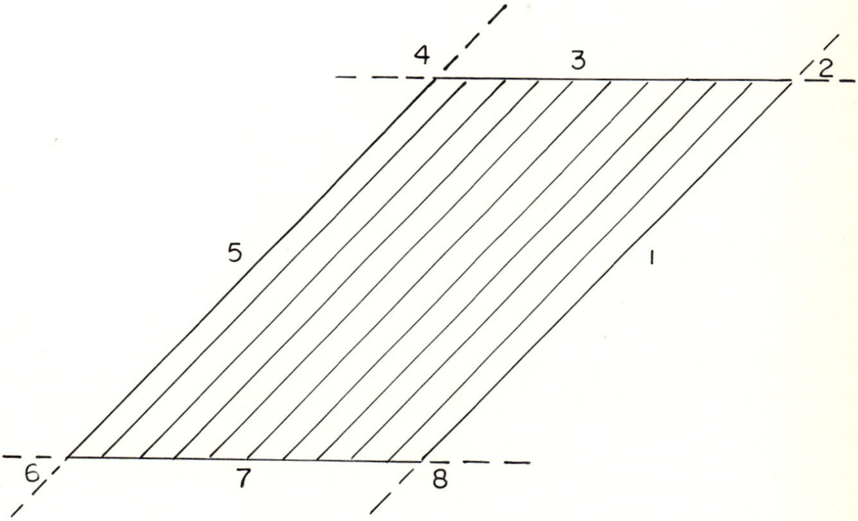

Figure 8.3

$$\left.\begin{array}{l} \beta_1 - \beta_5' \\ \beta_2 - \beta_6' \\ \beta_3 - \beta_7' \\ \beta_4 - \beta_8' \\ \beta_5 - \beta_1' \\ \beta_6 - \beta_2' \\ \beta_7 - \beta_3' \\ \beta_8 - \beta_4' \end{array}\right\} \qquad (8.1)$$

This means that a typical configuration has a Σ-type like the one in Figure 8.4. This is

<u>Case 8.3</u> (patterns on mosaics). G and $\mathcal{R}$ as defined above.

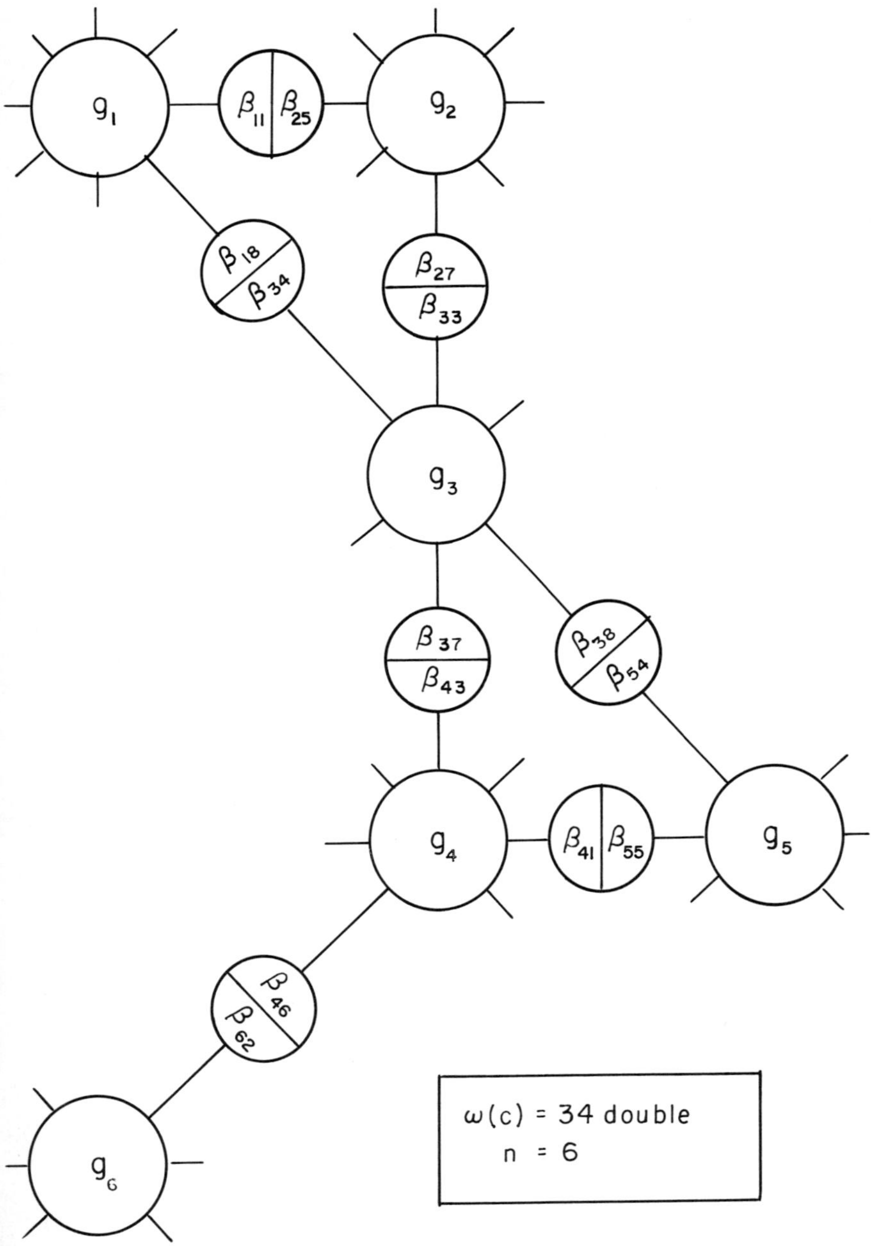
g1
β11 β25
g2
β18 β34
β27 β33
g3
β37 β43
β38 β54
g4
β41 β55
g5
β46 β62
g6
ω(c) = 34 double
n = 6

Figure 8.4

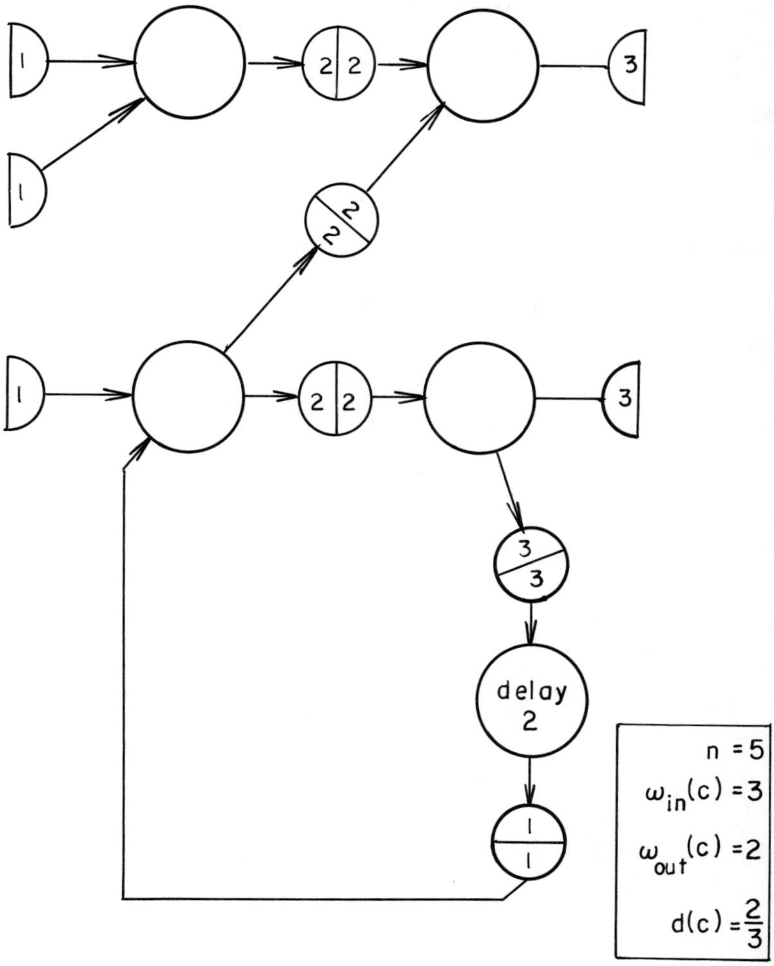

Figure 8.5

A case that we will return to several times and that will be studied in depth in Chapter 19 is made up of <u>formal neurons</u> as McCulloch-Pitts neurons, see e.g. Griffith (1971), or as linear operators.

<u>Case 8.4</u> (neural network configurations). The generators in

class G^{α} have in-arity n_{α} and out-arity m_{α}. The ratio $d_{\alpha} = m_{\alpha}/n_{\alpha}$ is called the <u>divergence</u> of the generator. The bond values are integers and for any $g \notin G^0$ $\beta_{out} = \beta_{in}+1$. G^0 consists of generators, the delay operator, for which $\beta_{out} \leq \beta_{in}-1$. The bond relation $\beta\rho\beta'$ says that $\beta \leq \beta'$. The connection type is arbitrary.

The ratio $d(c) = \omega_{out}(c)/\omega_{in}(c)$ is called the divergence of the configuration. A simple configuration of this type is shown in Figure 8.5.

9. Full Graph Structure.

In this section it will be assumed

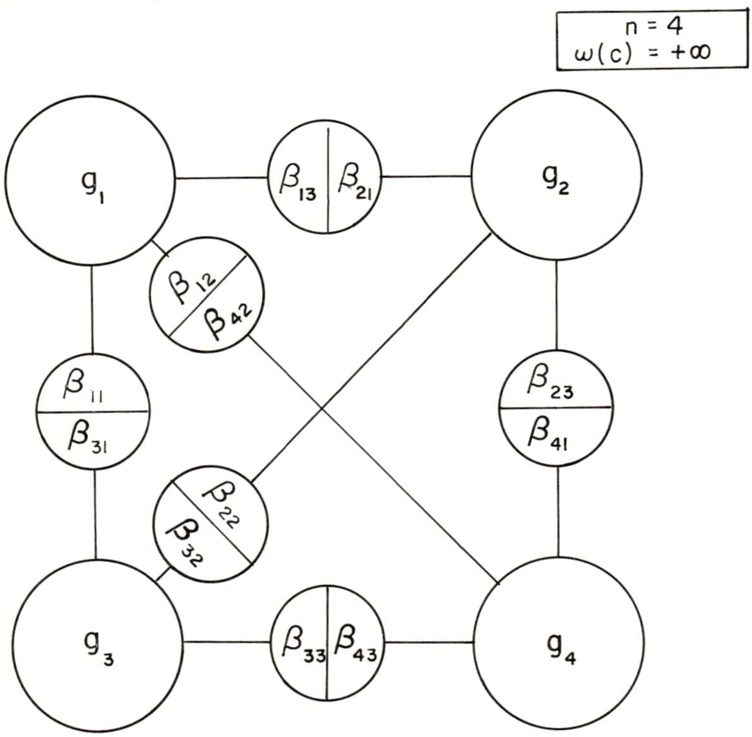

Figure 9.1

that $\omega(g) = \infty$ for all g, that ρ is symmetric and that for any $\sigma \in \Sigma_n$ all possible bi-directional arrows join any two distinct generators in the n-configuration. The typical 4-configuration hence looks as in Figure 9.1 where we have indicated only connected bonds and all arrows are bi-directional.

This case will be called Σ = FULL. A particularly simple case is when, for any generator g, all its bond values are the same. Let us now look at some cases where this connection type occurs.

Consider properties arranged into generator classes G^α, see Case 1.2.1, and with all the bond values of any g equal to its class index.

Case 9.1 (descriptions). With this G and Σ = FULL put ρ = DIFFERENT meaning that $\beta\rho\beta'$ if and only if $\beta \neq \beta'$. Such configurations are (partial) descriptions of objects.

The role of $\mathscr{R}$ in this case is to avoid contradictions and superfluous descriptors.

Now let G be as in Case 1.3.4, spheres as generators. We then get

Case 9.2 (excluded volume). With this set of generators and Σ = FULL, make the bond value of g equal to (x_1,x_2,x_3,r_α). The bond value carries all the information of the generator except for additional attributes such as "type of material" etc. Put $\beta\rho\beta'$ if

$$(x_1-x_1')^2 + (x_2-x_2') + (x_3-x_3')^2 > (r^\alpha+r^{\alpha'})^2, \qquad (9.1)$$

in other words if the two spheres exclude each other and do not overlap.

Two instances of contrast signs in the plane will also

be considered. The first one is

<u>Case 9.3</u> (joining contrast generators). Let the generators be continuous real valued functions defined on closed sets domain (g), with the bond values $\beta = (d,\gamma)$ where d is the domain of g and γ is the restriction of the function to the boundary ∂d. With the connection type Σ = FULL put

$$\beta\rho\beta' \leftrightarrow (d \cap d') \subset (\partial d) \cap (\partial d') \text{ and } \gamma \text{ and } \gamma' \text{ coincide on } (d \cap d'). \tag{9.2}$$

The second instance is similar.

<u>Case 9.4</u> (joining overlapping contrast generators). Let the generators be as in Case 9.3 except that the bond values now carry full information about the function. Same Σ but put

$$\beta\rho\beta' \leftrightarrow g \text{ and } g' \text{ coincide on the intersection } d \cap d'. \tag{9.3}$$

For future references we also mention the following cases with Σ = FULL

<u>Case 9.5</u> (patterns in spectrography). With $X = R'$ and G^α as in Case 1.3.21 with S as the direct product of translations on X and multiplication of measures with positive scalars. In each G^α a spectral measure should be uniquely specified by A,m,w where A is the total variation, m a location parameter and $\underline{w}$ a scale parameter. The bond relation says that $(m,w) \neq (m',w')$.

<u>Case 9.6</u> (polynomial patterns). Let $X = R'$ and G^α consist of all Ax^α, A real. S means multiplication with non-zero

scalars and ρ that $\alpha \neq \alpha'$.

Case 9.7 (orthogonal polynomial patterns). Same as previous case except that the generators are orthogonal functions.

A two-dimensional case will also be encountered later on.

Case 9.8 (star-shaped patterns). Let G consist of directed segments (x_0, y_0, x_1, y_1) and with the bond relation

$$x_0 = x_0',\ y_0 = y_0',\ \frac{y_1 - y_0}{x_1 - x_0} \neq \frac{y_1' - y_0'}{x_1' - x_0'}\ . \tag{9.4}$$

10. Probabilities on $\mathscr{C}(\mathscr{R})$.

Once $\mathscr{C}(\mathscr{R})$ has been defined, the question arises whether it is meaningful to treat all the regular configurations on an equal basis, or whether some of them should be given greater weight than others. In other words, how should one introduce measures on configuration space?

To do this we will have to specify the probability distribution in one way or another, or at least approximate it. If we have no knowledge about the random mechanism that gives rise to the measure, then we can only be empirical about it and count how many observed configurations we found, classifying them according to some attribute. It is more appealing, however, to try modelling the probability measure starting from first principles. As always in probabilistic modelling, one would attempt to express the stochastic elements in primary variables with simple distribution properties: stochastically independent, or equally distributed, or both.

This leads us to metric pattern theory: the combinatory aspect of the theory leads naturally to certain probabilistic consequences that we will begin to explore in this section.

Consider the space $\mathcal{C}_n$ and a simple random source of generators (see Section 3.4) so that an a priori distribution q is given on G. On the product space $G^n = G \times G \times \cdots \times G$ let q^n be the product measure with the marginal distribution q along each coordinate.

If $q^n\{\mathcal{C}_n(\mathcal{R})\} > 0$ we define the probability distribution $q^n_{\mathcal{R}}$ on the collection of regular configurations as the conditional measure

$$q^n_{\mathcal{R}}\{W\} = q^n\{W | \mathcal{C}_n(\mathcal{R})\} = \frac{q^n\{W \cap \mathcal{C}_n(\mathcal{R})\}}{q^n\{\mathcal{C}_n(\mathcal{R})\}} . \qquad (10.1)$$

Here we have indexed the n generators g_i appearing in the admissible n-configurations in a fixed manner that will be chosen according to what connection type is used. With this indexing we then look at $(g_1, g_2, \ldots, g_n)$ as a point in G^n and W is a subset in G^n.

Of course W must be in an appropriate σ-algebra but since the measure theoretic questions are easy to settle in most concrete cases we will meet, this will not be discussed further at this point.

Case 10.1 (excluded volume). Let X be R^3 with spheres as generators, obeying the restrictions of Case 9.2 and with the spheres restricted to X. Define q by letting x, y, z and r be independent, with x, y and z continuously distributed over R and r having a positive density over $(0,1)$. Then $q^n\{\mathcal{C}(\mathcal{R})\} > 0$ and (10.1) makes sense.

In the special case when the configurations have no internal structure so that no restrictions apply, we have simply $q^n\{\mathcal{C}_n(\mathcal{R})\} = 1$ so that $q^n_{\mathcal{R}} = q^n$.

Let us assume for the moment that G is denumerable, with the generators $g_1, g_2, g_3, \ldots$ and that $q(g_\nu) = q_\nu = 0$. The latter means no essential restriction since, if some $q_\nu = 0$, we would simply remove g_ν from the generator space. Also, let there be at least one admissible configuration of length n. Say that we consider condigurations of n generators with fixed internal structure specified through the connection σ and say that bond structure says that $A(g_i) = A(g_j)$ if i connects with j via σ and A is some given agreement function. To find $q^n_{\mathscr{R}}$ we calculate (10.1) for any given $c = (g_{i_1}, g_{i_2}, \ldots, g_{i_n})$. The numerator of this expression is equal to $q(g_{i_1}) \cdot q(g_{i_2}) \cdots q(g_{i_n})$ if $c \in \mathscr{C}_n(\mathscr{R})$ and 0 otherwise. But $c \in \mathscr{C}_n(\mathscr{R})$ if and only if $A(g_{i_\nu}) = A(g_{i_\mu})$ for any connected pair. This means that we have proved the following.

For restricted configuration spaces the probabilities are given by

$$q^n(c) = Kq_{i_1} q_{i_2}, \ldots, q_{i_n} \prod \text{truth value } \{A(g_{i_\nu}) = A(g_{i_\mu})\}, \tag{10.2}$$

where the product is over all the connections in the configuration and where K is a normalization constant to make $q^n(\mathscr{C}_n(\mathscr{R})) = 1$. In the special case where $A \equiv \text{TRUE}$ the above reduces to

$$q^n(c) = Kq_{j_1}^{n_1} q_{j_2}^{n_2} \cdots q_{j_\ell}^{n_\ell} \tag{10.3}$$

where $n_1, n_2, \ldots, n_\ell$ are the numbers in the decomposition of c and $g_{j_1}, g_{j_2}, \ldots$ the generators appearing in the decomposition.

Now turn to the case when q is a continuous measure over G (more than denumerably infinite). We then encounter the difficulty that $q^n(\mathscr{C}_n(\mathscr{R}))$ may be zero and (10.1) is meaningless. To clarify this situation we shall start with a special case that illuminates what can happen generally. Say that Σ = FULL and $A(g) = g$ for all g.

First, it is clear that in practice, agreement $g = g'$ must be expressed in terms of accuracy rather than exact equality. To calculate the conditional probabilities we must interpret the condition as satisfied with some tolerated error, and then let this tend to zero in some specified manner to get a limiting probability. Say that the generators are real numbers and that q is given through a continuous and positive density f on some finite interval (a,b) on the real line. We shall say that the generators $g_1, g_2, \ldots, g_n$ satisfy the agreement with the tolerance $\varepsilon > 0$ if

$$(g_1, g_2, \ldots, g_n) \in Q_\varepsilon \subset R^n. \tag{10.4}$$

It is natural to require that the sets Q_ε be monotonically decreasing as ε decreases to zero with the diagonal set as limit:

$$\lim_{\varepsilon \downarrow 0} Q_\varepsilon = \{g_1, g_2, \ldots, g_n \mid g_1 = g_2 = \cdots = g_n\}. \tag{10.5}$$

But this is not enough. We must also assume some similarity property and we shall do this by the condition that $(g_1+h, g_2+h, \ldots, g_n+h)$ belong to Q_ε if $(g_1, g_2, \ldots, g_n)$ does. The argument behind this choice goes as follows. Say that the values of g represent measurements such as locations where there is no preferred origin. It may then be natural to take the translations in $G = R$ as the similarity transformations and we would naturally ask for translation invariance as above.

On the other hand there is nothing unique about this choice. Say instead, that the measurements represent size, so that $G = R_+$. Here, there is a preferred origin, $g = 0$, and a more natural similarity would be the group of scale changes $g \to gk$ for $k > 0$. Then we would ask that if $(g_1, g_2, \ldots, g_n) \in Q_\varepsilon$ so should $(kg_1, kg_2, \ldots, kg_n)$, so that Q_ε would be formed from rays from the origin rather than from lines parallel to the diagonal as above. We shall treat the translation case here since it brings out the main difficulty clearly. The geometric interpretation of the condition is that Q_ε forms a cylinder about the diagonal with the same cross-section everywhere. Otherwise we would require different accuracy at different points of the diagonal and the result will be different. To calculate what corresponds to the expression in (10.1) we have to consider the conditional probability

$$\frac{\displaystyle\int_{Q_\varepsilon \cap \{\alpha < g_1 < \beta\}} f(g_1) f(g_2) \cdots f(g_n) dg_1 \cdots dg_n}{\displaystyle\int_{Q_\varepsilon} f(g_1) f(g_2) \cdots f(g_n) dg_1 \cdots dg_n} \tag{10.6}$$

where $a \leq \alpha < \beta \leq b$. We can write both numerator and denominator of (10.6) as

$$\int_{\text{interval}} f(g_1) dg_1 \int \cdots \int_{(g_1 - g_1^0) + Q_\varepsilon} f(g_2) \cdots f(g_n) dg_2 \cdots dg_n \tag{10.7}$$

where interval $= (\alpha,\beta)$ in the numerator and (a,b) in the denominator. We have used the notation Q'_ε for the subset

$$Q'_\varepsilon = \{g_2, g_3 \ldots, g_n \mid (g_1^o, g_2, \ldots, g_n) \in Q_\varepsilon\} \subset R^{n-1}. \tag{10.8}$$

The inner integral of (10.7) can be bounded from above and below by

$$(f(g_1) \pm \delta)^n m(Q'_\varepsilon) \tag{10.9}$$

where m is Lebesgue measure in R^{n-1} and δ can be made arbitrarily small if ε is chosen small enough. With these bounds for both numerator and denominator it follows that (10.6) has the limit, as $\varepsilon \to 0$,

$$\frac{\int_\beta^\alpha f^n(g)\,dg}{\int_a^b f^n(g)\,dg}. \tag{10.10}$$

We can immediately extend this to the case where n_1 generators are constrained to be equal, also n_2 others, and so on. We then get

<u>Theorem 10.1</u>. <u>For configurations of length n and with the given internal structure we have under the stated conditions that the induced probability distribution on</u> $\mathscr{C}_n(\mathscr{R})$ <u>is given by the marginal densities over the constrained generators</u>

$$\frac{f^{n_i}(g)}{\int_a^b f^{n_i}(g)\,dg}. \tag{10.11}$$

If the constraint classes are large, $n_i \gg 1$, then the

distribution (10.11) become more and more concentrated:

Corollary. *If, in addition to the conditions of Theorem 10.1, the function f has a unique maximum at $g = g'$, then the induced probability measure converges weakly to the one with all the mass in the element g'.*

This follows from well-known properties of generalized means (see, e.g., Hardy, Littlewood and Pólya (1934), p. 143).

This means that *increased algebraic dependence leads to higher probabilistic concentration*.

Another question that arises in this context is: when does the algebraic restriction have no effect on the marginal distributions? The answer is given easily.

Corollary. *Under the conditions of Theorem 10.1 the marginal distributions remain unchanged if and only if f takes only the values 0 and K', where K' is some constant.*

Proof. The marginal distribution is unchanged if and only if (see (10.11))

$$f(g) = \frac{1}{K} f^n(g) \tag{10.12}$$

where K is the integral in the denominator. Hence

$$f(g)[1 - \frac{1}{K} f^{n-1}(g)] = 0 \tag{10.13}$$

which implies either that $f(g) = 0$ or that $f(g) = K^{1/(n-1)} = K'$.

The probability measure on $\mathscr{L}_n(\mathscr{R})$ induced by q will be denoted $P = q/\mathscr{L}_n(\mathscr{R})$.

It is now possible to relax the conditions of Theorem 10.1 considerably. In particular, one can allow the support of the generators to be the whole real line, or, more interestingly, the whole or part of finite Euclidean space. In the latter case we may have agreement functions that involve only some of the coordinates of the vector. One condition that cannot be relaxed without corresponding change in the results is the translation invariance of the tolerance region Q_ε. We can use regions like

$$Q_\varepsilon = \{g_1, g_2, \ldots, g_n \mid \Sigma(g_i - \frac{1}{n}\Sigma g_j)^2 < \varepsilon\} \tag{10.14}$$

or

$$Q_\varepsilon = \{g_1, g_2, \ldots, g_n \mid \max_{i,j} |g_i - g_j| < \varepsilon\} \tag{10.15}$$

since they are translation invariant. If, however, we use other invariances the result in (10.11) will not hold. The case with scale changes can be cited as an example of this but we shall not pursue this any further here, nor shall we attempt to generalize the result to more general probability distributions.

Returning to the translation-invariant Q_ε we quote as an example what happens to normal distributions.

<u>Case 10.2</u> (constrained normal distributions). Let the generators $g_1, g_2, \ldots, g_n$ be real-valued and $N(m_i, \sigma_i^2)$ and belong to the same decomposition class. Then the marginal distribution of the admissible configuration is also normal.

<u>Proof</u>. We apply (10.11) but extended to infinite range and with different marginal distributions for the generators. This will lead, with essentially the same proof but with a uniformity argument added to it, to a similar expression but with the power f^n replaced by a product so that, in the present case, we get

$$\text{const. } \exp\left(-\frac{1}{2}\sum\frac{(g-m_i)^2}{\sigma_i^2}\right). \tag{10.16}$$

This can be rewritten in the standard form of a normal frequency function $N(m,\sigma^2)$ with the parameters

$$m = \frac{\sum \frac{m_i}{\sigma_i^2}}{\sum \frac{1}{\sigma_i^2}}, \quad \frac{1}{\sigma^2} = \sum \frac{1}{\sigma_i^2}. \tag{10.17}$$

Note that the resulting variance σ^2 is at most equal to all the original σ_i^2.

To see how the bond structure affects stochastic dependence we shall consider the case when Σ = LINEAR ORDER, ρ = EQUALITY, and the in- and out-bonds of a generator are real-valued and denoted by y and z. Q is given by a bivariate normal distribution with means zero, standard deviations one, and correlation coefficient ρ, $|\rho| < 1$.

For a $c \in \mathscr{C}_n(\mathscr{R})$, with bond values as indicated in Figure 10.1, we get the joint frequency function with the same sort of argument as in the proof of Theorem 10.1

$$f(x_1,x_2,\ldots,x_{n+1}) \propto \exp\left(-\frac{1}{2(1-\rho^2)} F_n\right) \tag{10.18}$$

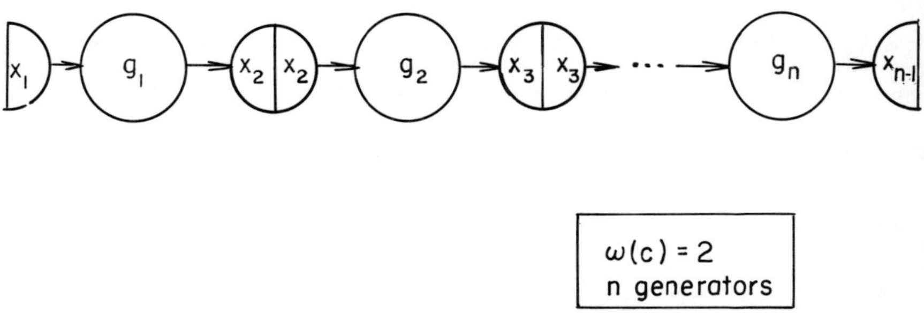

Figure 10.1

with the quadratic form

$$F_n = x_1^2 - 2\rho x_1 x_2 - 2\rho x_n x_{n+1} + x_{n+1}^2 + 2\sum_{i=2}^{n-1}[x_i^2 - \rho x_i x_{i+1}] \tag{10.19}$$

from which the second order moments can be calculated directly.

Before doing this however, let us look at P as defined in $\mathscr{C}_{n+\nu}(\mathscr{R})$, $\nu > 0$, which of course has the same form as in (10.18), (10.19) but with n replaced by $n + \nu$. Let us say that we are not interested in the last bond value $x_{n+\nu+1}$. Then we have to integrate out this variable from the joint frequency function which will lead to an expression like $F_{n+\nu-1}$ but where the coefficient 1 for $x_{n+\nu}^2$ is replaced by some constant c_1. Now we can repeat this procedure, and to get a recursive relation for the c_p's we write

$$\begin{aligned} &2x_{n+\nu-p}^2 - 2\rho x_{n+\nu-p}\cdot x_{n+\nu+1-p} + c_p x_{n+\nu+1-p}^2 \\ &= \left(2 - \frac{\rho^2}{c_p}\right) x_{n+\nu-p}^2 + \ell_p \left(x_{n+\nu+1-p} - \frac{\rho}{c_p} x_{n+\nu-p}\right)^2. \end{aligned} \tag{10.20}$$

But integrating out $x_{n+\nu-p}$ we get the recursion

$$c_{p+1} = 2 - \frac{\rho^2}{c_p}, \; p \geq 0; \quad c_0 = 1. \tag{10.21}$$

All the c_p will be in the interval [1,2] and since the right hand side of (10.21) is an increasing function of c_p they form a bounded, increasing sequence. With the usual fixed point argument it is seen that $\lim_{p\to\infty} c_p = c_\infty$ satisfying the equation

$$c_\infty = 2 - \frac{\rho^2}{c_\infty} \tag{10.22}$$

so that, since only the root greater than one is possible,

$$c_\infty = 1 + \sqrt{1-\rho^2}\,. \tag{10.23}$$

In the same way if we regard a configuration that is assumed to continue r steps to the left we can apply exactly the same argument because of the symmetry of the problem. This is so although the connection type Σ is not symmetric here, but because of the fact that the dependence between the in-bond and out-bond is symmetric.

Combining these results we get

Theorem 10.3. The measure induced on $\mathscr{C}_{n+p+r}(\mathscr{R})$ has the property that when we consider the regular configuration c such that $c' = c_r, c, c_p \in \mathscr{C}_{n+p+r}(\mathscr{R})$ has a joint density of the form (10.18) but with F replaced by

$$F_n^{p,r} = c_r x_1^2 - 2\rho x_1 x_2 - 2\rho x_n x_{n+1} + c_p x_{n+1}^2 + 2 \sum_{i=2}^{n-1} [x_i^2 - \rho x_i x_{i+1}] \tag{10.24}$$

where c_p is defined by (10.21). In the limit when $p \to \infty$, the value of c_p tends to $1 + \sqrt{1 - \rho^2}$.

Here we have considered c' as the concatenation of c_r, c, and c_p.

Rewriting (10.24) we get after some algebraic manipulations exploiting (10.21)

$$F_n^{p,r} = (c_r + c_{n+p-1} - 2)x_1^2 + \sum_{i=1}^{n} c_{n+p-i} [x_{i+1} - \frac{\rho}{c_{n+p-i}} x_i]^2 \tag{10.25}$$

which has some important consequences.

First it follows that our P has the Markov property: the linear connection type together with the bond relation ρ = EQUALITY leads to Markovian dependence.

Second we can use the stochastic independence of

$$\left.\begin{aligned} y_{i+1} &= x_{i+1} - \frac{\rho}{c_{n+p-i}}\, x_i ; \quad i = 1,2,\ldots,n \\ y_1 &= x_1 \end{aligned}\right\} \tag{10.26}$$

with the variances

$$\left.\begin{aligned} \operatorname{Var}(y_{i+1}) &= \frac{1-\rho^2}{c_{n+p-i}} ; \quad i = 1,2,\ldots,n \\ \operatorname{Var}(y_1) &= \frac{1-\rho^2}{c_r + c_{n+p-1}^{-2}} . \end{aligned}\right\} \tag{10.27}$$

In particular for the limiting case $p = r = \infty$ we get the simple result, using (10.22),

$$\begin{aligned} \operatorname{Var}(x_i) &= \frac{1-\rho^2}{c_\infty - 1} = \sqrt{1-\rho^2} \\ \operatorname{Corr}(x_i, x_{i+h}) &= \left(\frac{\rho}{c_\infty}\right)^h = \left(\frac{\rho}{1+\sqrt{1-\rho^2}}\right)^h, \quad 0 \le h \le n. \end{aligned} \tag{10.28}$$

This means that <u>the $\mathscr{R}$-structure has led to the decrease of the variance</u> from 1 to $\sqrt{1-\rho^2}$. Also <u>the correlation coefficient between two neighboring bond values</u>, $h = 1$, <u>is</u> $\rho/c_\infty < \rho$, <u>so it has also decreased</u>.

A systematic study of the influence of bond constraints on the stochastic dependence seems motivated by these results.

To illustrate the possibilities let us consider a more complex connection type. Let the generators be enumerated by two integers, k and ℓ, so that we can visualize them as situated at the lattice points in the plane. Rather than starting with finite configurations and applying a limiting argument as above we shall let the configuration be infinite from the beginning, all integral values of k and ℓ occur in the configuration. Say that all generators have the same bonds but of course the bond values may differ. For ρ we take EQUALITY just as before. Assume also that the connections are invariant with respect to translations. The arity of the generators will be denoted by $\omega = \omega(g)$, counting bi-directional arrows.

The generator may look as in Figure 10.2(a) and the connections as in (b) where the bonds have been left out in order not to clutter up the picture.

Let us assume that the bond values for a particular generator are normally distributed with means zero, standard deviations one, and completely symmetric correlations. The latter

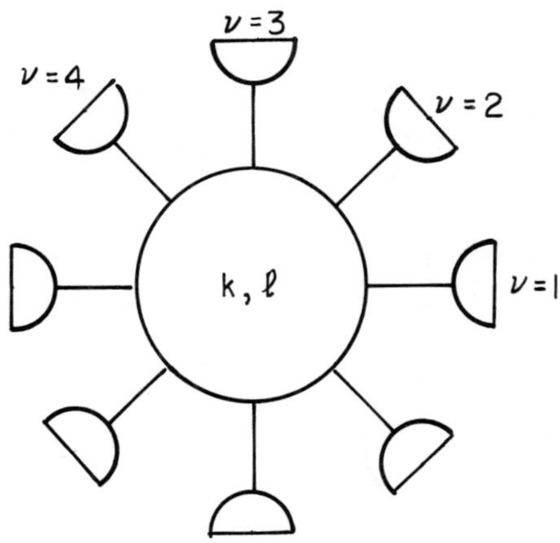

Figure 10.2(a)

$\omega(c)$ = 8 double arrows

means that all the correlation coefficients have the same value r. The correlation matrix is then

$$R = \left\{ \begin{matrix} 1 & r & r & \dots & r \\ r & 1 & r & \dots & r \\ \cdot & \cdot & \cdot & & \\ r & r & r & \dots & 1 \end{matrix} \right\}, \qquad (10.29)$$

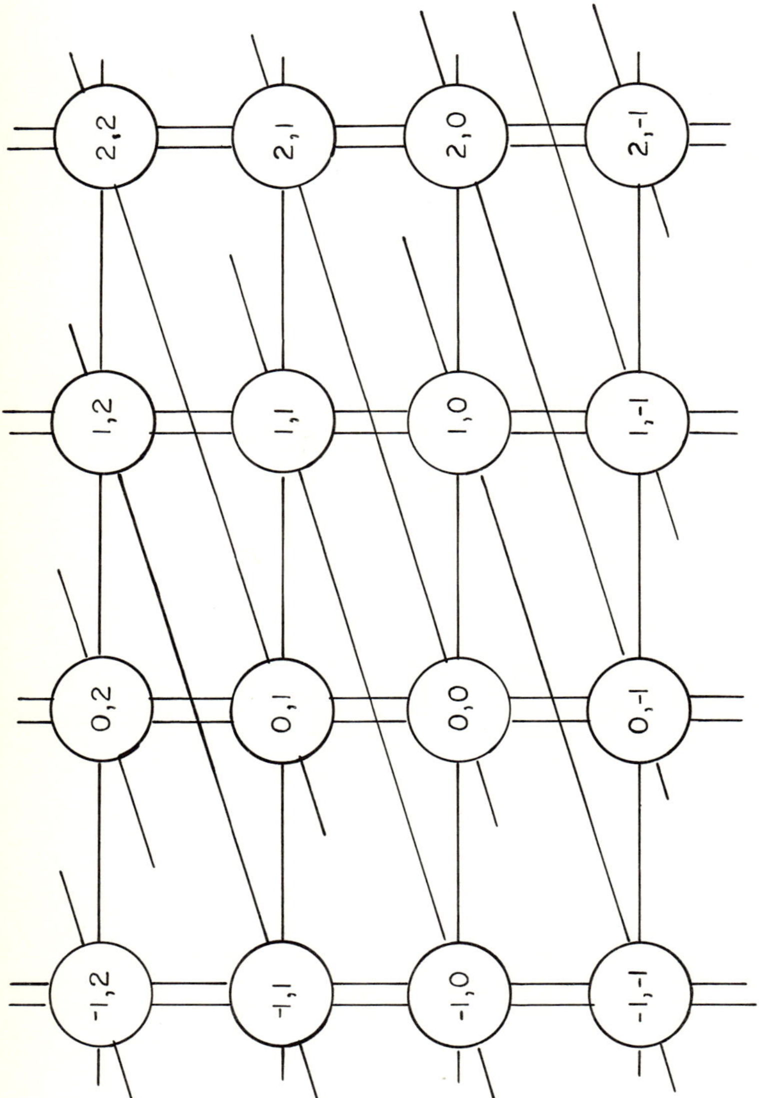

Figure 10.2(b)

a $\omega \times \omega$ matrix with the inverse

$$R^{-1} = \begin{Bmatrix} q_1 & q_2 & q_2 & \cdots & q_2 \\ q_2 & q_1 & q_2 & \cdots & q_2 \\ \cdot & \cdot & \cdot & \cdot & \cdot \\ q_2 & q_2 & q_2 & \cdots & q_1 \end{Bmatrix} \tag{10.30}$$

where

$$q_1 = \frac{1-2r+r\omega}{(1-r)(1-r+r\omega)}, \quad q_2 = \frac{-r}{(1-r)(1-r+r\omega)}. \tag{10.31}$$

In order that R be positive definite, as is necessary, it is well known that

$$1 > r > -\frac{1}{\omega-1}. \tag{10.31a}$$

Denote the bond values of the (k,ℓ) generator by $\beta_{k\ell}(\nu)$, where ν is an index as indicated in Figure 10.2(a). Because of the bonding we need only enumerate half the bonds (symmetry!), $\nu = 1,2,\ldots,p = \omega/2$. We consider the β's as forming a p-dimensional stationary Gaussian stochastic vector process with means zero and cross covariances $r_{k-k'\ell-\ell'}(\nu,\nu')$ which will now be calculated.

At this point, our argument is partly heuristic in that we do not start with finite configurations and then pass to the limit; this defect ought to be removed by a rigorous derivation.

In the quadratic form that will appear in the exponent we will have coefficients

$$q_{kk'\ell\ell'}(\nu,\nu') = \begin{cases} 2q_1 & \text{if } k = k',\ \ell = \ell',\ \nu = \nu' \\ q_2 & \text{if the bonds } (k,\ell,\nu) \text{ and } (k',\ell',\nu') \text{ are distinct but under the same generator} \\ 0 & \text{else} \end{cases} \tag{10.32}$$

Hence we have to solve the partial difference equation

$$\delta_{kk''}\delta_{\ell\ell''}\delta_{\nu\nu''} = \sum_{k',\ell',\nu'} r_{k-k'\ell-\ell'}(\nu,\nu')q_{k'k''\ell'\ell''}(\nu',\nu''). \tag{10.33}$$

But the cross covariances can be written as

$$r_{k-k'\ell-\ell'}(\nu,\nu') = \int_{-\pi}^{\pi}\int_{-\pi}^{\pi} \exp i[(k-k')u + (\ell-\ell')v] \; F_{\nu\nu'}(du,dv) \tag{10.34}$$

where $F(du,dv)$ is a matrix valued bounded measure on the square appearing as domain of integration in (10.34). The matrix values are non-negative symmetric matrices, see Doob (1953), p. 596.

On the other hand the function appearing on the left of (10.33) can also be written like this but with the special choice

$$F^{w}_{\nu\nu'}(du,dv) = \frac{1}{4\pi^2}\,\delta_{\nu\nu'}dudv. \tag{10.35}$$

Hence we get, by identifying the spectral measures,

$$\frac{1}{4\delta^2}\,\delta_{\nu\nu''}dudv = \sum c_{nm}(\nu,\nu')\exp i[nu+mv]F_{\nu'\nu''}(du,dv) \tag{10.36}$$

where the coefficients $c_{nm}(\nu,\nu')$ are given by

$$\left.\begin{array}{lll} c_{00}(\nu,\nu') = 2q_1 & \text{if} & \nu \neq \nu' \\ c_{nm}(\nu,\nu') = 2q_2 & \text{if } (n,m,\nu') \neq (0,0,\nu) \text{ but the} \\ & \text{bond } (n,m,\nu') \text{ borders on the} \\ & \text{generator } (0,0) \\ c_{nm}(\nu,\nu') = 0 & \text{else} \end{array}\right\} \tag{10.37}$$

Introducing the <u>Fourier transform of the connection type</u> through the matrix function

$$\phi_{\nu\nu'}(u,v) = \sum_{(n,m)\in\kappa} \exp i[nu+mv] \tag{10.38}$$

with κ = those (n,m) for which a bond (n,m,ν') borders on the generator $(0,0)$, we can solve (10.36) for the spectral measure

$$F(du,dv) = \frac{1}{4\pi^2}(2q_1I+q_2\phi(u,v))^{-1}dudv \tag{10.39}$$

where ϕ stands for the matrix function with element $\phi_{\nu\nu'}(u,v)$. This gives us the spectral measure and hence the cross-covariance sought.

We need a bound on the norm of the matrices $\phi(u,v)$ and get it from a standard matrix inequality

$$||\phi(u,v)|| \leq \max_{\nu} \sum_{\nu'} |\phi_{\nu\nu'}(u,v)| \leq 2(\omega-1) \tag{10.40}$$

But

$$\frac{|q_2|}{q_1} = \frac{|r|}{1-2r+r\omega} . \tag{10.41}$$

If $r > 0$ then

$$\frac{|q_2|}{q_1} = \frac{|r|}{1/(\omega-1)} . \tag{10.42}$$

On the other hand, if $r < 0$, using (10.31a), we get

$$\frac{|q_2|}{q_1} = \frac{-r}{1+(\omega-2)r} \leq \frac{-r}{1-\frac{\omega-2}{\omega-1}} = -r(\omega-1) < 1. \tag{10.43}$$

Hence, we can write

$$F(du,dv) = \frac{1}{8\pi^2 q_1}(I+\frac{z}{2}\phi(u,v))^{-1}dudv \tag{10.44}$$

where $z = q_2/q_1$. Since $|z| \leq 1/(\omega-1)$ we can define, for $r > 0$,

$$A(u,v) = (I+\frac{z}{2}\phi(u,v))^{1/2} \tag{10.45}$$

by the formal Taylor expansion of the square root function,

taking the positive branch at $z = 0$, for example. This matrix power series converges uniformly for real u and v and we can write

$$F = \left(\frac{1}{\pi\sqrt{8q_1}} \sum A(s,t) \exp i[su+tv]\right)^2 \tag{10.46}$$

which implies that we can write

$$\begin{aligned} \beta_{k\ell} &= (\beta_{k\ell}(1), \beta_{k\ell}(2), \ldots, \beta_{k\ell}(p)) \\ &= \sum_{s,t} A(s,t)\varepsilon_{k-s,\ell-t} \end{aligned} \tag{10.47}$$

where the ε-vectors and their components are all uncorrelated with variance $1/2q_1$.

The important part of the argument is the derivation of (10.44) and we sum it up as

<u>Theorem 10.4</u>. <u>For configurations as described above the covariance structure of the bond values</u> $\beta_{k\ell}(\nu)$ <u>is given by the spectral measure</u>

$$F(du,dv) = \frac{1}{8\pi^2 q_1} (I+\tfrac{z}{2}\phi(u,v))^{-1} du dv \tag{10.48}$$

<u>where</u> ϕ <u>is the Fourier transform of the internal structure defined by the connection type</u>; $r > 0$.

We must stress, however, that in contrast to the earlier case, this derivation is only heuristic since we have not proved convergence to a limiting distribution.

We now apply the same way of thinking to the bond structures appearing in finite-state languages and context-free languages, starting with the former.

Denote the absolute probabilities on G by $q_{ij}(x)$,

where i and j are states and x a terminal symbol. Then in $\mathcal{L}_n(\mathcal{R})$ the induced P-measure is given by

$$P_n(c) \propto \prod_{\nu=1}^{n} q_{i_\nu i_{\nu+1}}(x_\nu) \tag{10.49}$$

with obvious notation. Similarly, if c is embedded in $\mathcal{L}_{n+r+p}(\mathcal{R})$ as above we get a similar expression to (10.49) with $n+r+p+1$ bonds and we have to "sum out" the r first and the p last as well as the corresponding terminal symbols. Introducing the matrix

$$Q = \left\{\sum_x q_{ij}(x)\right\} \tag{10.50}$$

we can write the resulting probabilities

$$P_n^{p,r} \propto e^T Q^r e_{i_1} P_n(c) e_{i_{n+1}}^T Q^p e = a_{i_1}^{(r)} P_n(c) b_{i_{n+1}}^{(p)} \tag{10.51}$$

where the column vector e has 1's as coordinates, and e_s has 0's except for the sth coordinate which is one,

$$a_i^{(r)} = e^T Q^r e_i, \quad b_i^{(r)} = e_i^T Q^p e. \tag{10.52}$$

In $\mathcal{L}_n(\mathcal{R})$ we then have, if c' denotes the sub-configuration of c leaving out the last generator,

$$P_n(c|c') = \frac{q_{i_n i_{n+1}}(x_n)}{\sum_{i_{n+1}, x_n} q_{i_n i_{n+1}}(x_n)} \tag{10.53}$$

which depends only upon the last bond value i_n of c'. The connection type Σ = LINEAR ORDER induces the Markov property here just as it did in the previous case.

Similarly for context free languages the same argument shows that the induced probabilities can be regarded as obtained by successive choices of rewriting rules, the choices being conditionally independent, each one depending only upon the antecedent generator in the configuration. Hence, <u>the resulting measure belongs to a branching process caused by the connection type</u> Σ = TREE and ρ = EQUALITY. We now proceed to a more detailed examination of this case.

We shall use Greek letters, or sometimes numbers, for syntactic variables, the elements in N. Let us enumerate the set $\mathcal{R}$ of rewriting rules by an index r ranging from 1 to ρ. The set of rules rewriting ξ will be denoted by $\mathcal{R}_\xi$,

$$\mathcal{R} = \bigcup_{\xi \in N} \mathcal{R}_\xi . \tag{10.54}$$

As usual, we shall consider two derivations identical if they have the same generation tree.

<u>Case 10.3</u> (syntax-controlled probabilities). Given numbers p_r such that

$$1 > p_r > 0, \quad \sum_{r \in \mathcal{R}_\xi} p_r = 1 \quad \text{for any} \quad \xi \in N, \tag{10.55}$$

we associate the probability P(T) to any generation tree T:

$$P(T) = \prod_{r \in \Pi(T)} p_r \tag{10.56}$$

where $\Pi(T)$ is the set of rules used at the nodes of T counting multiple use of rules.

In this definition we have assumed that if some rule in $\mathcal{R}$ had a probability equal to zero it was removed from $\mathcal{R}$. Hence $p_r > 0$. Similarly, if there is only one derivation starting with $\xi \neq \sigma$ with positive probability (this probability must then be one) we delete ξ by replacing it on the right-hand side of all rules by its rewritten value. This means that $p_r < 1$ except possibly for those rules that rewrite σ.

To make this concrete, let us examine two examples, neither of them of independent interest but sufficient to indicate a difficulty that may otherwise be overlooked.

<u>Example 1</u>. Let us have only one terminal symbol a, and a single syntactic variable σ and the rules

$$\left.\begin{array}{ll} r = 1: & \sigma \to a \\ r = 2: & \sigma \to \sigma a \end{array}\right\} \tag{10.57}$$

so that $\rho = 2$. Assign the probability p to rule 1: $p_1 = p$, $0 < p < 1$. Then $p_2 = 1 - p_1 = 1 - p$. The general form of a terminating derivation consists of n applications of rule 2 followed by one applications of rule 1: $n > 0$. If this tree is denoted by T_n we get from (10.56)

$$P(T_n) = (1-p)^n p, \qquad n = 0,1,2,\ldots\ . \tag{10.58}$$

Since the set of all generation trees $\mathscr{T} = \bigcup_1^\infty T_n$, we get

$$P(\mathscr{T}) = \sum_0^\infty p(1-p)^n = 1. \tag{10.59}$$

Hence, we have a genuine probability distribution defined over the set $\mathscr{T}$ of terminating generation trees.

This is not always so, however. To show that, it is enough to take the same trivial set of grammatical strings $a\{a\}^* = \{a^m;\ m \geq 1\}$ as in Example 1 but generated by another grammar.

<u>Example 2</u>. Same as Example 1, except that we now have the two rules

$$\left.\begin{aligned} r = 1&: \quad \sigma \to a \\ r = 2&: \quad \sigma \to \sigma\sigma \end{aligned}\right\} \tag{10.60}$$

with probabilities p and $1 - p$ respectively. Look at e.g., T_{a^3}; this set consists of the two trees shown in Figure 10.3 to each of which we associate the probability $(1-p)^2p^3$. In

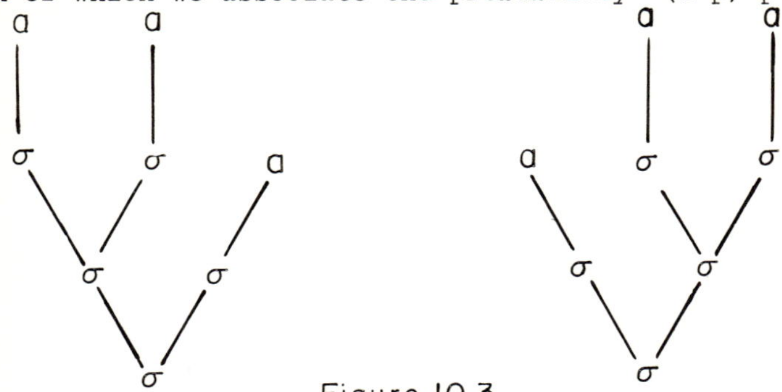

Figure 10.3

general, a derivation of a^i consists of i applications of rule 1 and of $i - 1$ applications of rule 2; the corresponding probability is hence $(1-p)^{i-1}p^i$. Adding all these probabilities, we get

$$P(\mathscr{T}) = \sum_{i=1}^{\infty} f_i(1-p)^{i-1}p^i \tag{10.61}$$

where f_i is the number of different generation trees leading to the string a^i. This function vanishes for $p = 0$ and for $p = 1$ we get the value $f_1 = 1$. It can be shown to be continuous in $[0,1]$ (but it is not analytic at $p = 1/2$), so that $P(\mathscr{T})$ will take on values strictly greater than zero and less than one if $0 < p < 1$. In order that we have a legitimate probability distribution over $\mathscr{T}$ we must, of course, require that $P(\mathscr{T}) = 1$; this example shows that this is not automatically true. We have to find conditions insuring that it is. To do this, a more sophisticated approach is needed.

It will be convenient to work with infinite generation trees. To the set $\mathscr{R}$ of rules we shall add the trivial rules $a \to a$ for each terminal. We associate the probability one to each such rule. This has the consequence that when we go on extending a generation tree <u>ad infinitum</u> a terminal symbol will never be rewritten as anything else but itself (with probability 1). We introduce the notion of level through Figure 10.4, which should be self-explanatory.

level 5	b		a	c		a	a
level 4	b		a	c		a	a
level 3	b		a	c		δ	a
level 2	b		σ		δ		a
level 1		ξ				η	
level 0			σ				

Figure 10.4

Denote by $\overline{\mathscr{T}}$ the set of all possible infinitely extended generation trees. Given a finite generation tree T, terminated or not, extending up to level ℓ, we introduce the subset $C(T)$ of $\overline{\mathscr{T}}$ consisting of all infinitely extended generation trees equal to T up to level ℓ. The set $C(T)$ is called a cylinder set with the base T of level ℓ. Now we introduce probabilities on the cylinder sets by assigning probabilities to their base according to (10.56), and extend the measure to the σ-algebra of subsets of $\overline{\mathscr{T}}$ generated from all cylinders. Although the measure obtained in this way is a legitimate probability measure, it is not quite what we want, since we are interested only in terminating derivations. In $\overline{\mathscr{T}}$ this corresponds to cylinders where, from some level on, all symbols are terminal words; this whole set we denote by F. We want to find criteria for $P(F) = 1$.

Consider an arbitrary generation tree and look at it at the ℓth level. Find the number $n_1(\ell)$ of occurrences on this level of the first syntactic variable, $n_2(\ell)$ of the second one, and so on. In this way we get a vector

$$n(\ell) = (n_1(\ell), n_2(\ell), \ldots, n_v(\ell)), \tag{10.62}$$

where v is the number of syntactic variables. We want to know if $n(\ell)$ eventually reduces (with probability one) to the zero vector; this means that the derivation has terminated.

We have now reduced our problem to one whose solution is known. Indeed, the vector $n(\ell)$ forms a multitype branching process (see Harris (1963), pp. 34-49). Enumerate the syntactic variables by the index $1,2,3,\ldots,r$, where 1 corresponds to the initial symbol. For the variable i consider the set $\mathcal{R}_i$ of rules and put

$$m_{ij} = \sum_{r \in \mathcal{R}_i} p_r n_j(r) \tag{10.63}$$

where $n_j(r)$ is the number of syntactic variables appearing on the right-hand side of rule r. Let M be the matrix

$$M = \{m_{ij};\ i,j = 1,2,\ldots,r\}. \tag{10.64}$$

We shall assume that M has no final classes (see Harris (1963), p. 46). It then follows from a theorem on branching processes that a derivation starting from an arbitrary syntactic variable will terminate eventually with probability one if and only if $\rho \leq 1$, where ρ is the positive eigenvalue of M that is greater in absolute value than any other eigenvalue.

We have then proved

Theorem 10.5. Under the given conditions the syntax-controlled probability model generates a legitimate probability measure over the set of terminating generation trees if $\rho \leq 1$ and M has no final class.

Returning to Example 2, we have just a single syntactic variable σ, so that $v = 1$, and the matrix M reduces to the real number

$$M = p \cdot 0 + (1-p)2 = 2 - 2p. \qquad (10.65)$$

Hence, if $p \geq 1/2$ the probability measure induced over the generation trees is legitimate.

From now on we shall assume that the conditions of Theorem 10.5 are satisfied, so that we are allowed to speak of the probability measure $P^{\mathscr{T}}$ over the set $\mathscr{T}$ of all generation trees, where $P^{\mathscr{T}}$ is induced by the syntax-controlled probabilities.

We use $P^{\mathscr{T}}$ to define a probability measure over the set of regular configurations generated by $\mathscr{T}$.

Definition 10.1. For an arbitrary grammatical string $I \in \mathscr{C}(\mathscr{R})$ we put

$$P(I) = \sum_{T \in T_I} P(T). \qquad (10.66)$$

This probability measure is denoted by P^{Φ}. T_I denotes the set of all generation trees of I.

This implies that a string is assigned positive probability if and only if it is grammatical. In Chapter 3, the I's will appear as the images in this particular context.

<u>Remark 1</u>. If the grammar is unambiguous we introduce P^{Φ} simply by looking up the unique derivation T of a given string and define the P^{Φ} probability of the sentence as $P^{\mathcal{T}}(T)$.

<u>Remark 2</u>. Starting from a given context-free language Φ we can sometimes generate it from many possible grammars G. These grammars, together with a specification of the probabilities of the productions involved, will give rise to different P^{Φ} measures. Given Φ we have a wide choice of possible probability distributions, not only with different numerical values but of different analytical forms. Going back to Remarks 1 and 2, we have in the first case the form

$$P(a^{i}) = p(1-p)^{i}; \quad i \geq 1 \tag{10.67}$$

and in the second case

$$P(a^{i}) = f_{i}p^{i}(1-p)^{i-1}; \quad i \geq 1. \tag{10.68}$$

<u>The grammar specifies the syntax, the</u> P^{Φ} <u>specifies the usage or style</u>.

It is not difficult to calculate the entropy, which we shall do now.

<u>Theorem 10.6</u>. <u>Introduce the entropies</u>

$$h_i = h(\xi_i) = -\sum_{r\in\mathcal{R}_{\xi_i}} p_r \log p_r; \quad i = 1,2,\ldots,r. \tag{10.69}$$

<u>If the eigenvalue</u> ρ <u>is strictly less than</u> 1 <u>then the entropy</u> H <u>of</u> P^{Φ} <u>is the first component of the vector</u>

$$[I-M]^{-1}h \tag{10.70}$$

where $h = (h_1, h_2, \ldots, h_v)$.

Proof. For each syntactic variable ξ_i, $i = 1,2,\ldots,v$, construct the sets $\mathscr{T}_i$ of generation trees rooted in ξ_i. On each $\mathscr{T}_i$ we induce the syntax-controlled probabilities as before, obtaining measures $P^{\mathscr{T}_i}$. For each such measure we calculate the entropy

$$H_i = H(\xi_i) = -\sum_{T\in\xi_i} P^i(T)\log P^i. \tag{10.71}$$

To show that these are finite we first work with the entropies H_i^ℓ associated with cylinders on level ℓ, write down a recursion relation between $H_i^{\ell+1}$ and H_i^ℓ using the well-known relation

$$H(X,Y) = H(X) + EH(Y|X). \tag{10.72}$$

Finally, we let ℓ tend to infinity, noting the monotonic dependence upon ρ. Convergence follows from $\rho < 1$. What is of interest from the present point of view is the form of the resulting relation

$$H(\xi_i) = h(\xi_i) + \sum_{r\in\mathscr{R}_{\xi_i}} p_r[H(\xi_{j_1}) + H(\xi_{j_2}) + \cdots] \tag{10.73}$$

where ξ_{j_1}, ξ_{j_2} are the syntactic variables appearing on the rewritten side of rule r with due regard to multiplicity. With the aid of the numbers $n_j(r)$ (see above), we can express (10.73) as

$$H_i = h_i + \sum_{r\in\mathscr{R}_{\xi_i}} p_r \sum_{j=1}^{n} n_j(r)H_j; \quad i = 1,2,\ldots,n. \tag{10.74}$$

Using (10.72), this becomes

$$H_i = h_i + \sum_{j=1}^{v} m_{ij} H_j \tag{10.75}$$

or in matrix notation

$$H = h + MH \tag{10.76}$$

so that

$$H = [1-M]^{-1} h; \tag{10.77}$$

the matrix $I - M$ is non-singular since $\rho < 1$. The entropy we are interested in is $h(\sigma)$ and since $\sigma = \xi_1$, the result stated in the theorem follows.

Using this theorem we can calculate the entropy without having to evaluate the probabilities $P^{\mathscr{T}}(T)$ explicitly. Take Example 2 again. Then $\rho = 2 - 2p$, so that if $p > 1/2$ we get from (10.77) directly

$$H(\sigma) = -\frac{1}{2(1-p)} [p \log p + (1-p) \log(1-p)]. \tag{10.78}$$

When we restrict ourselves to finite-state automata the question arises how these results apply in the more special setting.

<u>Theorem 10.7</u>. <u>Consider a right linear grammar and assume for each syntactic variable</u> ξ <u>there is at least one derivation with positive probability</u> $\xi \overset{*}{\Longrightarrow} x \in S^*$. <u>Then the induced probability measure is legitimate.</u>

<u>Proof</u>. For a right linear grammar every rule has the form $\xi \to u$ or $\xi \to u\eta$, where $\eta \in N$ and $u \in S^*$. This means that the numbers $n_j(r)$ above are either 0 or 1, so that

$$\sum_{j=1}^{v} m_{ij} \le 1; \quad i = 1,2,\ldots,v. \tag{10.79}$$

This implies that $\rho \leq 1$. On the other hand, there can be no final classes, since $\xi \stackrel{*}{\Longrightarrow} S^*$ is always possible with positive probability, by assumption. Appealing to Theorem 10.5, we have proved the theorem.

11. Mappings in Configuration Space.

In some applications one has to operate with more than one configuration space at a time and then we shall have to investigate certain natural mappings between them. Here, we shall only mention two types of such mappings: configuration homomorphisms and generator annihilation.

Consider the two configuration spaces $\mathscr{C}(\mathscr{R})$ and $\mathscr{C}'(\mathscr{R})$

$$\mathscr{C}(\mathscr{R}) = \langle G,S,\Sigma,\rho\rangle, \quad \mathscr{C}'(\mathscr{R}) = \langle G',S,\Sigma,\rho\rangle \tag{11.1}$$

where a map $h\colon G \to G'$ is given as bond invariant, see Section 1.1. This induces a map H from $\mathscr{C}(\mathscr{R})$ to $\mathscr{C}'(\mathscr{R})$ through the definition $H\colon c \to c'$ where $c' = (g_1', g_2', \ldots, g_n')$ with $g_i' = h(g_i)$ and structure (c) = structure (c'). Note that the latter statement makes sense since h leaves the bond structure of generators unchanged. The induced mapping H is a configuration homomorphism in the following sense.

Theorem 11.1. The mapping H has the homomorphism properties that if $c' = H(c)$, $c \in \mathscr{C}(\mathscr{R})$,

(i) $\operatorname{ext}(c) = \operatorname{ext}(c')$ and $c' \in \mathscr{C}(\mathscr{R}')$

(ii) $H(sc) = sH(c)$, all $s \in S$ and $c \in \mathscr{C}(\mathscr{R})$

(iii) if c_1, c_2 and $c_1 \sigma c_2 \in \mathscr{C}(\mathscr{R})$ then

$$H(c_1 \sigma c_2) = H(c_1) \sigma H(c_2) \in \mathscr{C}(\mathscr{R}'). \tag{11.2}$$

<u>Proof</u>. Since h leaves the bond structure unchanged for the individual generators it follows that $c \to c'$ implies $\mathrm{ext}(c) = \mathrm{ext}(c')$. If $c = (g_1, g_2, \ldots, g_n)$, $c \to c' = (g_1', g_2', \ldots, g_n')$ then, since $sg_i \to sg_i'$ and structure is not affected by s, the statement in (ii) follows. Finally, (iii) can be verified by using the bond invariance which makes the new configuration $H(c_1)\sigma H(c_2)$ regular.

As an example take the wave-form case, say for simplicity with $p = 1$, and say that the endpoints of the intervals are restricted to the set B of natural numbers. Following the construction at the end of Section 1.3 let X_0 be the set of all multiples k/p of $1/p$, where p is a fixed natural number. Since $B \subset X_0$ the construction defines a bond invariant mapping h to a set G' of new generators concretely defined as solutions of the given differential operators but restricted to X_0. According to the last theorem this mapping induces a homomorphism from the original configurations to the new ones. Such a homomorphism will come up naturally when studying images and their deformations.

Now assume for the moment that Σ is monotonic and introduce the annihilation operator $\mathscr{A}$ which takes a given configuration $c \in \mathscr{C}(\mathscr{R})$, deletes all the generators in it belonging to generator classes with index α in a certain given set A_0.

Since the result $A(c)$ has correct connection type due to the monotonicity of Σ and since now new connections are established and all the old ones remain true with respect to the bond relation ρ, it follows that $A(c) \in \mathscr{C}(\mathscr{R})$, so that we have a map $A\colon \mathscr{C}(\mathscr{R}) \to \mathscr{C}(\mathscr{R})$.

If $c = (g_1, g_2, \ldots, g_n)$, then $A(c) = (g_{i_1}, g_{i_2}, \ldots, g_{i_m})$, with g_i present in $A(c)$ if its generator index $\alpha(g_i) \notin A_0$. The structure of $A(c)$ is obtained from that of c by leaving out all connections to bonds of generators annihilated by $\mathscr{A}$.

It follows that $A(sc) = sA(c)$, property (ii) of Theorem 11.1, and if $c_1, c_2, c_1 \sigma \ell_2 \in \mathscr{C}(\mathscr{R})$ then

$$A(c_1 \sigma c_2) = A(c_1) \sigma' A(c_2) \tag{11.3}$$

where σ' is the combination operator obtained from σ by leaving out connections to and from generators in the S^α, $\alpha \in A$. Property (iii) in Theorem 11.1 is, therefore not strictly satisfied; only its modified form (11.3). Property (i) need not be true either since leaving out a generator may open up new external bonds that earlier were internal. The annihilation operator $\mathscr{A}$ is therefore not usually a homomorphism.

CHAPTER 3

THE IMAGES

1. The Image Algebra.

We shall now introduce the observables in the theory of patterns. Two distinct configurations c and c' from $\mathscr{C}(\mathscr{R})$ need not appear as different to the observer. Whether they do or not depends upon how the observer obtains his information about the configurations and how he then processes this information. We formalize this through an *identification rule* R and write $c \equiv c' (\text{mod } R)$ or cRc' if the c and c' are identified by the rule that expresses how the observer can distinguish between configurations. The following will be assumed in order that a relation be allowed as an identification rule.

Definition 1.1. *A relation* R *between configurations in* $\mathscr{C}(\mathscr{R})$ *is said to be an identification rule if*

(i) R *is an equivalence relation*

(ii) *if* cRc' *then* c *and* c' *have the same exter-*

nal bonds and external bond values

(iii) if cRc' then (sc)R(sc') for any $s \in S$

(iv) if $c = c_1 \sigma c_2$ and $c' = c_1' \sigma c_2'$ are regular and if $c_1 R c_1'$, $c_2 R c_2'$, then cRc'.

The equivalence classes of $\mathscr{C}(\mathscr{R})$ are called the images and generically denoted by I and the whole set of them is written as $\mathscr{T}$,

$$\mathscr{T} = \mathscr{C}(\mathscr{R})/R = \langle G,S,\Sigma,\rho\rangle/R. \tag{1.1}$$

To be more specific, we shall sometimes refer to the elements of $\mathscr{T}$ as the pure images, as opposed to the deformed images that will be introduced in Chapter 4. For the equivalence class I that contains a given configuration c we shall write I(c).

The set $\mathscr{T}$ is given an algebraic structure.

Theorem 1.1. The similarity transformations can be uniquely defined on $\mathscr{T}$ and combinations $I_1 \sigma I_2$ are uniquely determined for images I_1, I_2 when the connections in σ fit their external bonds. Then

(i) $s(I_1 \sigma I_2) = (sI_1)\sigma(sI_2)$ if the external bonds fit via σ

(ii) if $(I_1 \sigma_1 I_2)\ _2 I_3$ and $I_1 \sigma_1'(I_2 \sigma_2' I_3)$ are regular combinations they define the same image if $\sigma_1 \cup \sigma_2 = \sigma_1' \cup \sigma_2'$.

Proof: The obvious definition $sI = I(sc)$ is unique since if c_1 and $c_2 \in I$ then $I(sc_1) = I(sc_2)$, see Definition 1.1(iii). Similarly, the obvious definition $I_1 \sigma I_2 = I(c_1 \sigma c_2)$ if $I_1 = I(c_1)$, $I_2 = I(c_2)$, and the external bonds fit via σ

is also unique. This follows from Definition 1.1(iv). To prove (i) above it is sufficient to note that if I_1 and I_2 have external bonds fitting via σ, then the same is true about sI_1 and sI_2, see 2.2, so that the right hand side makes sense. The two sides then must be equal since we have equivalence modulo R between the two expressions we get by replacing I_1 and I_2 respectively by two of its configurations, see Theorem 1.2.1. The statement (ii) is proved in the same way.

For a given $\mathscr{L}(\mathscr{R})$ different identification rules will give rise to different image algebras. If R_1 and R_2 are two such rules and if R_1 is <u>finer</u> than R_2, in the sense that R_1-images are always contained in R_2-images but not always the converse, we shall write $R_1 < R_2$. In particular it is sometimes important that a rule can distinguish between configurations with just one generator. More precisely, considering a monatomic connection type, see 2.2, and an arbitrary pair of generators g_1 and g_2. The two configurations $\{g_1\}$ and $\{g_2\}$ are regular and we shall say that R <u>separates generators</u> if $\{g_1\} \equiv \{g_2\}$ (mod R) implies that $g_1 = g_2$.

Note that since all the configurations in an image have the same external bonds and bond values it makes sense to speak of these for the image itself.

The set $\mathscr{T}$ together with the similarity transformations and combinations via σ is called the <u>image algebra</u> and will also be denoted by $\mathscr{T}$ and can be represented by a 5-tuple

$$\mathscr{T} = \langle \mathscr{L}(\mathscr{R}), R \rangle = \langle G, S, \Sigma, \rho, R \rangle \tag{1.2}$$

A probability measure P on $\mathscr{L}(\mathscr{R})$ induces one on $\mathscr{T}$ by the relation, $E \subset \mathscr{T}$,

$$P(E) = P\{c \mid c \in \mathscr{L}(\mathscr{R}),\ I(c) \in E\} \tag{1.3}$$

To simplify the notation we use the same symbol P for the induced measure.

We shall encounter many types of identification rules later, but it may be illuminating to mention some simple ones already at this stage. A trivial one is by <u>equality between configurations</u> so that cRc' if and only if c = c'. Then of course we have $\mathscr{T} = \mathscr{L}(\mathscr{R})$.

Another rule R occurs when all regular configurations have arity zero and we define cRc' if and only if content(c) = content(c'): <u>identification by content</u>.

A more interesting rule is obtained for contrast generators combined as in Case 2.9:3 such that a function is defined uniquely in a subset of the background space. Two configurations are identified if they represent the same function and have the same external bonds: <u>identification as function</u>. It will often be the case that an image represents some function on a background space.

When we operate with two or more image algebras defined over the same $\mathscr{L}(\mathscr{R})$ it will be of particular importance to see if simple mappings take one into another. Especially we shall look for homomorphisms.

<u>Definition 1.2.</u> <u>A mapping</u> H <u>from one image algebra</u> $\mathscr{T}$ <u>to another</u> $\mathscr{T}'$ <u>with the same</u> S <u>and</u> $\mathscr{R}$ <u>is called a homomorphism if</u>

(i) $H(I_1) = H(I_2) \Rightarrow \text{ext}(I_1) = \text{ext}(I_2)$

(ii) <u>for any</u> $s \in S$ <u>we have</u> $H(sI) = sH(I)$

(iii) <u>if</u> I_1 <u>and</u> I_2 <u>match via</u> σ <u>then</u> $H(I_1 \sigma I_2) = H(I_1)\sigma H(I_2)$.

As a simple example of homomorphism we can mention $H_{s_o} : \mathcal{T} \to \mathcal{T}$ with $H_{s_o}(I) = s_o I$ where s_o is a fixed similarity transformation such that $ss_o = s_o s$, all s. When S is a group this means that s_o should be a central element. Others will appear later.

If the mapping H is also one-one we speak of an <u>isomorphism between image algebras</u>. In the example, if S forms a group H_{s_o} will be an isomorphism.

Let us consider an image algebra with monatomic connection type, see Section 2.2. Then for any two generators g_1 and g_2 the corresponding configurations $c_1 = \{g_1\}$ and $c_2 = \{g_2\}$ are regular. It can happen, although this is certainly not necessary, that there exists a generator g such that $g \equiv (c_1 \sigma c_2) \bmod R$. If, moreover, R separates generators, then g is uniquely determined and we shall write

$$g = g_1 \sigma g_2. \tag{1.4}$$

In this way pairs of generators can <u>collapse</u> into a single one and this procedure may be repeated. A consequence is the following:

<u>Theorem 1.2</u>. <u>Consider a configuration</u> $c \in I$ <u>involving two generators</u> g_1 <u>and</u> g_2. <u>If</u> g_1 <u>and</u> g_2 <u>are connected in</u> c <u>via</u> σ <u>and if</u> $g_1 \sigma g_2 = g$, <u>then</u> c <u>is</u> R-<u>equivalent to the configuration</u> c' <u>obtained from</u> c <u>by replacing the subconfiguration</u> $g_1 \sigma g_2$ <u>by</u> g.

Proof: To illustrate what is happening look at Figure 1.1, where the subconfiguration c' of c is indicated as what is inside the dotted line in (a) and is shown in isolation in (b), and is R-equivalent to $\{g\}$. Replacing $g_1 \sigma g_2$ by $\{g\}$ we get the configuration as shown in (c). That this statement is true is seen as follows. Consider the two configurations

$$\begin{cases} c = (g_1 \sigma g_2)\sigma' c'' \\ c' = \{g\}\sigma' c'' \end{cases} \tag{1.5}$$

In the figure c'' corresponds to everything outside the dotted line in (a). We have of course the reflexive property $c'' \equiv c''(\text{mod } R)$, but we have also assumed that $(g_1 \sigma g_2) \equiv \{g\}$ (mod R). Combining this with (iv) in Definition 1.1 it follows that $c \equiv c'$ (mod R) as claimed.

This opens up the possibility of reducing configurations. Given an image I and one of its configurations, we will try to collapse two of its generators into a single one and repeat this until no further reduction is possible: the resulting configuration is an irreducible configuration modulo R.

In general this result is not unique, but we shall encounter cases occasionally where this is true.

For a given image I we will typically have many configurations belonging to the equivalence class and observing I we will not be able to single out any particular configuration. This is the segmentation problem: find an irreducible configuration belonging to I.

For a given image algebra $\mathcal{T}$ with the identification rule R characterizing the observer, another observer with

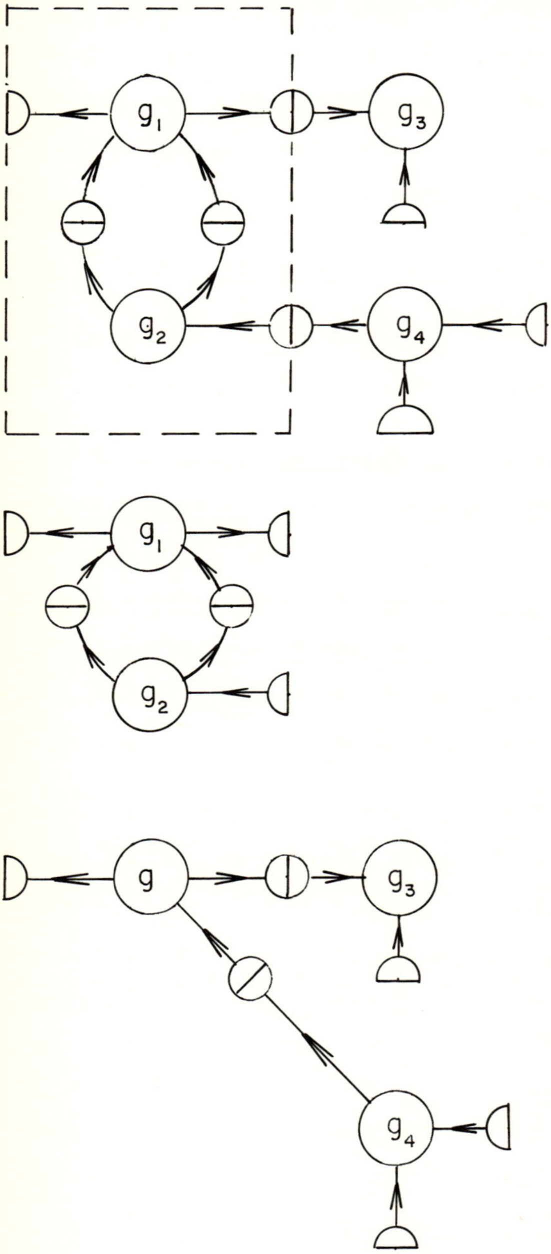

(a) $n = 4$

$\omega_{out}(c) = 1$

$\omega_{in}(c) = 3$

(b) $n = 2$

$\omega_{out}(g_1 \sigma g_2) = 2$

$\omega_{in}(g_1 \sigma g_2) = 1$

(c) $n = 3$

$\omega_{out}(c') = 1$

$\omega_{in}(c') = 3$

Figure I.1

less powerful information gathering devices will have a rule $R' > R$ (see earlier in this section). This will give rise to another partition of $\mathcal{L}(\mathcal{R})$ coarser than the first. In general we can characterize the relation between the partitions and corresponding mappings by

<u>Theorem 1.2.</u> (i) <u>Let</u> H <u>be a homomorphism</u> $\mathcal{T} \to \mathcal{T}'$ <u>and define</u> R' <u>in</u> $\mathcal{L}(\mathcal{R})$ <u>through</u>

$$c_1 R' c_2 \quad \text{iff} \quad H[I(c_1)] = H[I(c_2)] \tag{1.6}$$

<u>Then</u> R' <u>is an identification rule cruder than</u> R, $R < R'$, <u>and the resulting image algebra</u> $\mathcal{T}(\mathcal{R}')$ <u>is isomorphic to</u> $\mathcal{T}'$.

(ii) <u>On the other hand if</u> R' <u>is an identification rule cruder than</u> R, $R < R'$, <u>then the natural mapping</u> M <u>that takes an</u> R-<u>image</u> I <u>into an</u> R-<u>image</u> I' <u>if</u> $I \subset I'$ <u>is a homomorphism from</u> $\mathcal{T}$ <u>to</u> $\mathcal{T}' = \mathcal{T}(\mathcal{R}')$.

<u>Proof of (i)</u>: Obviously R' is reflexive, symmetric, and transitive, hence an equivalence relation in $\mathcal{L}(\mathcal{R})$ and condition (i) in Definition 1.1 holds.

If $c_1 R' c_2$ so that $H[I(c_1)] = H[I(c_2)]$ it follows that $\text{ext}[I(c_1)] = \text{ext}[I(c_2)]$ so that $\text{ext}(c_1) = \text{ext}(c_2)$, condition (ii) holds.

Again let $c_1 R' c_2$. Then

$$\begin{aligned} H[I(sc_1)] &= H[sI(c_1)] = sH[I(c_1)] = \\ &= sH[I(c_2)] = H[sI(c_2)] = H[I(sc_2)] \end{aligned} \tag{1.7}$$

so that $(sc_1)R'(sc_2)$ and condition (iii) holds.

Finally if $c = c_1 \sigma c_2$ and $c' = c_1' \sigma c_2'$ are regular in $\mathscr{C}(\mathscr{R})$, and $c_1R'c_1'$, $c_2R'c_2'$ then

$$\begin{cases} H[I(c_1)] = H[I(c_1')] \\ H[I(c_2)] = H[I(c_2')] \\ H[I(c)] = H[I(c_1)]\sigma H[I(c_2)] \\ H[I(c')] = H[I(c_1')]\sigma H[I(c_2')]. \end{cases} \tag{1.8}$$

It then follows that $H[I(c)] = H[I(c')]$ so that $cR'c'$, property (iv), so that R' is an identification rule. The induced mapping between $\mathscr{T}(\mathscr{R}')$ and $\mathscr{T}'$ is now one-one and we have an isomorphism.

Proof of (ii): Property (i) of Definition 1.2 holds since if $M[I_1] = M[I_2]$ so that I_1 and I_2 are contained in the same R'-image I' then $\text{ext}(I_1) = \text{ext}(I') = \text{ext}(I_2)$.

Also, if $M[I] = I'$, then for any $c \in I \subset I$ we must have $sc \in sI \subset sI'$ so that M takes sc into sI', property (ii).

Finally, if I_1 matches I_2 via σ and if

$$\begin{cases} M[I_1] = I_1' \\ M[I_2] = I_2' \end{cases} \tag{1.9}$$

then I_1' matches I_2' via σ since the external structure of elements in I_1' is the same as that of I_1 and similarly for the second image. Further any $c_1\sigma c_2$ with $c_1 \in I_1$, $c_2 \in I_2$ will be mapped into $I_1'\sigma I_2'$ which completes the proof.

When we discuss symmetries of individual images, we shall assume that the similarity transformations s form a group S. Consider the subset

$$S(I) = \{s | sI = I\} \subset S \tag{1.10}$$

Then $S(I)$ is a subgroup of S and is called the <u>covering group of the image</u> I. It expresses the <u>full symmetry</u> of I, which is often too strong.

Instead let us form S from two groups of similarity transformations of G, S_1 and S_2 with unit elements e_1 and e_2 respectively and such that $s_2 S_1 s_2^{-1} \subset S_1$ for any $s_2 \in S_2$. Put

$$S = \{s_1 s_2 | s_1 \in S_1,\ s_2 \in S_2\}. \tag{1.11}$$

Then S is a group. Indeed, it has a unit element $e = e_1 e_2$ and multiplication is associative, and closed under multiplication, since for any $s_1, s_1' \in S_1$ and $s_2, s_2' \in S_2$ there is a $s_1'' \in S_1$ such that

$$s_1 s_2 s_1' s_2' = s_1 s_1'' s_2 s_2' \tag{1.12}$$

which can be written as $s_1''' s_2'''$ with $s_1''' \in S_1$ and $s_2''' \in S_2$. The inverse of $s_1 s_2$ is of course $s_2^{-1} s_1^{-1}$ which can be written as $s_1'''' s_2^{-1}$, $s_1'''' \in S_1$.

Then S_1 is a normal subgroup of S. Indeed for any $s_1^o \in S_1$ and $s = s_1 s_2 \in S$ we have

$$s s_1^o s^{-1} = s_1 s_2 s_1^o s_2^{-1} s_1^{-1} = s_1 s_1' s_1^{-1} \in S_1 \tag{1.13}$$

for some $s_1' \in S_1$.

As before we introduce the subgroups $S(I)$, $S_1(I)$ and $S_2(I)$ of S, S_1 and S_2 respectively that leave a given image I invariant. Again $S_1(I)$ is a normal subgroup of $S(I)$ and the factor group $S(I)/S_1(I)$ will express the S_2-symmetries of the image.

Having defined the image algebra we shall formalize the notion of a pattern in accordance with the discussion in Chapter 2.

<u>Definition 1.3</u>. <u>A pattern family</u> $\{\mathcal{P}_r\}$ <u>is a collection of disjoint and</u> S-<u>invariant subsets</u> $\mathcal{P}_r$ <u>of the image algebra</u> $\mathcal{T}$. <u>Each of the subsets</u> $\mathcal{P}_r$ <u>is a pattern</u>. <u>Further, a diffuse pattern family</u> $\{\mathcal{P}_r\}$ <u>is a collection of</u> S-<u>invariant probability measures</u> $\mathcal{P}_r$ <u>over</u> $\mathcal{T}$. <u>Each of the measures</u> $\mathcal{P}_r$ <u>is a diffuse pattern</u>.

The cardinality of the pattern index r is arbitrary.

Three special cases should be mentioned. If $r = 1$, we have only one pattern class identical to the whole image algebra, and there is only one pattern. This will occur when we study structurally homogeneous situations.

An intermediate case is when each pattern class $\mathcal{P}_r$ is generated from a <u>prototype</u> I_r such that different I_r are not S-equivalent and

$$\mathcal{P}_r = \{I | I = sI_r,\ s \in S\} \tag{1.14}$$

If S forms a group then for a given image I the corresponding pattern class $\mathcal{P}_r$ can be uniquely determined since these classes do not overlap. Also if $I = sI_r = tI_r$ we have

$$t^{-1}sI_r = I_r \tag{1.15}$$

so that $t^{-1}s$ belongs to the covering group of I_r, see (1.10). In other words, knowing the images enables us to identify <u>s</u> in the relation $I = sI_r$ to the extent that we know to what left coset of the covering group it belongs.

As mentioned earlier one may operate with two or more groups of similarity transformations simultaneously. If one of them degenerates to contain just the identity transformations the corresponding pattern classes (1.14) consist of individual images. In this third case, which is quite frequent, we can use the terms image and pattern interchangeably.

Given an image algebra $\mathcal{I}$ we shall speak of <u>derived images</u> via a mapping $T: \mathcal{I} \to$ some image algebra $\mathcal{I}'$ if to each $I \in \mathcal{I}$ corresponds one in $\mathcal{I}'$, $I' = TI$, where T need not be a homomorphism. This can happen when one image I describes an activity and I' represents the result. See, for example, the spatial images derived from motion images in Section 5.8.

Similarly an image can be derived from two (or more) images: $I''' = I_1 \times I_2$, I_1 and $I_2 \in \mathcal{I}$. The interesting case is when the "$\times$" operation means something more than just one of the usual binary operators in $\mathcal{I}$, say that I_1 interferes with I_2 in some sense. The Moiré patterns that will be discussed in 3.5 are derived from pairs of images.

2. <u>Abstract Sequence Images</u>.

In this section we shall consider configurations made up of abstract generators from a finite set G.

Let us begin with the case when the generators are letters used in English, including space, and configurations are formed with Σ = LINEAR. Given a simple random source over G, the probability measure Q over $\mathcal{C}_n$ is well defined and configurations can be generated according to Q.

We shall identify regular configurations by their

external bonds plus their strings of terminal symbols. Note that the image is not just the terminal string, it also has external bonds although the latter will not always be indicated explicitly.

A celebrated experiment due to Shannon (1949) produced strings like

OCRO HLI RGWR NMIELWIS EU LL NBNESEBYA TH EEI ALHENHTTPA OOBTTVA NAH BRL.

It shows little relation to real English.

If the generators are of the form (a) in Figure 2.1, signifying transitions from one letter to another, with ρ = EQUALITY we know that a simple random source over G defines a Markov chain measure P over $\mathscr{L}_n(\mathscr{R})$, see Section 2.10. A generator here has two bonds, and the values ℓ and ℓ' are letters. Shannon's experiment produced strings like

ON IE ANTSOUTINYS ARE T INCTORE ST BE S DEAMY ACHIN D ILONASIVE TUCOOWE AT TEASONARE FUSO TIZIN ANDY TOBE SEACE CTISBE,

slightly more similar to English.

Choosing the generators as in (c) in the figure and with connection type as in (d) one can show that this induces a Markov chain probability measure P of order $p = 2$, and a string can look like

IN NO IST LAT WHEY CRATICT FROURE BIRS GROCID PONDENOME OF DEMONSTURES OF THE REPTAGIN IS REGOACTIONA OF CRE.

The "words" now look more like words in English. Taking the

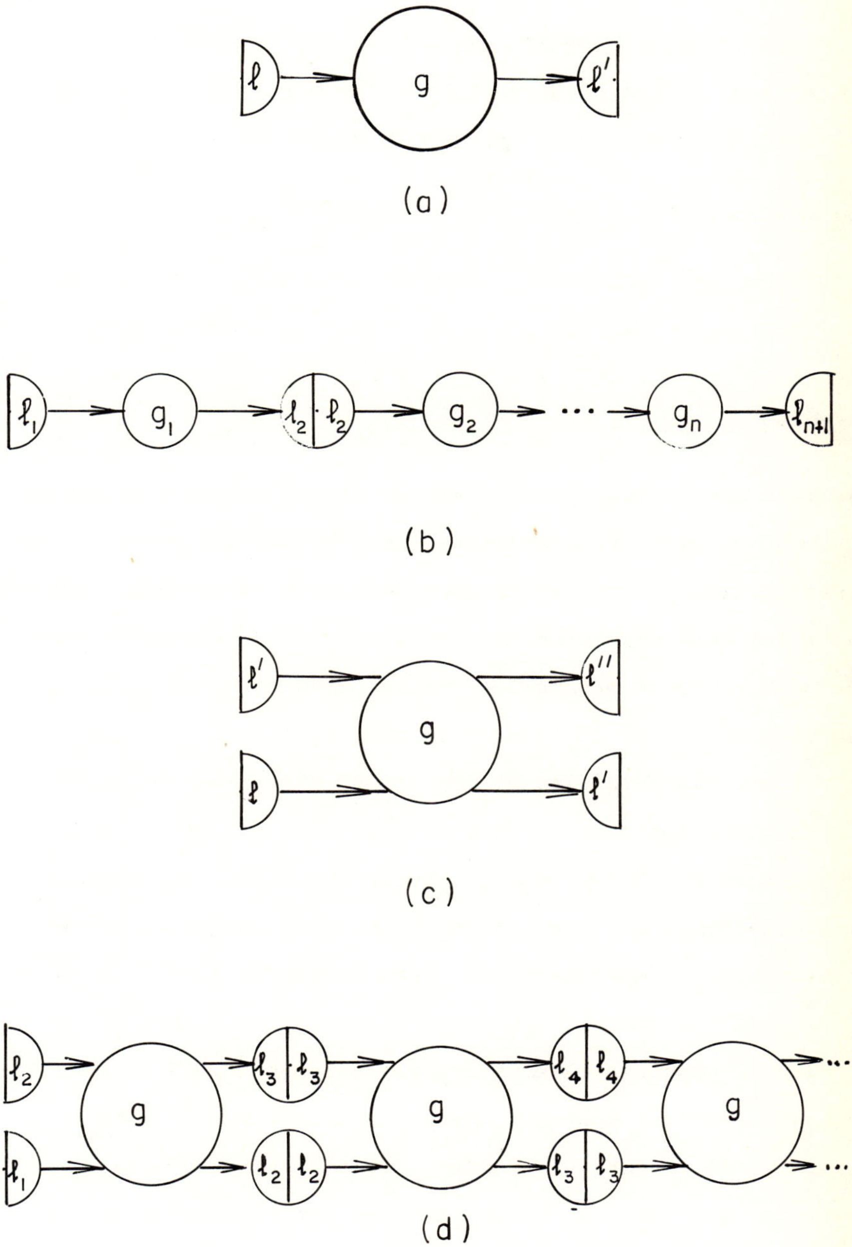

ℓ
g
ℓ'
(a)
ℓ₁
g₁
ℓ₂
ℓ₂
g₂
...
gₙ
ℓₙ₊₁
(b)
ℓ'
ℓ''
g
ℓ
ℓ'
(c)
ℓ₂
ℓ₁
g
ℓ₃
ℓ₃
ℓ₂
ℓ₂
g
ℓ₄
ℓ₄
ℓ₃
ℓ₃
g
...
...
(d)

Figure 2.1

generators as words rather than letters Shannon got for the simple random source

REPRESENTING AND SPEEDILY IS A GOOD APT OR COME CAN DIFFERENT NATURAL HERE THE A IN CAME THE TO OF TO EXPERT GRAY COME TO FURNISHES THE LINE MESSAGE HAD BE THESE

and similarly for a first order Markov chain

THE HEAD AND IN FRONTAL ATTACK ON AN ENGLISH WRITER THAT THE CHARACTER OF THIS POINT IS THEREFORE ANOTHER METHOD FOR THE LETTERS THAT THE TIME OF WHO EVER TOLD THE PROBLEM FOR AN UNEXPECTED.

The last string even has substrings that seem to be meaningful. Increasing the order even more, we can expect more and more lifelike strings to be produced by the stochastic source. The price paid for the better realism is the complexity of the model. If $n = \text{card}(G)$ then a pth-order Markov chain has to specify n^{p+1} transition probabilities. In order that these probabilities be consistent they have to satisfy n^p linear relations, so that we have effectively $n^p(n-1)$ free parameters in the model. It is clear that the exponential dependence upon p in this expression makes high-order Markov sources impractical and other types of sources will be needed.

One possible alternative was discussed in Section 2.10: syntax-controlled probabilities. To illustrate this, consider the finite-state automaton with six states, initial state 1 and terminal state 6. The wiring diagram of the automaton is given in Figure 2.2.

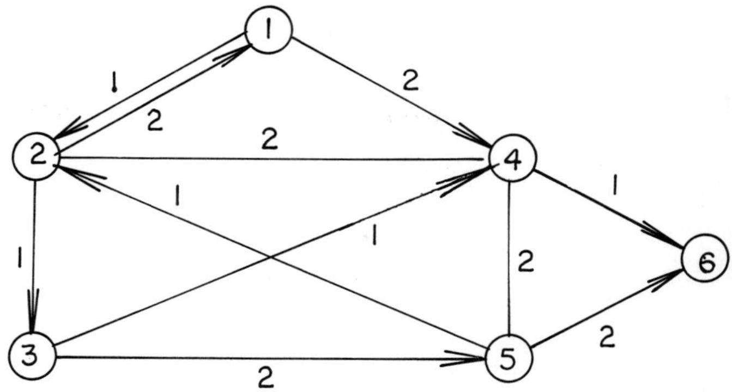

Figure 2.2

The corresponding rules of the right linear grammar have five syntactic variables $\sigma = v_1, v_2, v_3, v_4, v_5$ with the rewriting rules that play the role of generators here

$v_1 \to 2\ v_4$	
$v_1 \to 1\ v_2$	$v_3 \to 2\ v_5$
$v_2 \to 2\ v_4$	$v_4 \to 1$
$v_2 \to 1\ v_3$	$v_4 \to 2\ v_5$
$v_2 \to 2\ v_1$	$v_5 \to 1\ v_2$
$v_3 \to 1\ v_4$	$v_5 \to 2$

Table 2.1

For simplicity let Q be defined by assigning equal probabilities to all the generators; we can calculate $P = Q/\mathscr{C}(\mathscr{R})$ as in Section 2.10.

Simulating the automaton with a pseudo-random number

generator, the following strings were obtained (see Table 2.2 where they are given in the order in which they were produced).

1 1 2 2
1 1 1 1
2 1
1 1 1 2 1 1 1 2 2
2 2 1 1 2 2
1 2 1
1 1 2 2
2 2 1 1 2 1 1 2 2
1 2 2 1 2 2 1 1 1 2 2
1 2 2 1 1 2 1 1 1 2 1 1 2 1 1 2 1 2 1
2 1
1 1 2 2
1 1 2 1 2 2 2
2 1
1 1 2 2
2 1
1 1 2 1 2 1
1 2 1
2 1
2 2 1 2 2 1 1 1 2 2
1 1 2 1 1 1 2 1 1 1 2 2
1 1 1 2 1 2 1
2 1
1 1 2 2
2 2 2
2 2 2
2 2 1 2 1
1 1 1 1
1 1 1 1
2 2 2
2 1
2 1
1 1 1 1
2 1
1 1 2 1 1 2 1 2 2 1 2 1
1 1 2 2
2 2 1 2 2 1 1 1 2 1 2 2 1 1 1 2 2
1 1 2 1 1 1 1
1 2 2 1 1 2 1 2 2 2
1 1 2 1 1 2 2
2 2 1 1 2 2
1 2 2 2
2 2 1 2 2 1 1 1 1
1 2 1
2 2 2
1 1 2 1 1 1 2 1 1 2 1 1 2 2
2 1
2 2 2
1 2 2 2

Table 2.2

Leaving this example, for a general finite state language the configurations can be represented as $c = (g_1, g_2, \ldots, g_n)$ where the generator g_k is a transition from state i_k to i_{k+1} writing a terminal symbol x_k. It is then natural to use the identification rule cRc' if c' is another admissible configuration with the same number of generators, the same external bonds and with the same terminal string. It can easily be verified that R has the properties of an identification rule, see Definition 1.1.

This means that only the first and last state of c together with all its terminals can be seen by the observer. In other words he cannot identify the derivation unless the grammar is unambiguous. To get $P(I)$ for a given image we enumerate the syntactic variables by $1,2,3,\ldots,v$, and consider all the rules

$$\begin{cases} i \to xj, & x \in S, \quad j \in N \\ i \to x, & x \in S \end{cases} \tag{2.1}$$

making up the set $\mathcal{R}_i$. Denoting the corresponding probabilities by $p_{ij}(x)$ and $r_i(x)$ respectively, we get

$$\sum_{x \in \Gamma_t} \sum_{j=1}^{v} p_{ij}(x) + \sum_{x \in \Gamma_t} r_i(x) = 1; \quad i = 1,2,\ldots,v \tag{2.2}$$

or with a matrix-vector notation,

$$\begin{cases} P(x) = p_{ij}(x); \quad i,j = 1,2,\ldots,v \\ r(x) = \mathrm{col}[r_1(x), r_2(x), \ldots, r_v(x)] \end{cases} \tag{2.3}$$

and

$$P = \sum_{x \in \Gamma_t} P(x), \qquad r = \sum_{x \in \Gamma_t} r(x). \tag{2.4}$$

We also introduce the row vector $d = (1,0,0,\ldots,0)$. Then (2.3) can be written as $Pe + r = e$, where e is the column vector with all v components equal to 1. We assume that the conditions of Theorem 2.10.7 hold.

<u>What is the probability that</u> P <u>attributes to a given string image</u>? We shall carry out the derivation for images that start in the initial state and end in the final state, grammatical strings. The general case can be handled in exactly the same way. We get directly if the terminal string is $x = (x_1, x_2, \ldots, x_n)$

$$P(x) = \Sigma p_{1i_1}(x_1)p_{i_1i_2}(x_2)\cdots p_{i_{n-2}i_{n-1}}(x_{n-1})r_{i_{n-1}}(x_n). \quad (2.5)$$

This can be written as

$$P(x) = d\, P(x_1)P(x_2)\ldots P(x_{n-1})r(x_n). \quad (2.6)$$

In order that we have a legitimate probability distribution, we shall assume that the final state can be reached from any state. From this attractive matrix-form representation of the probabilities we can derive closed expressions for many of the quantities we need for analysis in part III (see Appendix).

What is <u>the probability of obtaining a string of the form</u>

$$\left\{\begin{array}{l} \alpha \text{ unspecified terminals, followed by } x_1\text{, followed by} \\ \beta \text{ unspecified terminals, followed by } x_2\text{, etc.} \\ \ldots \\ \gamma \text{ unspecified terminals?} \end{array}\right. \quad (2.7)$$

To get this we just have to sum (2.6) over all unspecified terminal symbols and get the probability

$$dP^{\alpha}P(x_1)P^{\beta}P(x_2)\ldots P^{\gamma-1}r$$
$$= dP^{\alpha}P(x_1)P^{\beta}P(x_2)\ldots P^{\gamma-1}(1-P)e \tag{2.8}$$

where we have used the relation $r = e-Pe$.

A special case of (2.6) is when we ask what happens when all the terminals, say L in number, are left unspecified. We then get from (2.8) the <u>sentence length distribution</u>

$$d(P^{L-1} - P^{L})e, \quad l = 1,2,\ldots \tag{2.9}$$

Now let us calculate the probability that a string <u>starts with a given string</u> $y = y_1, y_2, \ldots, y_n$. We get the sum

$$\begin{aligned}&P(x=y) + P(x=y \text{ plus one terminal}) + P(x=y \text{ plus two terminals}) + \ldots\\ &= dP(y_1)\ldots P(y_{h-1})r(y_h) + dQr + dQPr + dQP^2r + \ldots\\ &= dP(y_1)\ldots P(y_{h-1})r(y_h) + dQ(1-P)^{-1}r\\ &= dP(y_1)\ldots P(y_{h-1})r(y_h) + dQe\end{aligned} \tag{2.10}$$

where

$$Q = P(y_1)P(y_2)\ \ldots\ P(y_h) = P(y). \tag{2.11}$$

Finally, <u>what is the expected number of occurrences of a substring</u> y? Denote by $n_y(x)$ the number of occurrences of the substring y in a string x; $0 \le n_y(x) \le 1 + \text{length of } x - \text{length of } y$. Enumerating the possibilities that give rise to positive values of $n_y(x)$, we get

$$\begin{aligned}En_y(x) &= \sum_{x\in\phi} n_y(x)P^{\phi}(x) = P(x=yu) + P(x=u_1yu)\\ &\quad + P(x=u_2yu) + \ldots\end{aligned} \tag{2.12}$$

Here u denotes an arbitrary string, possibly empty, while u_k denotes an arbitrary string of k terminal symbols.

Using (2.6), we can express (2.12) as

$$\begin{aligned}
En_y(x) &= dP(y_1)\dots P(y_{h-1})r(y_h) + dQ[1-P]^{-1}r \\
&\quad + dP\, P(y_1)\dots P(y_{h-1})r(y_h) + dPQ[1-P]^{-1}r \\
&\quad + dP^2P(y_1)\dots P(y_{h-1})r(y_h) + dP^2Q[1-P]^{-1}r + \dots \\
&= d[1-P]^{-1}P(y_1)\dots P(y_{h-1})r(y_h) + d[1-P]^{-1}Q[1-P]^{-1}r \\
&= d[1-P]^{-1}P(y_1)\dots P(y_{h-1})r(y_h) + d[1-P]^{-1}Qe. \qquad (2.13)
\end{aligned}$$

Summing up, we have proved

Theorem 2.1. *If the final state can be reached with positive probability from any state, then*

(i) $P(x = x_1x_2\dots x_n) = dP(x_1)\dots P(x_{n-1})r(x_n)$

(ii) $P(x = u_\alpha x_1 u_\beta x_2 \dots u_\gamma) = dP^\alpha P(x_1)P^\beta P(x_2)\dots P^{\gamma-1}(1-P)e$

(iii) $P(x \text{ has length } L) = d(P^{L-1} - P^L)e$

(iv) $P(x=x_1x_2\dots x_h u) = dP(x_1)\dots P(x_{h-1})r(x_h)+dQ[1-P]^{-1}r$

(v) $En_y(x) = d[1-P]^{-1}\{P(x_1)\dots P(x_{h-1})r(x_h) + Qe\}$

The behavior predicted by (iii) is illustrated by Table 2.3 that gives the observed sentence length distribution for the simulation experiment mentioned.

For context free languages on the other hand we have seen in 2.10 how to calculate the probability of a configuration, which in this context means a left-most derivation of a string. Such configurations have tree connection type. We know how to guarantee that P is a legitimate probability measure on $\mathscr{C}(\mathscr{R})$, see Theorem 2.10.5.

sentence length L	observed in %	predicted in %
1	0	0
2	22	25
3	16	25
4	24	19
5	20	6
6	6	8
7	8	6
8	0	3
9	6	2
10	4	2
11	2	1
12	4	1
13	0	1
14	2	0
15	0	0
16	0	0
17	2	0
18	0	0
19	2	0
20	0	0

Table 2.3

The natural identification rule here is to identify two configurations if the derivation starts from the same syntactic variable and if they have the same sequence of symbols at the bottom nodes. This statement guarantees that the external structures are the same. It can be verified that this is indeed an identification rule according to Definition 1.1.

For a given CF grammar we can have several Q-(and hence P-) specifications. It is natural to speak of the probabilities appearing in Q, preferably normalized for each rewriting rule, as _stylistic parameters_. They describe how the various syntactic operations are used, some more often than others: linguistic performance as contrasted to linguistic competence.

A small computational experiment starting from the

example in 4.6 has been carried out as an illustration. With Q as defined as in Table 2.4 it gave the following set of twenty terminal strings

```
(~Z)
(V∧(Z∧V))
(~V)
(((Z∧V)∧(~(~U)))∨(~V))
((Y∨(X∧Y))∧((((X∧(X∧(~U)))∨Y)∨((((~((~(V∨(~Z)))∨(X∧(V∨(Y∨
    (X∧U))))))∨Z)∧U)∧((~((V∨V)∨((((~V)∧(((Y∨Z)∧X)∧(~(~Y)))))∨
    (Y∧X))∨U)))∨(U∧(Z∨(~V)))))) (Y∧Y)))
U
V
(X∨(~(~Y)))
X
(Z∨(((~((~Y)∨(X∧(~X))))∧V)∨((((~V)∧U)∧(X∧V))∧Z)))
(~(U∧( X∧V)))
V
(~Y)
X
V
(~Y)
Y
((U∨(Y∨(~V)))∨(U∧V))
((U∧U)∨Y)
(U∧((~(Z∨(~(V∨Y))))∨((Z∧U)∨(Z∧U))))
```

$Q(\sigma \to X) = 1/10$	$Q(\sigma \to (\sigma \vee \sigma)) = 1/6$
$Q(\sigma \to Y) = 1/10$	$Q(\sigma \to (\sigma \wedge \sigma)) = 1/6$
$Q(\sigma \to Z) = 1/10$	$Q(\sigma \to (\sim \sigma)) = 1/6$
$Q(\sigma \to U) = 1/10$	
$Q(\sigma \to V) = 1/10$	

Table 2.4

One of the strings is long with a high degree of nesting. Now we change three stylistic parameters to

$$\begin{aligned} Q(\sigma \to (\sigma\vee\sigma)) &= 3/10 \\ Q(\sigma \to (\sigma\wedge\sigma)) &= 1/10 \qquad (2.14) \\ Q(\sigma \to (\sim \sigma)) &= 1/10 \end{aligned}$$

leaving the others as they were. The following twenty

terminal strings were generated

Z
(X∨(X∧(~(Z∨(Z∨V)))))
(~(((((X∨Z)∨(Y∨Y))∨((U∧Z)∨(~(~(V∨(V∨V))))))∧V)∨((Y∧(V∨X))∨Y)))
Z
V
((X∨V)∨((U∧(V∧Z))∨(~(((Z∨(((~(Y∨Z))∨U)∨X))∨(Y∧((V∨(~Z))∨(X∨
  ((U∨U)∨(Y∨(((~V)∧(~(V∨V)))∨V)))))))∨(~Y)))))
Y
(V∨Y)
(~((~V)∧V))
(U∧((U∨(Y∨Y))∧Y))
X
(V∨Z)
U
(V∨(~X))
((Y∧((X∨Y)∧((((((U∨V)∨Z)∨Y)∧(U∧((~Z)∨(Z∨Y))))∨((((Y∧(V∨X))∨
  ((X∨Z)∧Y))∨((((V∨Z)∨Z)∧Z)∧V))∧(~Z)))∨((~(Y∨((Y∨(U∨(Y∧
  ((Y∧U)∨(Y∨Y)∨((U∧((Z∨Y)∨X))∨(Y∨(V∨(~Z))))))))))∨(Z∨U))))
  ∨(~(U∨Y))))))∨(X∨Y))
(U∨U)
(~U)
(Y∧(~U))
(((V∧U)∧(~(((U∨Y)∨Y)∨Y)))∨(V∨Z))
((Y∧(~X))∨Y)

Already a cursory inspection shows that the strings tend to be longer and with more occurrences of the symbol "∨". The reason is that the rule $\sigma \to (\sim\sigma)$ with little nesting has been less frequent than the rule $\sigma \to (\sigma\vee\sigma)$. The style has changed in a noticeable way in this single example; in a more complex one a more systematic analysis would be needed and we shall return to this in Parts III and IV.

Formal language patterns are here represented by diffuse patterns, where the syntax corresponds to the image algebra $\mathscr{T}$ and style is given through the probability measure over $\mathscr{T}$. Style is then characterized by the stylistic parameters and by the additional variables, the stylistic variables, that are needed to arrive at an acceptable approximation to the linguistic performance of the group of language users being studied.

One can also ask for more concise measures of style. One simple criterion is Yule's coefficient of repetition, see 1.4, but applied to images rather than generators. A more informative entity is the <u>rank function</u> to be introduced below.

Let us take the finite state case and say that we have generated n sentences, grammatical strings. Denote by $V^{(n)}$ the number of different sentences.

The <u>type-token ratio</u> $\psi_n = V^{(n)}/n$ expresses how often sentences are repeated: if ψ_n is small repetitions occur often and vice versa. It was found in some simulation experiments that the type-token ratio was quite small but it was not clear whether this was just due to random fluctuations, or whether this is an intrinsic property of these diffuse patterns.

To find out whether this was indeed the case two methods could be used. One was numerical and based on Theorem 2.1. Say that a deterministic finite state automaton generates sentences over a lexicon with words $x,y,z,\ldots$ and variables $1,2,\ldots v$ using the rewriting rules

$$i \to xj, \quad i \text{ and } j \text{ variables}, \quad x \in \text{lexicon} \tag{2.15}$$

with probabilities $p_{ij}(x)$, and the terminating rewriting rules

$$i \to x, \quad i \text{ variable}, \quad x \in \text{lexicon}, \tag{2.16}$$

with probabilities $r_i(x)$. Then the probability of a string of words

$$S = x_1, x_2, \ldots, x_\ell \tag{2.17}$$

has a probability given by the attractive expression

$$P(S) = dP(x_1)P(x_2) \dots P(x_{\ell-1})r(x_\ell) \tag{2.18}$$

with the matrices

$$P(x) = \{p_{ij}(x); \quad i,j = 1,2,\dots,p\} \tag{2.19}$$

and the vectors

$$\begin{cases} r(x) = \text{col}[r_1(x),r_2(x),\dots,r_v(x)] \\ \\ \quad d = (1,0,0,\dots,0) \in R^v. \end{cases} \tag{2.20}$$

Enumerating all the grammatical sentences in some order $S_1, S_2, \dots$ we could study the behavior of

$$p_\nu = P(S_\nu), \quad \nu = 1,2,\dots \tag{2.21}$$

all of which will be positive.

When doing this numerically one faces the difficulty that many strings will be ungrammatical and hence have probability zero and not contribute to the rank function, see below. This means that we will have to go through a very large number of strings and evaluate their probabilities before we have a fair number of legal sentences.

To avoid this we chose instead an analytical approach. For this purpose the expression (2.18) could perhaps be used but it seems less promising when one realizes that the matrices $P(x)$ need not commute, causing trouble when studying the asymptotic properties. Instead we decided to use the Laplace transform of logarithmic rank functions as will be explained below.

Introduce the integer valued <u>rank function</u>

$$\psi(x) = \#\{\nu | P(S_\nu) \geq x\}, \qquad 0 < x \leq 1. \tag{2.22}$$

This function is a non-increasing step function and is unbounded since we exclude the trivial case when the number of grammatical sentences is finite.

The rank function is the appropriate tool for studying the question posed and we first give some preliminary results that can be found in greater detail in Bouchard (1967).

Theorem 2.2. *The rank function* ψ *has the following properties.*

(a) $\psi \in L(0,1)$ *and*

$$\int_0^1 \psi(x)dx = 1 \tag{2.23}$$

(b) *The expected value of the type-token ratio is*

$$\psi_n = \frac{m_n}{n} = E(\frac{V^{(n)}}{n}) = \int_0^1 (1-x)^{n-1}\psi(x)dx \tag{2.24}$$

and $\lim_{n\to\infty} \psi_n = 0$ *with convex behavior*, $\Delta^2\psi_m \geq 0$.

(c) *The normalized type-token ratio* ψ_n/m_n *converges in the mean to* 1.

Proof: (a) The rank function is non-negative and, integrating by parts,

$$\int_0^1 \psi(x)dx = \sum_{\nu=1}^{\infty} p_\nu = 1. \tag{2.25}$$

(b) We have

$$m_n = E[V^{(n)}] = \sum_{\nu=1}^{\infty} P(S_\nu \text{ is represented in a sample of } n \text{ sentences})$$

$$= \sum_{\nu=1}^{\infty} [1-(1-p_\nu)^n] = \int_0^1 [1-(1-x)^n]\psi(dx)$$

$$= n \int_0^1 (1-x)^{n-1}\psi(x)dx. \tag{2.26}$$

Since ψ is integrable $\psi_n \to 0$ follows from dominated convergence in the last term of the above relation. Convexity is directly established since

$$\begin{aligned}\Delta^n\psi_n &= \psi_{n+2}-2\psi_{n+1}+\psi_n = \int_0^1 [(1-x)^2-2(1-x)+1](1-x)^{n-1}\psi(x)dx \\ &= \int_0^1 x^2(1-x)^{n-1}\psi(x)dx > 0.\end{aligned} \tag{2.27}$$

It is also clear from looking at the second sum in (2.26) that $m_n \uparrow +\infty$ as $n \to \infty$.

(c)

$$\mathrm{Var}(\frac{V^{(n)}}{m_n}) = \frac{1}{m^2}\sum_{\nu=1}^{\infty} \mathrm{Var}(e_\nu) + \frac{1}{m^2}\sum_{\nu\neq\mu} \mathrm{Cov}(e_\nu,e_\mu) \tag{2.28}$$

where e_ν is the indicator variable for sentence S_ν appearing. But

$$\begin{aligned}&\frac{1}{m_n^2}\sum_{\nu=1}^{\infty} \mathrm{Var}(e_\nu) = \\ &\frac{1}{m_n^2}\sum_{\nu=1}^{\infty} [1-(1-p_\nu)^n](1-p_\nu)^n \leq \frac{m_n}{m_n^2} = \frac{1}{m_n} \to 0.\end{aligned} \tag{2.29}$$

But the covariances cannot be positive since for $\nu \neq \mu$

$$\begin{aligned}\mathrm{Cov}(e_\nu,e_\mu) &= P(e_\nu = e_\mu = 1) - P(e_\nu=1)P(e_\mu=1) \\ &= 1-(1-p_\nu)^n-(1-p_\mu)^n+(1-p_\nu-p_\mu)^n \\ &\quad -[1-(1-p_\nu)^n][1-(1-p_\mu)^n] \\ &= (1-p_\nu-p_\mu)^n-(1-p_\nu-p_\mu+p_\nu p_\mu)^n \leq 0.\end{aligned} \tag{2.30}$$

Hence $\mathrm{Var}(\frac{V^{(n)}}{m_n}) \to 0$ so that $\frac{V^{(n)}}{m_n} \to 1$ in the mean.

The rank function has some of the properties of a density. It is interesting to see what the mean of this density is. We have

$$\int_0^1 x\psi(x)dx = -\frac{1}{2}\int_0^1 x^2\psi(dx) = \frac{1}{2}\sum_1^\infty p_\nu^2$$

so that the mean is proportional to Yule's coefficient of repetition.

As an illustration let us see what happens with the type-token ratio for Zipf's law

$$p_\nu = \text{const.}\ \ \nu^{-\alpha}, \quad \alpha > 1 \tag{2.31}$$

so that

$$\psi(x) \sim cx^{-\beta}, \quad \beta = 1/\alpha, \quad 0 < \beta < 1 \tag{2.32}$$

and

$$\begin{aligned} \frac{\psi_n}{n} &\sim c\int_0^1 x^{-\beta}(1-x)^{n-1}dx = c\,\frac{\Gamma(1-\beta)\Gamma(n)}{\Gamma(n+1-\beta)} \\ &\sim c\Gamma(1-\beta)\,\frac{1}{n^{1-\beta}} \end{aligned} \tag{2.33}$$

with a power law decrease with exponent $1-\beta$ between 0 and 1.

It should be noted that this sort of discussion of type-token ratio is usually about words rather than about sentences. Since an abduction machine (see Part IV) is confronted with unknown syntax it must of course be exposed to whole sentences, not just single words, which motivates the difference.

Rather than just illustrating the problem by examples we shall explore it systematically by developing an analytical method that seems to be a powerful tool for handling questions of this type.

The grammatical sentences generated by a finite state grammar can also be obtained from a deterministic finite state automaton with a starting state, say corresponding to the variable 1, and a final state corresponding to the terminating rewriting rules.

We are interested in the behavior of the rank function $\psi(x)$, and especially its behavior at $x = 0$. We shall use the <u>logarithmic rank function</u>

$$\phi(y) = \psi(x), \quad y = -\log x. \tag{2.34}$$

It does not seem to have been observed that this function and its Laplace transform is the natural tool for studying <u>stylistic features</u> of stochastically generated languages according to the syntax-controlled model.

Style on the simplest level is characterized by the probabilities of rules $i \to x$, $x \in$ Lexicon. This will just tell us how often one uses different words for given parts of speech or other syntactic constructs.

Style of a more sophisticated sort is described by the probabilities of rules $i \to xj$. Here one would expect to find more meaningful descriptors that would be more powerful in singling out characteristic linguistic performance.

The particular stylistic quantity that we study in this paper is <u>flexibility</u> which is related to the rank function for small arguments. It tells us how extensive use of the language is being made. If $\psi(x)$ is large for small x then $E[V^{(n)}]$ is large, see (2.24), and vice versa.

Consider a variable i and the rewriting rules that lead to it

$$\begin{cases} j \rightarrow ai & \text{with probability} \quad \pi_1 \\ k \rightarrow bi & \text{"} \qquad \text{"} \qquad \pi_2 \\ \ell \rightarrow ci & \text{"} \qquad \text{"} \qquad \pi_3 \\ \text{etc.} \end{cases} \tag{2.35}$$

Denote by Σ_i the set of strings that can be generated by starting from state 1 and ending in i, all the time following the grammatical rules. Since we use a deterministic automaton all the sets Σ_i are different: any particular string can only end up in one state.

Denote by $\psi_i(x)$ the rank function associated with the probabilities generated by the branching decisions that led to the strings in Σ_i. We also use the logarithmic rank functions $\phi_i(y)$.

Considering the rules in (2.35) we note that the strings in Σ_i are made up of strings $a\Sigma_j$, $b\Sigma_k$, $c\Sigma_\ell$, etc. each with probabilities that are obtained from the ones for Σ_j, Σ_k, Σ_ℓ, etc. multiplied with $\pi_1, \pi_2, \pi_3 \ldots$ This means that

$$\psi_i(x) = \psi_j\left(\frac{x}{\pi_1}\right) + \psi_k\left(\frac{x}{\pi_2}\right) + \psi_\ell\left(\frac{x}{\pi_3}\right) + \ldots \tag{2.36}$$

since no aggregation of probabilities occurs due to the deterministic property of the system. If one of the states on the right in (2.35) is the initial state 1 then we get an additional term in (2.36)

$$\delta\left(\frac{x}{\pi}\right), \ \delta(u) = 1 \ \text{ for } \ u \leq 1, \ \delta(u) = 0 \ \text{ for } \ u > 1, \tag{2.37}$$

because the start always is at 1. The value of π is the branching probability from 1 to i.

For i = terminal state, which corresponds to the

terminating rewriting rules, we have $\psi = \psi_i$ so that if we can solve the system (2.36), for $i = 1,2,\ldots,v-1$ we get ψ as desired.

Let us express the system of equations in the logarithmic rank functions so that for all real y

$$\phi_i(y) = \phi_j(y+c_1) + \phi_k(y+c_2) + \phi_\ell(y-c_3) \tag{2.38}$$

with the negative constants $c_1 = \log \pi_1$, etc. We need a bound on the ϕ's and we get it by letting μ be the minimum of all branching probabilities used to generate grammatical sentences. Then for an arbitrary i

$$\psi_i(x) = \#\{(j_1,j_2,\ldots,j_n) \mid P(\text{states } j_1,j_2,\ldots,j_n) > x\}$$
$$\leq \{(j_1,\ldots,j_n) \mid \mu^r > x\} \leq \sum_{r=1}^{\frac{-\log r}{-\log \mu}} v^r \sim \frac{v^{\frac{-\log r}{-\log \mu}+1}}{v-1}. \tag{2.39}$$

Hence the rank functions have the bounds

$$\psi_i(x) = O(e^{\frac{-\log r}{-\log \mu}\log v}) \tag{2.40}$$

so that

$$\phi_i(y) = O(e^{y\frac{\log v}{-\log \mu}}). \tag{2.41}$$

Because of the at most exponential increase the ϕ_i have Laplace transforms

$$\lambda_i(z) = L\phi_i = \int_{y=0}^{\infty} e^{-zy}\phi_i(y)dy \tag{2.42}$$

for sufficiently large z, $z > \log v/(-\log \mu)$, and we shall only consider such values of z. It is well known that the Laplace transform determines the transformed function uniquely if it is given for $z > z_o$.

Then the system (2.38) is transformed into

$$\lambda_i(z) = \lambda_j(z)e^{c_1 z} + \lambda_k(z)e^{c_2 z} + \lambda_\ell(z)e^{c_3 z} + \ldots \tag{2.43}$$

or, if the initial state 1 is involved on the right, with the additional term

$$\int_0^\infty \delta\left(\frac{e^{-y}}{\pi}\right)e^{-zy}\, dy = \frac{e^{z \log \pi}}{z} = \frac{\pi^z}{z} \ . \tag{2.44}$$

We can therefore look at (2.43) for fixed z as an inhomogeneous system of linear equations, whose coefficients are exponential functions e^{cz}, $c < 0$, and whose constant terms are of the form $e^{z \log \pi/z}$ or zero.

The general element of the matrix $M(z)$ to be inverted in (2.44) is

$$m_{ij} = e^{c_{ij}z} - \delta_{ij}, \quad c_{ij} = -\log p_{ij} \tag{2.45}$$

for the pairs (i,j) corresponding to a possible transition $p_{ij} > 0$, and

$$m_{ij} = -\delta_{ij} \quad \text{else.}$$

The absolute value of the determinant $D(z)$ of M will for large values of z approach 1 since $-M$ will get close to the identity matrix. Hence $M(x)$ can be inverted and the determinant, a polynomial in the functions $e^{c_{ij}z}$, will appear in the denominator of the solution. This proves the following result that will be used to answer the question raised in the first section.

<u>Theorem 2.3</u>. <u>The syntax-controlled probability model for a finite state grammar generates a diffuse language pattern whose logarithmic rank functions have Laplace transforms that can be written as</u>

$$\lambda(z) = \frac{1}{z} \sum_{i=1}^{v-1} e^{z \log \pi i} R_i(z) \tag{2.46}$$

<u>where</u> $R_i(z)$ <u>are rational functions of exponential functions</u> e^{cz}, $c < 0$.

To study the different behaviors that are possible one should express $R_i(z)$ in partial fractions and apply an elementary calculus of residues argument for the Laplace inversion. It should be noted that a pole for the rational function gives rise to an infinite number of poles for $\lambda(z)$ periodically spaced on vertical lines in the complex plane. This is caused by the fact that ϕ is piecewise constant and does not tend smoothly to infinity as $y \to 0$. It is illuminating to look at the Laplace transform of $[y]$ to see what happens here. The asymptotic behavior of ϕ and ψ is not influenced by this however.

Consider first the machine with two words a and b, all branching probabilities equal to $1/2$ and $k+1$ states, starting in 1 and ending in $k+1$, see Figure 2.3(a). Then

$$\begin{cases} \lambda_1 = \alpha\lambda_1 + \beta \\ \lambda_2 = \alpha\lambda_1 + \alpha\lambda_2 + \beta \\ \lambda_3 = \alpha\lambda_2 + \alpha\lambda_3 \\ \quad \cdots\cdot \\ \lambda_k = \alpha\lambda_{k-1} + \alpha\lambda_k \end{cases} \tag{2.47}$$

with $\alpha = 2^{-z}$, $\beta = \alpha/z$. Hence

$$\begin{cases} \lambda_1 = \dfrac{\beta}{1-\alpha} \\ \lambda_2 = \dfrac{\beta}{(1-\alpha)^2} \\ \lambda_3 = \dfrac{\alpha\beta}{(1-\alpha)^3} \\ \cdot \quad \cdot \quad \cdot \quad \cdot \\ \lambda_k = \dfrac{\alpha^{k-2}\beta}{(1-\alpha)^k} \end{cases} \tag{2.48}$$

so that

$$\lambda = \alpha\lambda_k = (\tfrac{1}{2})^{kz} \frac{1}{z} \frac{1}{[1 - (\tfrac{1}{2})^z]^k} . \tag{2.49}$$

The behavior at $z = 0$ is of the form

$$\lambda(z) \sim \frac{\text{const}}{z^{k+1}} \tag{2.50}$$

so that the standard Tauberian theorem (see Feller, II, Chapter VIII) gives

$$\phi(y) \sim \text{const } y^k \tag{2.51}$$

and the rank function

$$\psi(x) \sim \text{const } (-\log x)^k . \tag{2.52}$$

This is not very large for x close to zero compared to Zipf law behavior. It therefore seems to support the conjecture that the style of these languages is characterized by low flexibility.

To show that this is not always so we shall choose a case where the topology of the wiring diagram of the automaton is quite different.

Consider the cascade like wiring diagram in Figure 2.3(b). The lexicon has two words, a and b, and there are

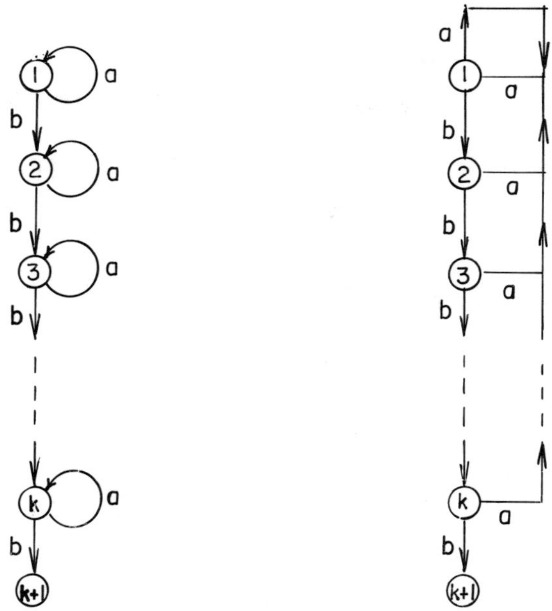

(a) Figure 2.3 (b)

k+1 states with start in 1 and stop in k+1. The Laplace transform of the logarithmic rank functions will then satisfy

$$\begin{cases} \lambda_1 = (\frac{1}{2})^z \lambda_1 + (\frac{1}{2})^z \lambda_2 + \cdots + (\frac{1}{2})^z \lambda_k + (\frac{1}{2})^z \frac{1}{z} \\ \lambda_2 = (\frac{1}{2})^z \lambda_1 + (\frac{1}{2})^z \frac{1}{z} \\ \lambda_3 = (\frac{1}{2})^z \lambda_2 \\ \cdots\cdot \\ \lambda_k = (\frac{1}{2})^z \lambda_{k-1} \end{cases} \tag{2.53}$$

where we have assumed all the branching probabilities to be 1/2. Hence

$$\lambda_1 = \beta \frac{1-\alpha^k}{1-2\alpha+\alpha^{k+1}}, \quad \alpha = (\tfrac{1}{2})^z, \; \beta = \frac{\alpha}{z} \tag{2.54}$$

and finally

$$\lambda(z) = (\tfrac{1}{2})^{kz} \frac{1}{z} \frac{1}{1-(\frac{1}{2})^{z}-(\frac{1}{2})^{2z}-\cdots-(\frac{1}{2})^{(k+1)z}} \tag{2.55}$$

The behavior of ϕ depends upon the value of the positive root ξ of

$$1 - x - x^{2} \ldots - x^{k+1} = 0 \tag{2.56}$$

where $1/2 < \xi \leq 1$ and is close to $1/2$ if k is large. Then $\lambda(z)$ will be regular in the half plane for which

$$\mathrm{Re}(z) > \frac{-\log \xi}{\log 2} = \gamma, \; 0 < \gamma < 1. \tag{2.57}$$

The behavior of ϕ is then governed by

$$\lambda(z) \sim \frac{\text{const.}}{(\frac{1}{2})^{z} - (\frac{1}{2})^{\gamma}} \tag{2.58}$$

close to $z = \gamma$. This means that $\phi(y)$ behaves as

$$\phi(y) \sim Ae^{y\gamma} \tag{2.59}$$

and

$$\psi(x) \sim \frac{A}{x^{\gamma}}, \quad 0 < \gamma < 1 \tag{2.60}$$

and we can make γ as close to one as we wish by making k large. We can therefore conclude that the flexibility of the syntax-controlled performance model can be very large.

3. Other Abstract Images.

Let the generators still be abstract symbols whose meanings are expressed directly without need of an information carrying medium, no background space. We shall now look at a few cases when the connection type Σ is not LINEAR or TREE.

Case 3.1. Colored graphs. Let the generators be enumerated as $G = \{1,2,3,\ldots\}$ and with arbitrary connections in the configurations, ρ = TRUE. Identify two configurations $c = (g_1,g_2,\ldots,g_n)$ and $c' = (g_1',g_2',\ldots,g_m')$ if

(i) $n = m$

(ii) there exists a permutation p of the n first natural numbers such that if p: $i \to i'$, $j \to j'$ then g_i and g_j are connected by structure (c) iff $g_{i'}$ and $g_{j'}$ are connected by structure c'. This R is called <u>identification by permutation</u>.

One can think of the g's as colors attached to the nodes of the graph of connections.

In Figure 3.1 the configurations c and c' are identified as belonging to the same image I (use the permutation $(1,2,3,4) \to (2,1,4,3)$, but c'' is not R-equivalent to any of them).

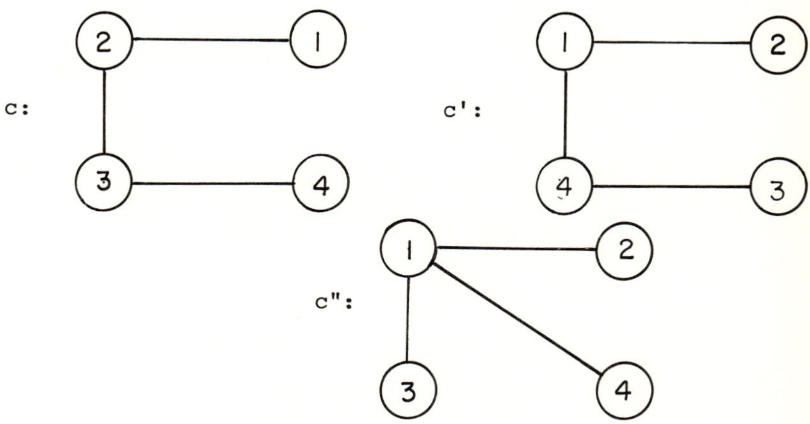

Figure 3.1

Very complex images can occur in the study of <u>kinship relations</u>, dear to anthropologists as well as to population geneticists. We shall look in particular at two aspects that complement each other: inbreeding and taboos against incest.

Consider the family tree in Figure 3.2, showing some of the ancestors of the famous bull Comet. The signs here are only two, male and females:

$$S = \{♂, ♀\}, \tag{3.1}$$

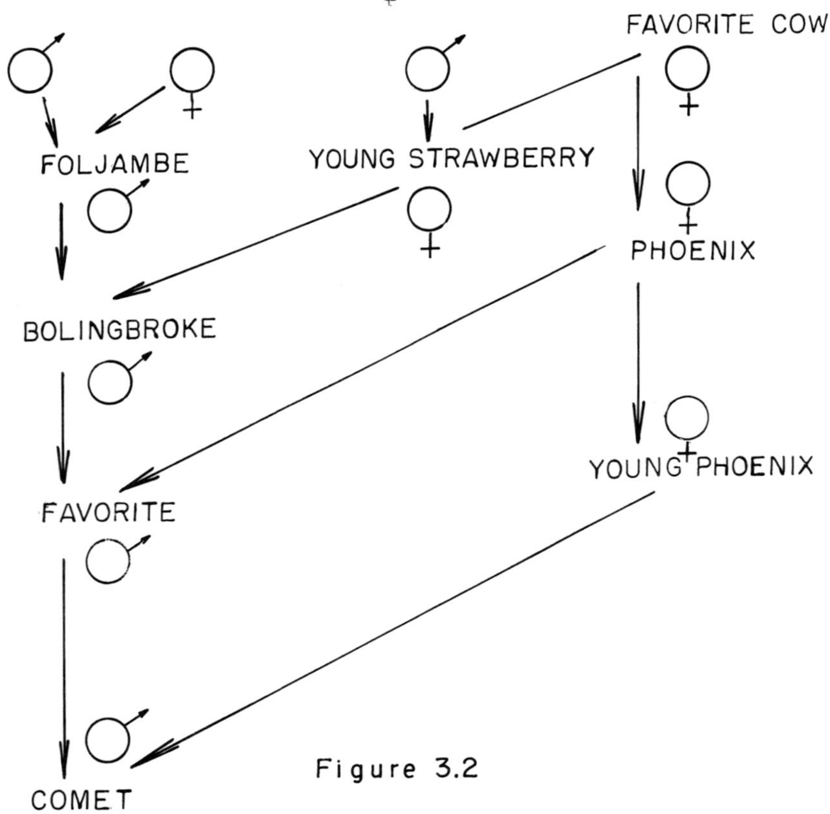

Figure 3.2

and their different occurrences are indicated by names. This small image does not show much regularity, but due to the inbreeding there are some multiple relationships; Comet is grandson to Pheonix both maternally via Young Phoenix and paternally via Favorite. He is also his own cousin and nephew to his mother.

The choice of connection type Σ depends upon the mating system, more about this later. Anyway it is clear that loops are ruled out always. Also it is clear that this sort of family tree does not have tree structure since an individual may descend from two individuals with a common ancestor.

To decide what constitutes a regular configuration we have to determine two things: <u>the mating rules and how complete the graph should be</u>.

In human populations mating is usually restricted by taboos preventing some incestuous relations. These vary widely and we have to choose the restrictions accordingly for each particular case. There may also be generative rules specifying preferential marriage between groups.

Completeness, on the other hand, should mean that no parent-child relation is left out in the graph. For any finite configuration, however, there will always be some elements with no or only one parent belonging to it, and others for which none or only some of the children belong to the configuration.

As a minimum set of restrictions we assume that

1. To each element of c are associated one or no male parent and one or no female parent.
2. All parent-child relations between elements in c are included.

Random sources of admissible configurations are of interest both in genetics and anthropology and are easiest to define recursively. Given a set of generators, we need the following specifications.

1. A random mechanism for mating between the generators following the restrictions $\mathcal{R}$.
2. A random mechanism for the number and sex of the resulting offspring.
3. A mechanism determining how generations follow upon each other, say by giving a distribution for the length of life and the fertile part of it.

Once we have a configuration the question arises how to describe it functionally. To characterize inbreeding and kinship in family patterns the following reasoning was introduced by Wright (1921) and made more precise by Malécot (1948). Consider two genes a and b in a population. It can happen that they are alike because they have arisen as a result of the reproductive process from <u>one</u> gene occurring in the ancestry of the two carriers: then a and b are said to be <u>identical by descent</u> and one writes $a \equiv b$. On the other hand, it can happen that they are just alike but that they do not result from the same chromosome in any common ancestor: then a and b are called <u>identical in state</u>.

The <u>coefficient of inbreeding</u> F_x of an individual x is defined as the probability that the two genes of the individual x at a particular locus are identical by descent, so that if x has genotype ab, then $F_x = P(a \equiv b)$.

For two individuals x and y the <u>coefficient of kinship</u> r_{xy} is defined as the probability that a random gene (at a particular locus) of x is identical by descent to a random gene of y (at the same locus), so that if x has genotype ab and y genotype cd

$$r_{xy} = \frac{1}{4} P(a \equiv c) + \frac{1}{4} P(a \equiv d) + \frac{1}{4} P(b \equiv c) + \frac{1}{4} P(b \equiv d). \tag{3.2}$$

Some simple relations exist between F and r. First, if x has the genotype ab (at some locus) then its coefficient of kinship with itself is given by

$$\begin{aligned} r_{xx} &= \frac{1}{4} P(a \equiv a) + \frac{1}{4} P(a \equiv b) + \frac{1}{4} P(b \equiv a) + \frac{1}{4} P(b \equiv b) \\ &= \frac{1}{2} (1+P(a \equiv b)) = \frac{1}{2} (1+F_x), \ F_x = 2r_{xx} - 1. \end{aligned} \tag{3.3}$$

On the other hand, crossing x and y we consider an offspring denoted by $x \times y$. If x is ab and y is cd then $F_{x\times y}$ is $\frac{1}{4} P(a \equiv c) + \frac{1}{4} P(a \equiv d) + \frac{1}{4} P(b \equiv c) + \frac{1}{4} P(b \equiv d)$ so that (see equation (3.2)):

$$F_{x\times y} = r_{xy}. \tag{3.4}$$

A more general relation is obtained as follows. Consider the individuals x,y,z with genotypes ab, cd, ef. Then an offspring $y \times z$ will have one of the genotypes ce, cf, de, df, each case with probability one-quarter. Therefore

$$\begin{aligned} r_{x,y\times z} &= \frac{1}{4}[\frac{1}{4} P(a \equiv c) + \frac{1}{4} P(a \equiv e) + \frac{1}{4} P(b \equiv c) + \frac{1}{4} P(b \equiv e)] \\ &\quad + \frac{1}{4}[P(a \equiv c) + \frac{1}{4} P(a \equiv f) + \frac{1}{4} P(b \equiv c) + \frac{1}{4} P(b \equiv f)] \\ &\quad + \frac{1}{4}[P(a \equiv d) + \frac{1}{4} P(a \equiv c) + \frac{1}{4} P(b \equiv d) + \frac{1}{4} P(b \equiv c)] \\ &\quad + \frac{1}{4}[P(a \equiv d) + \frac{1}{4} P(a \equiv f) + \frac{1}{4} P(b \equiv d) + \frac{1}{4} P(b \equiv f)] \\ &= \frac{1}{8} [P(a \equiv c) + P(a \equiv d) + P(a \equiv e) + P(a \equiv f) \\ &\quad + P(b \equiv c) + P(b \equiv d) + P(b \equiv e) + P(b \equiv f)] \\ &= \frac{1}{2} (r_{xy} + r_{xz}). \end{aligned} \tag{3.5}$$

With these relations it is easy, at least in principle, to calculate successively the kinship coefficients in a given family tree, and then $F_x = 2r_{xx}-1$,

As an example, see Figure 3.3 and consider the offspring $7 = 3\times 6$ of an uncle-niece marriage where 1,2, and 5 are not inbred and not related to each other. To find its inbreeding coefficient we apply (3.4) and (3.5) repeatedly.

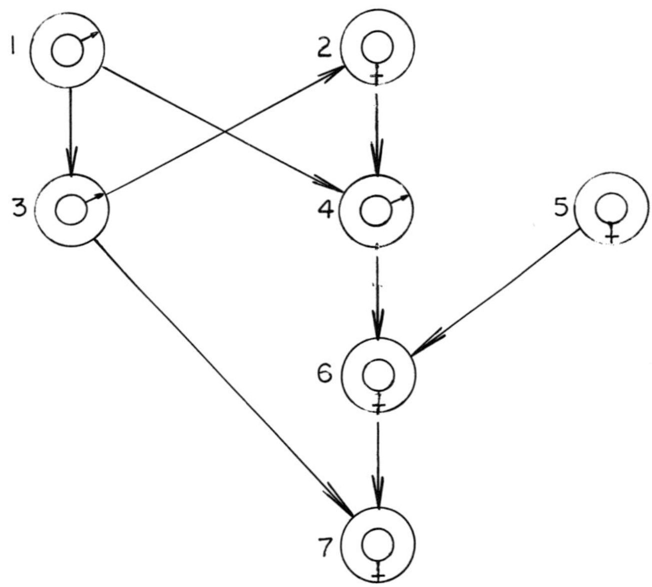

Figure 3.3

We get $F_1 = F_2 = F_5 = 0$ and $r_{11} = r_{22} = r_{33} = 1/2$, while $r_{12} = 0$. Hence

$$\begin{aligned} r_{34} &= r_{3\ 1\times 2} = \frac{1}{2} r_{31} + \frac{1}{2} r_{32} = \frac{1}{2} r_{1\ 1\times 2} + \frac{1}{2} r_{2\ 2\times 1} \\ &= \frac{1}{4} [r_{11} + 2r_{12} + r_{22}] = \frac{1}{4}. \end{aligned} \tag{3.6}$$

Proceeding recursively down the tree, we finally arrive at the inbreeding coefficient $F_7 = 1/8$.

On the other hand, if we have a regular mating system the recursive relation (3.5) leads to a difference equation with constant coefficients. As an example consider double-first-cousin mating as in Figure 3.4 with only three generations indicated. The individuals 1 and 2 are assumed to be full sibs, as are 3 and 4.

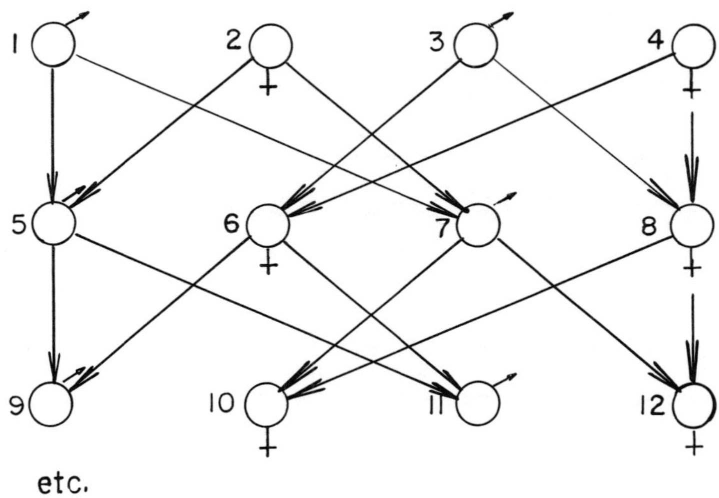

Figure 3.4

Introducing the notation F_t for the inbreeding coefficient in generation number t, we have $F_2 = r_{12}$ and $F_3 = r_{56}$. Now

$$F_4 = r_{9,10} = \frac{1}{4} (r_{57} + r_{58} + r_{67} + r_{68}) \tag{3.7}$$

and $r_{57} = r_{68}$, $r_{58} = r_{67}$, so that

$$F_4 = \frac{1}{2}(r_{57} + r_{58}). \tag{3.8}$$

But

$$\begin{aligned} r_{57} &= \frac{1}{4}(r_{11} + r_{22} + 2r_{12}) \\ &= \frac{1}{4}\left[\frac{1}{2}(1+F_1) + \frac{1}{2}(1+F_1) + 2F_2\right] = \frac{1}{4}[1+F_1+F_2] \end{aligned} \tag{3.9}$$

and $r_{58} = F_3$, so that

$$F_4 = \frac{1}{8} + \frac{1}{8}F_1 + \frac{1}{4}F_2 + \frac{1}{2}F_3. \tag{3.10}$$

For general generation number t we get similarly the linear difference equation with constant coefficients of order three

$$F_{t+3} = \frac{1}{8} + \frac{1}{8}F_t + \frac{1}{4}F_{t+1} + \frac{1}{2}F_{t+2} \tag{3.11}$$

and this can be solved by the standard method.

Thus we can describe the individuals in terms of inbreeding and by collecting statistics also average inbreeding, and so on.

Looking at the same sort of configuration, but from another perspective, we can ask: given the restrictions on mating, what structural properties do the resulting configurations have? A celebrated case is that of the Murngin system (see Notes). The population is divided into eight groups 1,2,3,...,8, with males and females in each, so that we have in all 16 generations:

$$S = \{1♂, 1♀, 2♂, 2♀, \ldots, 8♀\}. \tag{3.13}$$

The exchange relations are given in Table 3.1 consisting of two sets of rules, a and b. A male in 1 can marry a female from 3 (see (a)), or in 4 (see (b)), a male in 2 can marry a female in 4 (see (a)) or in 3, etc. There are

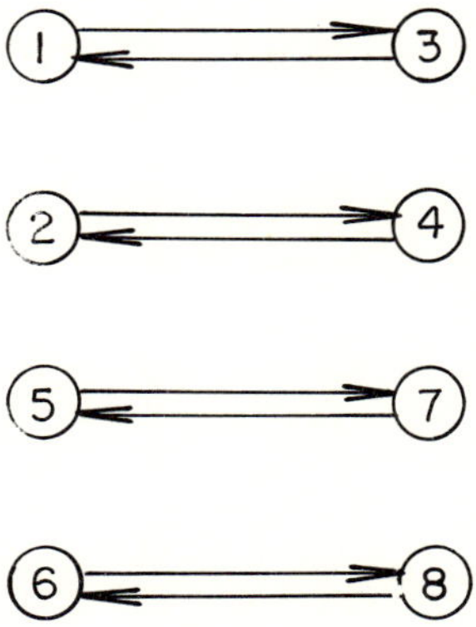

Table 3.1a

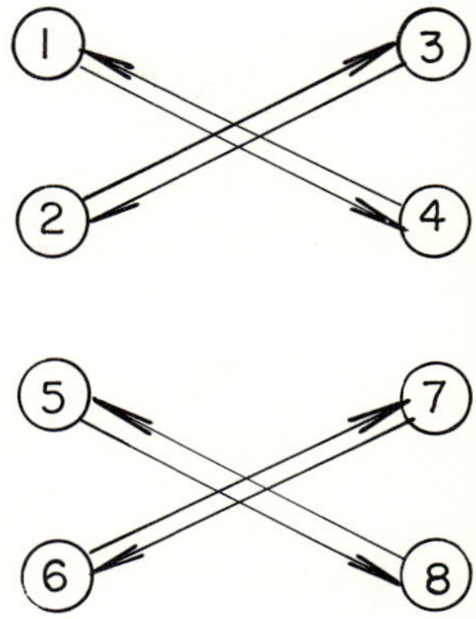

Table 3.1b

Mother's class	Children's class
1	6
2	5
3	8
4	7
5	1
6	2
7	3
8	4

Table 3.2

sixteen such types of marriage and we enumerate them 1,2,...,16. The offspring of a union depends only upon what group the mother belongs to, as indicated in Table 3.2.

One regular configuration is shown in Figure 3.5.

For an individual in class i there are two choices depending upon whether rules (a) or (b) are applied. Assume

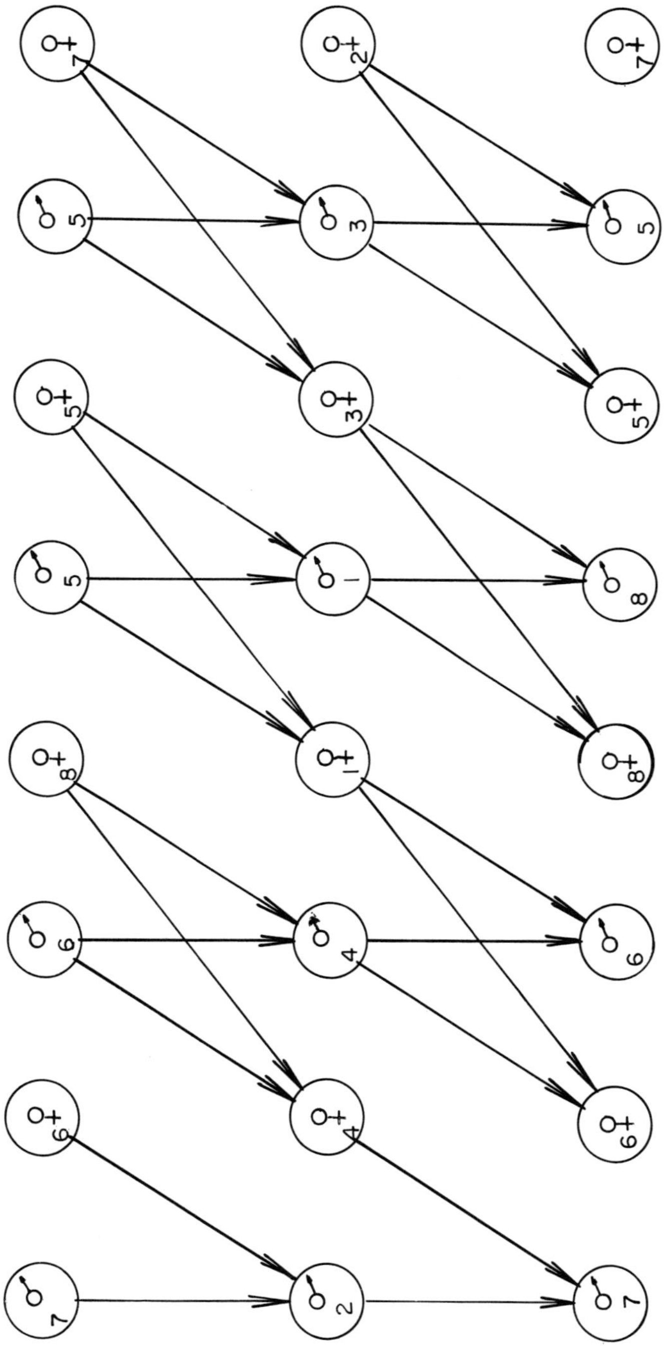

Figure 3.5

that this choice depends only upon the individual's sex and the type j of the marriage of its parents. If we characterize an individual by (i,j) we get a set of permutations

$$\begin{cases} f\colon (i,j) \to (i',j') & \text{for females} \\ m\colon (i,j) \to (i'',j'') & \text{for males} \end{cases} \qquad (3.13)$$

The resulting permutation group will simplify if one imposes the particularly interesting condition that a male should always be allowed to marry his mother's brother's daughter. Consider a brother and sister from a j marriage and an i mother. Then the brother will have to marry in group i'' and the sister in i'. If the sister's marriage results in a son and the brother's in a daughter, then the condition that these two should be allowed to marry implies $fm = mf$: the two permutation operators should commute with each other. In this way one can proceed to study the algebraic structure of $\mathscr{L}(\mathscr{R})$. A reader interested in a more detailed examination of this case is recommended to consult the references in the Notes.

A simplifying assumption often used in population studies is that the population evolves through disjoint generations, one generation consisting of individuals of the same age. Enumerating the generations by an index $t = 1,2,3,\ldots$ we get configurations as in Figure 3.6 with a highly irregular mating mechanism. We have left out the identifiers, only show $\alpha = ♂$ or ♀, and also left out the bonds on the arrows. The successive generations γ_t now play the role of macrogenerators. It is of interest to note that the γ_t form a Markov process if the mating is

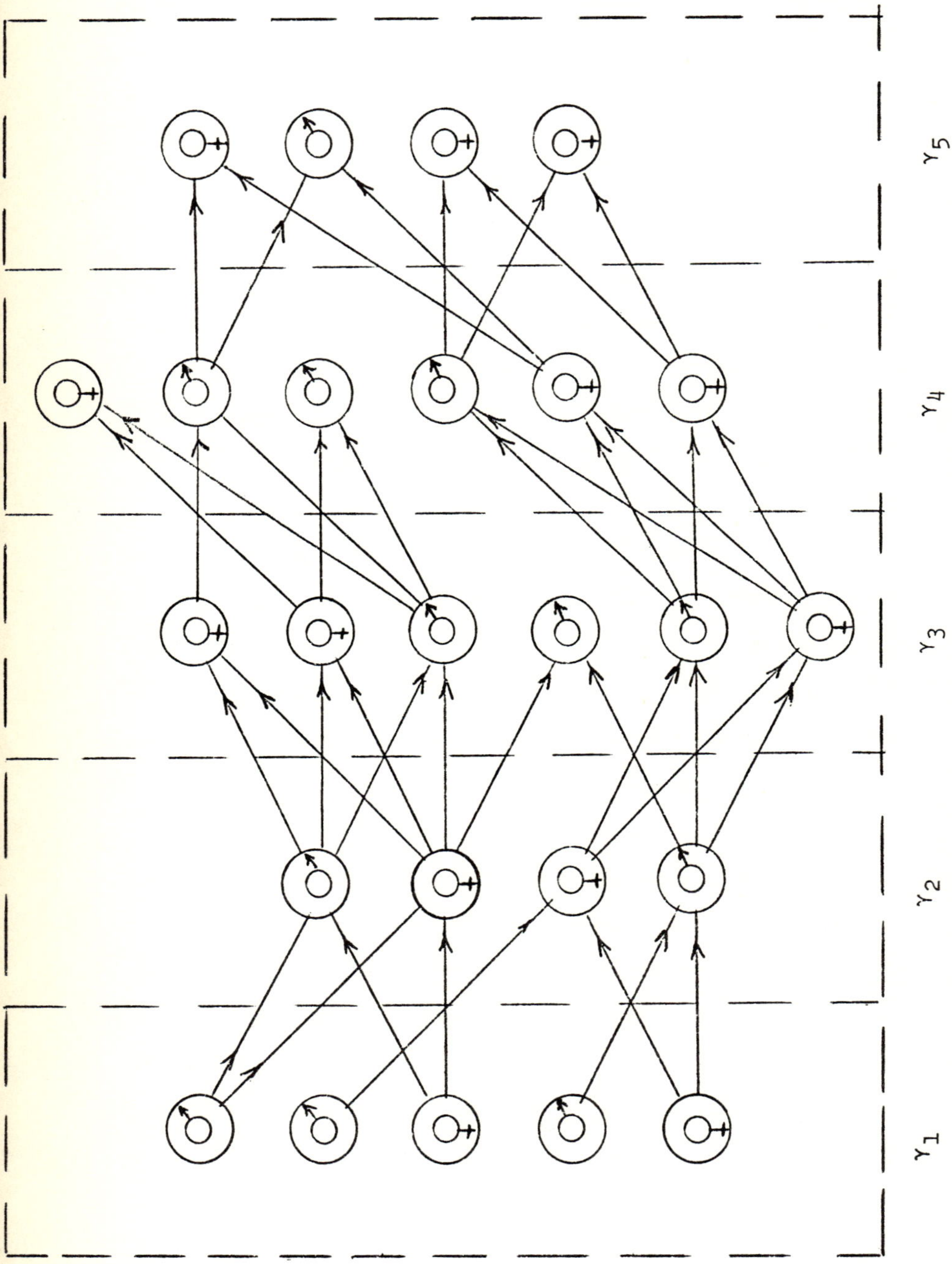

Figure 3.6

random and the number of offsprings from a mating has a distribution independent of the ancestry of the couple.

In this section the generators have been abstract. We shall postpone the discussion of similar looking configurations embedded in a background space to Section 5.5 and Part IV.

4. Temporal and Other One-Dimensional Patterns.

With generators defined against the real line as background space we shall assume throughout this section that the connection type is LINEAR. This represents the direction of the arrow of time.

Let us start by examining sequences of integers treated as numbers, not as abstract symbols. We are all familiar with quizzes from intelligence tests where a short sequence is given and one is asked to continue it. Consider the three sequences

I: 7, 10, 13, 16, 19

II: 3, 9, 27, 81, 243

III: 2, 6, 12, 20, 30

One is supposed to recognize I as an arithmetic series of the form $x_t = 3t+4$ satisfying the recursive relation $x_t = x_{t-1}+3$ so that the continuation would be 22, 25, ...

The sequence II can be looked at as a fragment of the geometric series $x_t = 3^t$ satisfying the recursion $x_t = 3x_{t-1}$. The continuation would be 729, 2187,... .

Finally, III can be regarded as an arithmetic series of order 2, $x_t = t^2+t$, satisfying $x_t = 2x_{t-1}-x_{t-2}+2$, with the continuation 42, 56, ...

Strictly speaking, a problem like this has, of course, no unique solution, but this is not really the question. The underlying idea, not always spelled out, is that one wants to generate the sequence and its continuation by a simple combination of familiar arithmetic operations. This naturally fits into the formalism of patterns.

Indeed, let the generators have $\omega_{in} = \omega_{out} = 1$ with the bonds β and β' and an attribute a. A function $f: Z \to Z$ links the bond values together, $\beta' = f(\beta)$. With the bond relation chosen as EQUAL we get the sort of numerical sequence patterns we are looking at.

Different function classes, $f \in \mathscr{F}_r$, will give rise to different pattern classes: arithmetic ones, arithmetic of order two, geometric, etc. Here it is helpful to view the generators as macro-generators of lower level generators, and hence consider the numerical sequence patterns as derived from operator patterns.

Case 4.1. (Arithmetic sequence patterns). The sequences are generated recursively from an operator configuration, see Case 2.7.1, where all $\nu = \mu = p$ and all bond values are Z or subsets of it. The resulting sequence configurations then consists of p-tuples of integers linked together with linear connection type. A function $\phi: Z^p \to Z$ is used to identify images, so that two sequence configurations $c = (g_1, g_2, \ldots, g_m)$, $c' = (g_1', g_2', \ldots, g_n')$ are identified modulo R iff $n = m$ and $\phi(g_k) = \phi(g_k')$ for all k.

The function ϕ could be for example the first component of the p-tuple.

In the present context the operators would accept as inputs and produce as outputs integers. As an example we could use as generators the binary operators ADD and MULTIPLY with $\omega_{in} = 2$, $\omega_{out} = 1$, and unary operators ADD(a) and MULTIPLY(a) with $\omega_{in} = \omega_{out} = 1$. Then III above will be derived from the configuration made up of the generators ADD, ADD(2), MULTIPLY, MULTIPLY(2), MULTIPLY(-1), and $p = 2$. The bond values of the lower level generators are of course domains and ranges, in this case $= Z$ and subsets of it.

Let us now examine some patterns appearing in the manufacturing of <u>yarns</u>. The yarn is made up by fibers and we shall assume that they are arranged parallel to the main axis of the yarn and are identical except for their length and location. We shall not go into the interesting question of the spatial structure of the yarn such as the twist it gets during the spinning process.

The generators will be line segments $g = (x,y)$, $x < y$, and we shall let $\omega_{in}(g) = \omega_{out}(g) = 1$ each bond value equal to the <u>front end</u> y of the fiber g. The length is another attribute, $\ell = y-x$. Since we shall deal only with yarns made of one sort of fiber we need only one generator class, $\alpha = 1$. The bond relation $g \rho g'$ will be chosen as $y \leq y'$.

<u>Case 4.2</u>. (<u>Thin sliver patterns</u>). With G and $\mathscr{R}$ as above and with S as the translation group the resulting configuration space $\mathscr{L}(\mathscr{R})$ will be used to form sliver patterns by employing the identification rule that two configurations are identified if the number $I(x)$ of fibers are equal for them and this holds for any location x.

Of course we could have used the <u>trailing end</u> x as

the bond value. Many yarns do not have the same structure in both directions, so that front ends and trailing ends can very well behave differently. This may be due to the <u>drafting process</u> and should not be overlooked when trying to synthesize these patterns.

Let us consider a configuration $c = (g_1, g_2, \ldots, g_n) \in \mathscr{C}_n(\mathscr{R})$ and think of it as <u>a long sliver</u>. As long ago as 1945 it was pointed out by Martindale that it is not possible to control the location of the fibers exactly and that it is therefore necessary to describe the structure of the sliver in probabilistic terms.

In $\mathscr{C}_\infty(\mathscr{R})$ we introduce a measure as follows. The right endpoints y_i form a homogeneous renewal process with the continuous distribution function F for the differences $y_{i+1} - y_i$. All the lengths $\ell_i = y_i - x_i$ are independent between themselves and of the x's and have a distribution function P with mean m_L.

The observer will see the contrast image $I(y) = \#\{\text{fibers at location } y\}$. This function is piecewise constant except at front- and trailing ends where it jumps -1 or $+1$ respectively. It will be characterized by

$$\begin{cases} m(y) = EI(y) \\ r(y,y+h) = \mathrm{Cov}[I(y), I(y+h)] \end{cases} \tag{4.1}$$

Two extreme patterns are shown in Figure 4.1. In (a) both F and P have only one point of increase and we get the highly <u>regular sliver</u>. On the other hand if F is exponential, and if P is continuous, so that the y's form a Poisson process, we get the <u>random sliver</u>. Of course inter-

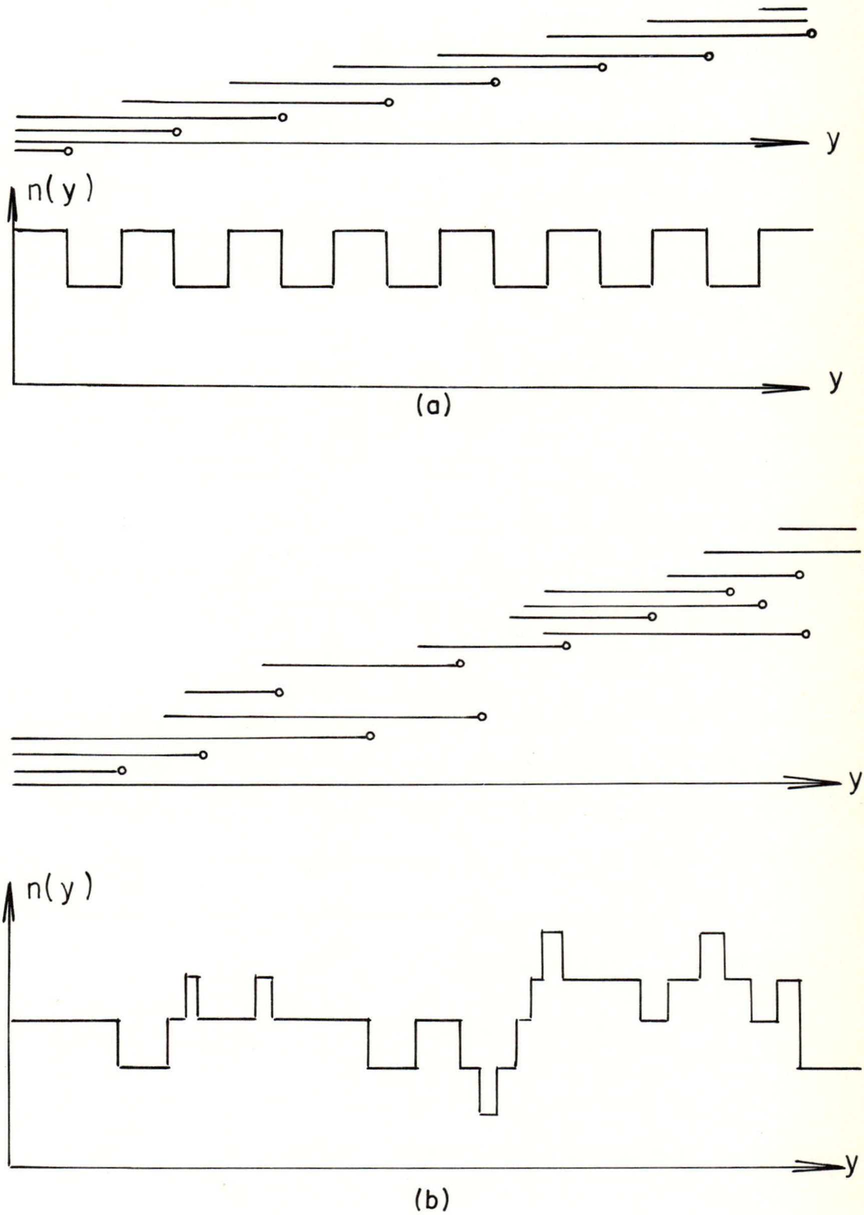

Figure 4.1

mediate patterns are also possible.

With these assumptions and with $G = 1-F$ it is clear that

$$m(y) = \sum_i EP(\ell_i > y_i - y,\ y_i > y) \tag{4.2}$$

or, if we define $G(y) = 0$ for $y < 0$,

$$m(y) = \sum_i EG(y_i - y) \tag{4.3}$$

and a standard argument in renewal theory gives us the rather obvious result,

$$m(y) = \lambda \int_0^\infty G(z)dz = \lambda m_L. \tag{4.4}$$

Similarly for the covariance function we get first for $h \geq 0$

$$E[I(y)I(y+h)] = \sum_{i,j} EP\{\ell_i > y_i - y, y_i > y,\ \ell_j > y_j - y - h,\ y_j > y+h\}. \tag{4.5}$$

The non-diagonal terms of this double sum give the expression

$$\sum_{i \neq j} E[G(y_i - y)G(y_j - y - h)] \tag{4.6}$$

which can be evaluated with the same method to be $\lambda^2 m_L^2$. The diagonal terms have to be treated separately since the same ℓ_i appears in the two factors. Doing this we get

$$\sum_i EG(y_i - y)e(y_i > y+h) \tag{4.7}$$

where $e(\cdot)$ is the indicator function of the event inside parenthesis. Hence (4.7) reduces to

$$\lambda \int_h^\infty G(z)dz = \lambda H(h); \quad H(x) = \int_x^\infty G(x)dx \tag{4.8}$$

so that the covariance function can be expressed as

$$r(y,y+h) = (8) + (6) - \lambda^2 m_L^2 = \lambda H(h). \tag{4.9}$$

In particular the variance becomes λm_L.

The spectral density is then

$$\phi(\mu) = \frac{1}{2\pi}\int_{-\infty}^{\infty} e^{-ih\mu} r(y,y+h)dh = \frac{1}{\pi}\int_0^{\infty} \cos(h\mu) r(y,y+h)dh \tag{4.10}$$

or, after integrating by parts twice,

$$\phi(\mu) = \frac{\lambda}{\pi}\int_0^{\infty} \frac{1-\cos \mu h}{\mu^2} F(dh). \tag{4.11}$$

These patterns can be used to derive new patterns corresponding to the <u>drafting process</u>. The sliver is moving along its axis at uniform velocity and under one roller, B. Somewhere between B and another roller F the fibers are suddenly accelerated to a higher velocity. Just where the change occurs can vary in an irregular manner, possibly due to the interactions between adjacent fibers. To get some feeling for the way the derived patterns may look we shall consider a highly idealized model.

Consider a long random sliver with the front end coordinates of the fibers y_i. A fiber g_i enters the back roller B at speed v_1 and accelerates to speed v_2 at the location ξ_i, see Figure 4.2, with $D = v_2/v_1 > 1$, and $0 \leq \xi_i \leq \ell$. This implies that the drafted sliver has front end points y_i that can be written, up to an additional constant, as

$$y_i = Dx_i + \eta_i \tag{4.12}$$

where $\eta_i = (1-D)\xi_i$.

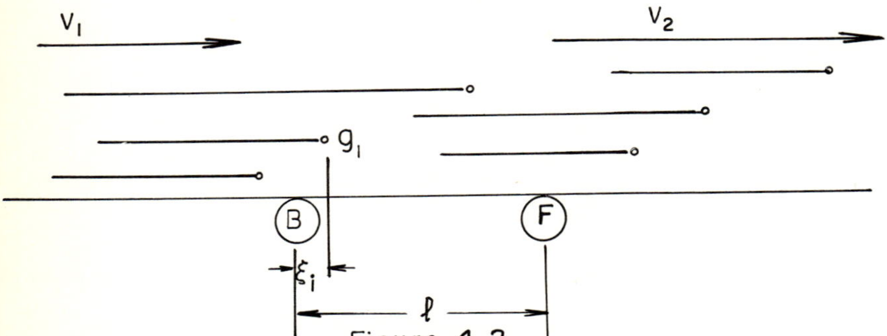

Figure 4.2

Assume that all the η's have the same frequency function f which of course has to vanish outside a finite interval. We shall determine the structure of the derived pattern, in this case the properties of the fiber image $I_o(y)$ as depending upon the location y.

It is easy to find the mean since we can write

$$\begin{aligned} m_o(y) &= E[I_o(y)] = \sum_i EP\{y \leq Dx_i+\eta_i \leq y+\ell_i\} \\ &= \sum_i EG(Dx_i+\eta_i-y) \end{aligned} \tag{4.13}$$

which is evaluated, as before, to be

$$m_o(y) = \lambda \int_{-\infty}^{\infty} G_f(Dx-y)dx = \frac{\lambda}{D} m_L \tag{4.14}$$

where we have written

$$G_f(u) = \int_{-\infty}^{\infty} G(u+v)f(v)dv \tag{4.15}$$

A could be expected (4.14) means simply that the <u>average</u>

width of the sliver has decreased by a factor $1/D$ due to conservation of mass.

To get the second order properties of the derived sliver pattern we note that for $h > 0$

$$E[I_o(y)I_o(y+h)] = \sum_{i,j} EP\{y \le Dx_i+\eta_i \le y+\ell_i, \\ y+h \le Dx_j+\eta_j \le y+h+\ell_j\} = S_1 + S_2 \tag{4.16}$$

where S_1 is the sum over $i = j$ and S_2 the rest.

Then

$$\begin{aligned} S_1 &= \sum_i EP\{y+h \le Dx_i+\eta_i \le y+\ell_i\} = \\ &= \sum_i E[G(Dx_i+\eta_i-y)e(Dx_i+\eta_i \ge y+h)] = \\ &= \lambda \int_{-\infty}^{\infty} \int_{Dx+v-y>h} G(Dx-y+v)f(v)dvdx = \frac{\lambda}{D} H(h). \end{aligned} \tag{4.17}$$

and S_2 is calculated just as before.

Summing up, we see that the derived pattern has the same mean and covariances as the primary one except that they have been reduced by the factor $1/D$. The coefficient of variation for the derived pattern is $\sqrt{D}$ times larger than the primary one. This is in accordance with the empirically established fact that the drafting process tends to increase the relative irregularity of the sliver pattern.

We have found the somewhat puzzling result that except for the reduction by the factor $1/D$ the covariance function is unchanged by the drafting process. Its length scale is not affected, for example. Nor is its spectrum more concentrated as it ought to be to account for the drafting waves that have been observed empirically. The model is not detailed enough to account for this phenomenon. Attempts to do this can be found in Foster (1951) and Sudarsana Rao (1962).

It should be pointed out that the derived pattern has changed in a less obvious manner than that measured by second order properties. The new positions of the front ends do no longer form a Poisson process: dependence has been introduced and exponential inter-point distribution does no longer apply.

In yarn patterns the background represents a space dimension. For the other one dimensional patterns to be discussed in this section the background will be the time axis.

In the case of speech patterns one can study them at different stages of their production. One can think, for example, of speech on a rather abstract level as consisting of a linear chain of atomic units, say phonemes, which are linked together according to rules and frequencies that characterize the speaker and his language. On the other hand one can measure some physical quantity, such as sound pressure variations, that expresses the acoustic effects of speech. Still other levels exist, but for the present discussion these two will suffice.

How is the acoustic wave pattern derived from the string of phonemes? A simple, but too naive, approach would assume that to each phoneme there would correspond a well defined acoustic wave form. In other words a configuration $c = (g_1, g_2, \ldots, g_n)$ of chained phonemes would give rise to the derived configuration $c' = (w(g_1), w(g_2), \ldots, w(g_n))$, where $w(g)$ stands for the wave form associated with a phoneme g. This would lead to a homomorphism $\mathscr{C}_n \to \mathscr{C}'_n$ based on a simple rule of substitution $w\colon G \to G'$, with G' consisting of the possible wave forms associated with phonemes.

The wave form $w(g)$ need not be completely specified by g. The bond values may be left open and be determined later by continuity conditions. Or, only the power spectrum of $w(g)$ is specified, not the whole function. Anyway, these are minor modifications and change little in principle.

It is known, however, that this approach does not work. The reader is referred to the influential 1967 paper by Lieberman, Cooper, Shankweiler, and Studdert-Kennedy for a discussion of the speech code and the role of context for acoustic cues. The coding is not just a cipher, $G \to G'$, but a more complex mapping $\mathscr{C} \to \mathscr{C}'$.

Lieberman et al. show how the coding differs from a straightforward cipher in several ways. A spectral component of a speech wave segment, a formant, can be directly influenced by the sound element that comes after it. Also there are no well defined boundaries in the speech wave dividing it into intervals of phonemic size. They point out too that "spectrograms are, in fact, extremely difficult to read." In general, their findings indicate that the mapping of a string of phonemes into an acoustic wave is not as direct as one may have hoped.

Many attempts have been made to account for observed features of the acoustic wave in terms of changes in the geometry of the vocal tract and in the spectrum of the glottal sound source. Based on the more or less idealized equations for the waves produced computer simulations have been carried out and analytic studies have also been attempted by several researchers. As far as the author knows the latter have been based on quasi-stationary considerations (instantaneous spectra). For a review see Flanagan (1972). Here we shall sketch the analysis of a highly idealized system of pattern synthesis, but one which need not be limited to the quasi-stationary case. The reader should be warned that this is only a capability study in the spirit of the introduction.

Let us consider a mechanical system consisting of ℓ harmonic oscillators coupled to each other with a coupling constant s, and with damping f. We shall assume that the damping coefficient f is very small. With appropriate choice of unit the system equations can be written

$$\begin{cases} \ddot{x}_1+f\dot{x}_1 = s(x_2-2x_1)+F(t) \\ \ddot{x}_k+f\dot{x}_k = s(x_{k+1}-2x_k+x_{k-1}), \quad k = 2,3,\ldots,\ell-1 \\ \ddot{x}_\ell+f\dot{x}_\ell = s(x_{\ell+1}-2x_\ell+x_{\ell-1}). \end{cases} \tag{4.18}$$

Here F is the driving force characterized by its mean value m and spectral density f. The system is bound to $x_{\ell+1}$ which is fixed so far.

We are of course allowing F to change character, actually with changes at certain time points $\{t_\nu\}$. More interestingly, <u>we will allow the geometry to change</u> suddenly by letting ℓ increase or decrease. If $\ell \to \ell+1$ (only for $\ell < n$) the number of equations increases by one; if $\ell \to \ell-1$, which is only allowed for $\ell > 1$, then we lose one equation. The maximum number of ℓ is n so that we always have $1 \leq \ell \leq n$ and x_{n+1} is defined as $x_{n+1} = L$, a fixed number. When a new equation is added we shall start the new variable at its old value and derivative zero. Conversely, when we drop one equation the old variable will be locked at its old value and with derivative zero.

The quantity we shall be interested in is the wave $x_1(t)$ and its spectral properties, but to get it we must of course deal with the entire x-vector.

It will be convenient to use an n-dimensional state vector $x = \text{col}(x_1, x_2, \ldots, x_n)$. For $t_v < t < t_{v+1}$ the system equations will consist of (4.18) plus

$$\dot{x}_{\ell+1} = \dot{x}_{\ell+2} = \ldots = \dot{x}_n = 0 \tag{4.19}$$

and we shall assume that at any time, also at the transition points t_v, the column vector $y = (x_1, \ldots x_n, \dot{x}_1, \ldots \dot{x}_n)$ is a continuous function. At $t = 0$ we start the system at rest, $y(0) = 0$.

Denoting the static solution of (4.18) by $\bar{x}_k$ when $F \equiv 0$ we get by direct calculation

$$\bar{x}_k = \frac{k}{\ell+1} x_{\ell+1} \tag{4.20}$$

so that it is reasonable to introduce the differences $u_k = x_k - \bar{x}_k$ into (4.18). They satisfy equations that look formally just as (4.18) except that the term corresponding to $u_{\ell+1}$ should be put equal to zero.

We can then write the equations, for t between two transition points, as

$$\ddot{u} + f\dot{u} = M_\ell u + \xi_\ell(t) \tag{4.21}$$

or in standard form

$$\dot{y} = A_\ell y + \eta_\ell(t)$$

with

$$[\eta_\ell(t)]_k = \begin{cases} -su_{\ell+1} & \text{if } k = \ell \\ F(t) & \text{if } k = 1 \\ 0 & \text{else} \end{cases} \tag{4.22}$$

and the matrix A_ℓ is partitioned into blocks

$$A_\ell = \left\{ \begin{array}{cc|cc} \multicolumn{2}{c|}{\multirow{2}{*}{0_n}} & I_\ell & 0 \\ & & 0 & 0_{n-\ell} \\ \hline sM_\ell & 0 & -fI_\ell & 0 \\ 0 & 0_{n-\ell} & 0 & 0_{n-\ell} \end{array} \right\} \tag{4.23}$$

where 0_r denotes an $r \times r$ zero matrix, I_r an $r \times r$ identity matrix and the $\ell \times \ell$ matrix

$$M_\ell = \left\{ \begin{array}{cccccc} -2 & 1 & 0 & \cdots & & \\ 1 & -2 & 1 & \cdots & & \\ 0 & 1 & -2 & \cdots & & \\ & \cdots & & 0 & 1 & -2 \end{array} \right\} \tag{4.24}$$

If t_1 and t_2 are two successive transition points one gets

$$y(t) = \exp-(t-t_1)A_\ell y(t_1) + \int_{t_1}^{t} \exp(t-s)A_\ell \eta_\ell(ds) \tag{4.25}$$

We shall use (4.25) later on, but let us first examine the simpler problem of the quasi-stationary behavior of the solution.

Let F be a process with mean zero and a spectral density $\phi(\lambda)$. If its spectral process is denoted by ζ and if we denote the spectral density matrix and spectral process of the stationary vector solution $u = u(t)$ by f and z respectively we have by direct calculation

$$[-\lambda^2 I_\ell + if\lambda I_\ell - sM_\ell]z = \begin{pmatrix} \zeta \\ 0 \\ 0 \\ \vdots \\ 0 \end{pmatrix} \tag{4.26}$$

Hence the density matrix f is (T = complex transpose)

$$f = N^T(\lambda) \begin{Bmatrix} \phi & 0 \dots & 0 \\ 0 & 0 \dots & 0 \\ \dots 0 & & 0 \end{Bmatrix} N(\lambda) \tag{4.27}$$

with

$$N(\lambda) = [-\lambda^2 I_\ell + if\lambda I_\ell - sM_\ell]^{-1}. \tag{4.28}$$

The eigenvalues λ_j of M_ℓ are all negative and can be written

$$\lambda_j = 2(\cos \pi \frac{j}{\ell+1} - 1), \quad j = 1,2,\dots\ell \tag{4.29}$$

so that with some orthogonal matrix $\mathcal{O}$

$$M_\ell = \mathcal{O}^T \Lambda \mathcal{O}, \quad \Lambda = \operatorname{diag}[\lambda_1, \lambda_2, \dots \lambda_\ell] \tag{4.30}$$

and

$$N(\lambda) = \mathcal{O}^T \Lambda(\lambda) \mathcal{O} . \tag{4.31}$$

Here $\Lambda(\lambda)$ is a diagonal matrix with elements

$$\lambda_j(\lambda) = [-\lambda^2 + if\lambda - s\lambda_j]^{-1} \tag{4.32}$$

which means that the spectral density of $u_1(t)$ can be written as

$$f_{11}(\lambda) = \left| \sum_{j=1}^{\ell} \frac{c_j}{-\lambda^2 + if\lambda - s\lambda_j} \right|^2 \phi(\lambda), \tag{4.33}$$

an expression that can be given a direct physical interpretation. The denominator in the j'th fraction will become small when

$$\lambda = \pm \sqrt{2s(1-\cos \frac{j}{\ell+1})} \tag{4.34}$$

so that this frequency component will be strong unless $\phi(\lambda)$ is small. This will be the more pronounced the smaller the loss factor f is.

Consider two sorts of driving forces F. If it forms white noise, so that ϕ is flat up to high frequencies, then the values in (4.34) will be stressed and the spectrum of the u_1-process will have a great number of peaks if ℓ is large, a smaller number if ℓ has been decreased. On the other hand, if the driving force has a narrow spectrum then only a few peaks will appear in the spectrum of the generated wave $u_1(t)$ even if ℓ is fairly large.

More generally this sort of pattern can be described as follows.

<u>Case 4.3</u>. (<u>Stationary regime patterns</u>). Given an image algebra of abstract sequences $\alpha_1, \alpha_2, \alpha_3, \ldots$ we use as generators for the derived temporal patterns spectral measures F_α. These generators have in- and out-bonds s_k, t_k as endpoints of intervals (see early part of this section) and $\mathcal{R}$ stipulates that the string of α's belongs to an abstract sequence image algebra, and that $t_k = s_{k+1}$. The image is the resulting stochastic process, in other words we deal with diffuse patterns, and the subprocesses over disjoint intervals are treated as independent and stationary.

One should also study what happens if the independence assumption is modified by a continuity condition at the endpoints of the intervals. This would require an analysis of the type carried out in (2.10) but so far this has not been attempted.

This quasi-stationary approximation implies that we can

think of the primary pattern as made up of generators characterized by ℓ (the size of the vibrating system), ϕ (the spectrum of the driving force), and the bond values $t_\nu, t_{\nu+1}$ denoting the endpoints of the time interval. The derived pattern is of regime type with the spectrum given by (4.33) which really means that we only have a generator mapping $G \to G'$ as mentioned above. This is a cipher, not a code and hence is not consistent with facts; a more detailed analysis is needed.

Let us look into the transient behavior of the vibrating system. It will then be necessary to examine (4.25) more carefully.

To find the eigenvalues of the $(2n) \times (2n)$ matrix A_ℓ write an eigenvector z as

$$z = \text{col}[u,a,v,b] \tag{4.35}$$

where the vectors u and v are ℓ-dimensional and a and b of dimension $n-\ell$. If μ is the corresponding eigenvalue we must have

$$\begin{cases} v = \mu u \\ 0 = \mu a \\ sM_\ell u - fv = \mu v \\ 0 = \mu b \end{cases} \tag{4.36}$$

One way to satisfy (4.36) is to put $\mu = 0$ and let a and b be arbitrary $(n-\ell)$-vectors. In the opposite case, $\mu \neq 0$, we get from the first and third equations

$$sM_\ell u = (\mu f+\mu^2)u \tag{4.37}$$

so that μ must be a root of the equation (see (4.29))

$$\mu^2 + f\mu = s\lambda_j \tag{4.38}$$

so that

$$\mu = -\frac{f}{2} \pm \sqrt{\frac{f^2}{4} + s\lambda_j} \quad . \tag{4.39}$$

Note that the real part is always negative.

Going back to the general solution (4.25) we can get $\exp A_\ell(t-s)$ by expressing A_ℓ in Jordan's canonical form. The exponential function of A_ℓ is of similar form (see e.g., Hale (1969), p. 99) and it is seen that an element of this matrix can be written as a linear function of $\exp(t-s)\mu$, μ as in (4.39), except for possibly one $\lambda = 0$ of algebraic multiplicity greater than one. This would contribute also a term $(t-s)\exp(t-s)\mu$. The algebraic multiplicity of $\mu = 0$ does not contribute any terms as can be seen by observing the way the zero-blocks appear in (4.23).

This was for a particular regime with fixed value of ℓ. When applying (4.25) successively we will get the general solution in terms of matrices $\exp A_\ell(t-s)$ multiplied together. For small f the μ's will be complex and we shall write them as $\rho e^{i\gamma}$.

We now get the interesting phenomenon that, when we cross regime boundaries, terms will appear in the general solution $u_1(t)$ with <u>sum and difference frequencies</u>, $\gamma_r \pm \gamma_s$. They will die out more quickly since their modulus is the product $\rho_r\rho_s$ but are nevertheless present in the transient. The absolute effect will of course, again, depend upon how the spectrum of the driving force $F(t)$ behaves over the regimes in question.

Here we have still used the same G as above, but the result is not just a generator mapping $G \to G$: Indeed, the transient behavior involves interaction at least between adjacent generators and possibly even over longer time spans: it is a real code. Whether this analysis has any relevance for actual speech synthesis is not clear but it indicates a possibility that may be worth keeping in mind.

Temporal patterns with high degree of complexity also occur in electro-physiological recordings: electrocardiograms (EKG), electroencephalograms (EEG), electromyograms (EMG). The mathematical modelling of such phenomena started a long time ago. Many of the attempts employed expansions in orthogonal functions, especially Fourier representations. If one chooses a phenomenological approach it becomes really a question of approximation theory and one has to decide on what features the approximation should be based to achieve an economical and accurate representation.

If we limit ourselves to linear features in the stationary case the selection of the main features can be made by using McClure's definitive result as will be discussed in Chapter 13. However, if one requires that the model should be based on physiological principles and anatomical evidence a more detailed examination of the pattern structure will be necessary. We shall outline a recent attempt to do so, Rosenberg et al. (1972) for EKG.

In EKG measurements one records the skin potentials obtained from electrodes placed at points on the body surface. With nine points one would record twelve differences, or leads, denoted by I, II, III, aVl, aVr, aVF and $V_1, V_2, V_3, V_4, V_5, V_6$. The leads have different characteristics. Although the time functions are not strictly periodic, it is useful to talk of "periods" even though they may not be well defined. A "period" is described in terms of its maxima, minima and the arcs that connect them using the notation indicated schematically in Figure 4.3. In addition to the P wave, T wave, U wave, and the QRS complex, one also speaks of the PQ and ST segments. The physiological interpretation of the electric activity of the heart goes as follows, in drastic abbreviation. An impulse starts in the sino-auricular node situated in the wall of the right auricle (see Figure 4.4).

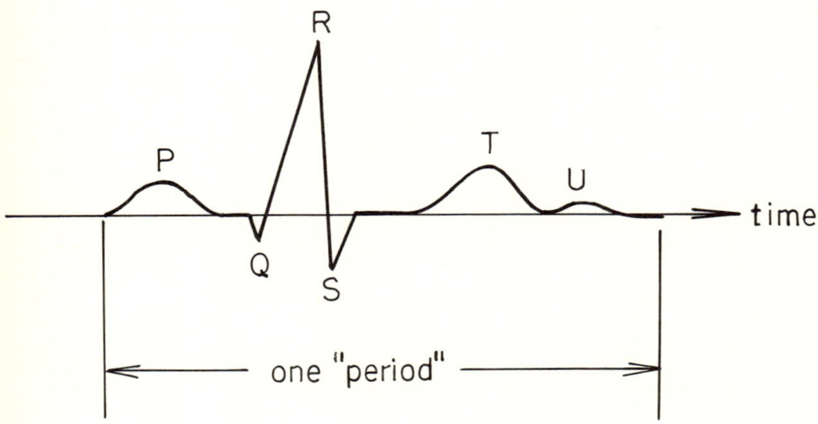

Figure 4.3

The wave passes through the walls of the atria until it reaches the atrio-ventricular node close to the septum, after which a brief delay of the electrical activity takes place. Then the bundle of His is activated, and both branches of the bundle propagate the impulse of the right and left ventricles. Finally the wave spreads through the ventricular walls. The

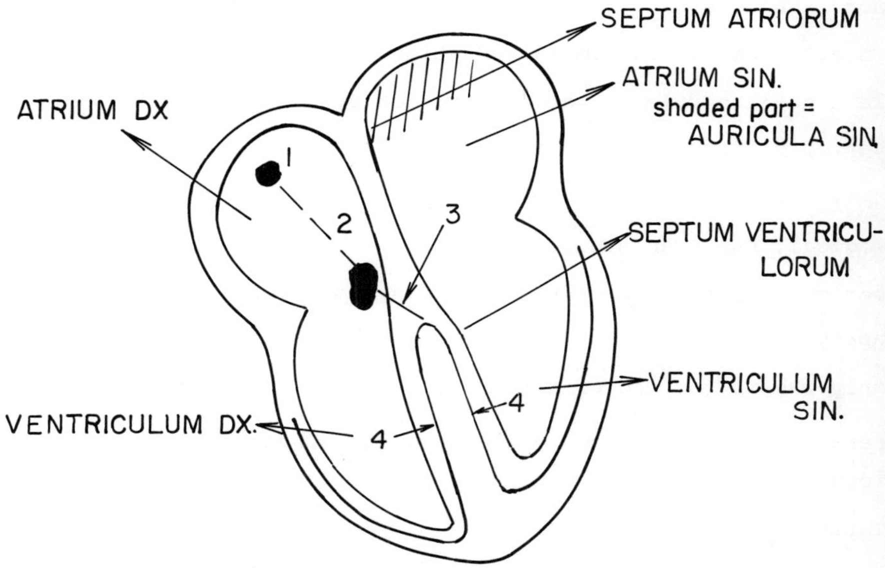

1. PACEMAKER (sinoauricular node of Keith & Flack)

2. ATRIOVENTRICULAR NODE OF HIS-TAWARA

3. BUNDLE OF HIS

4. RIGHT AND LEFT BUNDLE BRANCH

Figure 4.4

velocity of propagation is quite different in different parts of the heart, being highest in the bundle of His and the branches.

During the depolarization of the auricles the P wave is generated. After the short interval PQ depolarization of the ventricles begins, resulting in the QRS complex. The repolarization which occurs afterwards gives rise to the

T wave.

Of course, this description gives only a superficial account of the mechanism, leaving out such possibilities as more than one R or S wave, and, in particular, anomalies in all their overwhelming variation. We have intentionally left out the role of the propagation system leading to observed differences in skin potentials.

The question of how to model the EKG, normal or pathological, has been studied for many years. The more or less periodic appearance of EKG recordings makes it tempting to apply Fourier analysis. Following Oscar Wilde's maxim - the only way to get rid of a temptation is to yield to it - many researchers have indeed yielded to it. Their attempts have resulted in less than complete success and it is worth speculating on why this is so.

If we only aim for data compression, Fourier methods may certainly be of help. To understand the patterns - both their regularity and even more the anomalies - a less phenomenological attitude is called for, based on the known properties of the biological system. Due to the high complexity of the system, with its many and different components, it is unlikely that a simple model would do the job. Instead one needs a model which describes at least the main components separately as well as their connections.

The model suggested by Rosenberg et al. is of this type. They use 13 components connected as in Figure 4.5 where the abbreviations are explained in Table 4.1. The diagram of the network should be compared to the above sketch of the electrical generation of the EKG.

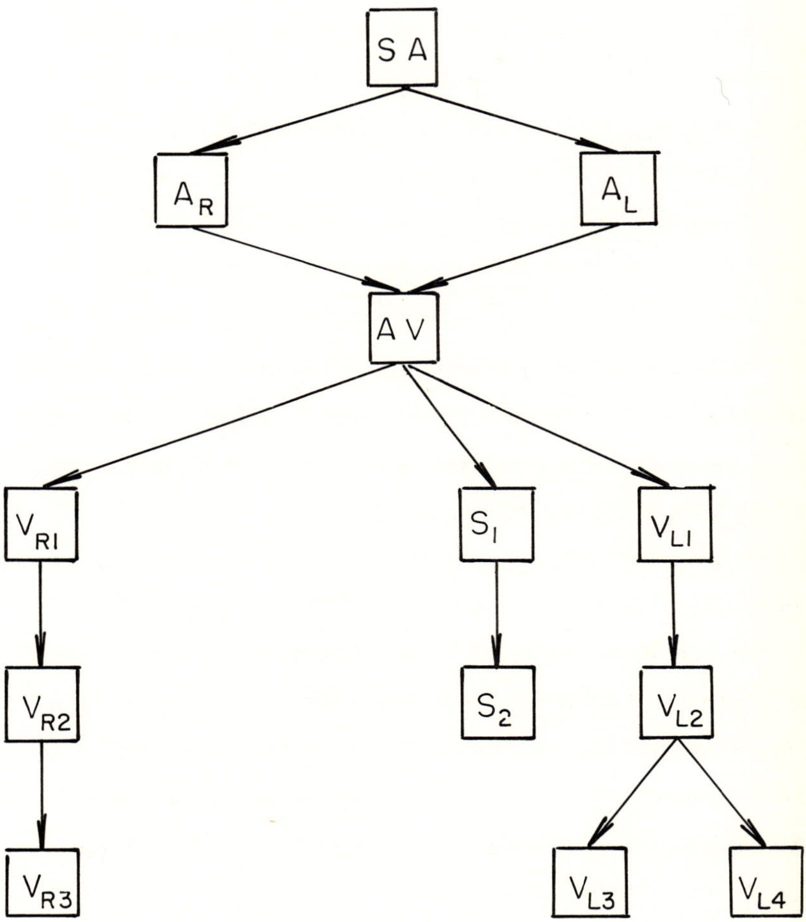

Figure 4.5
(from Rosenberg 1972)

The 13 components are nonlinear devices based on mechanical analogues that need not concern us at the moment. Solving the system equations numerically with parameters chosen to fit empirical evidence about frequency and conduction properties of heart and body tissue, simulated EKGs appear as in Figure 4.6.

SA	sino-auricular mode
A_R, A_L	right and left atria
S_1, S_2	septum
V_{R1}, V_{R2}, V_{R3}	right ventriculum
AV	atrio-ventricular mode
$V_{L1}, V_{L2}, V_{L3}, V_{L4}$	left ventriculum

Table 4.1

It is too early to judge the success of this model definitively, but it certainly generates patterns that appear to be in close qualitative agreement with normal EKGs. The model can also simulate some anomalies.

In this way the patterns that model idealized electrical heart activity are generated from configurations with non-trivial internal structure and that function as mathematical machines. The regularities of the pattern mirrors the values of the attributes of the generators occurring in the configuration. Varying the values of the attributes, we can obtain different images, resulting in a family of pattern classes.

To formalize the above similarly to Case 4.3 let us again start with sequence patterns of abstract symbols. To each value we associate an ordinary differential equation of order p: $L^\alpha y = 0$, where the differential operator L^α does not involve time. We assume that on any interval (s,t) the solution of $L^\alpha y = 0$ is unique given $y(s), y'(s), \ldots y^{(p-1)}(s)$.

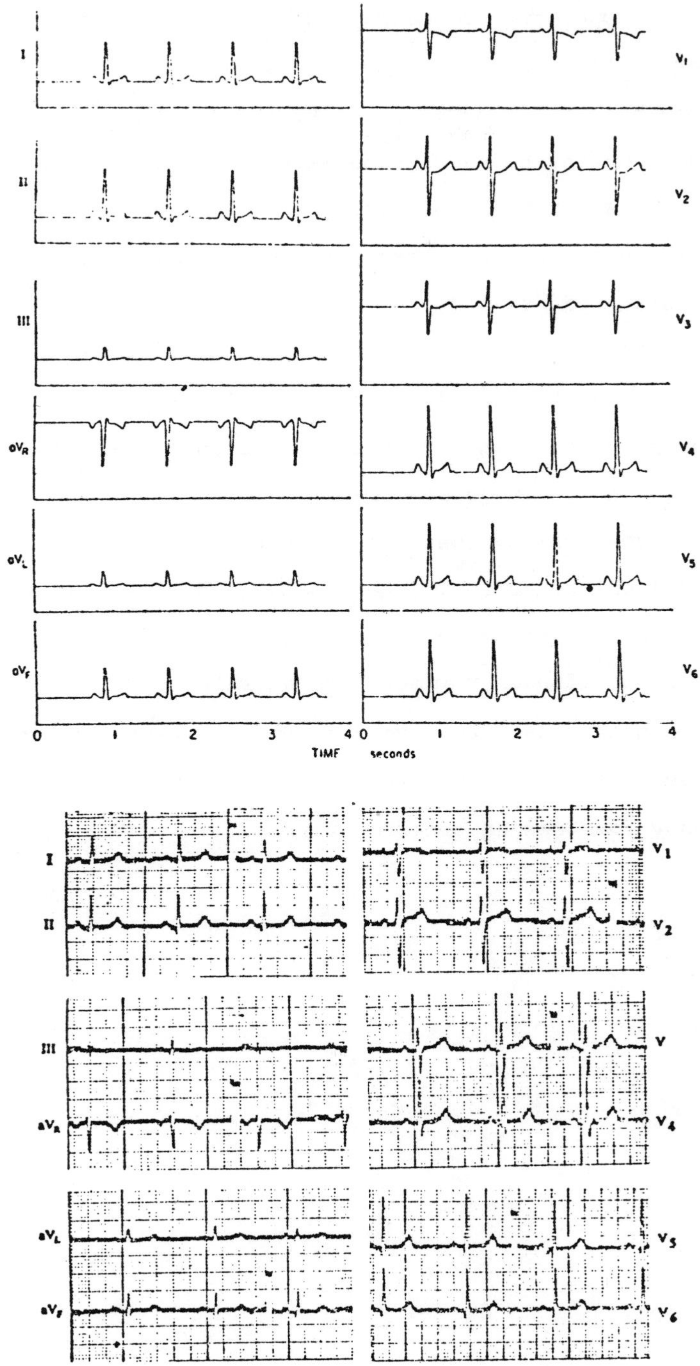

Figure 4.6

(from Rosenberg (1972))

Case 4.4. (Operator regime patterns). The configuration structure will be chosen as before except that we now also include the bond relation that the vector $(y,y',,\ldots y^{(p-1)})$ be continuous on the boundaries between regimes. Images are identified as functions of time.

The generators of these regime patterns have α as their index, and s and t (endpoints of regimes) together with the values of $(y,y',\ldots y^{p-1})$ as their bond values. This is similar to Case 4.1 but it should be noted that in the present case we insist that the abstract α-sequence be admissible in its image algebra.

A numerical synthesis of this pattern is shown in Figure 4.7 which has four generators, each of which is a second order difference operator driven by a spike input at the beginning of each regime. The attributes of the generators have been chosen to give the temporal patterns at least some resemblance to an EKG trace. Note that this was done using only four second order operators.

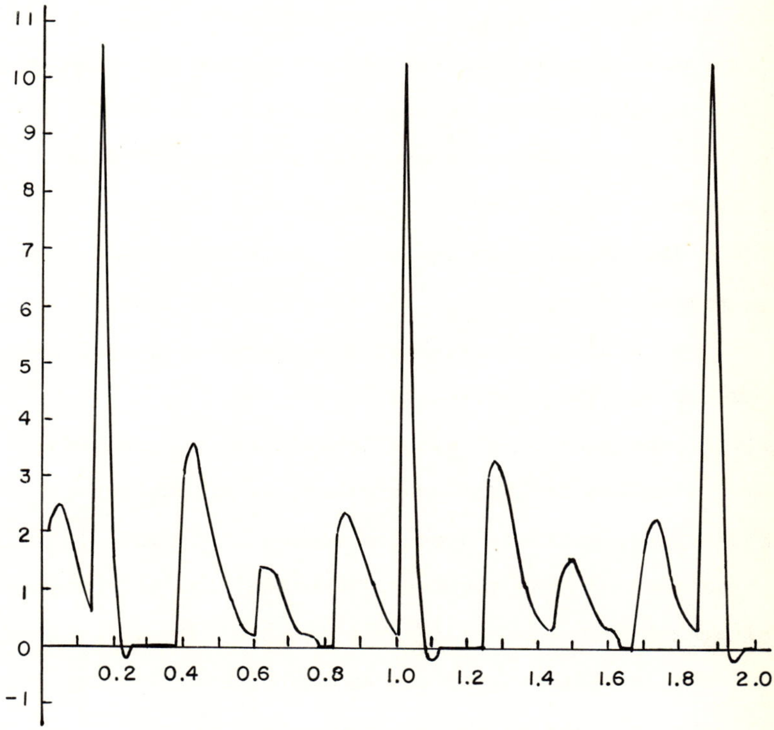

PLOT OF AN UNDISTORTED ELECTROCARDIOGRAM

Parameters Yielding the Above Plot:

Degree of Generator 1 = 2;	Time Constant = 2;	Period = 48
" " " " 2 = 2;	" " " " = 1;	" " = 8
" " " " 3 = 2;	" " " " = 2;	" " = 56
" " " " 4 = 2;	" " " " = 2;	" " = 24

Figure 4.7

5. Plane Patterns.

With the background space equal to the plane, a configuration made up of point generators (see Section 1.3) gives us the simplest sort of plane patterns: the point patterns. Here we shall identify configurations as sets. The image is given as a finite or denumerably infinite set of points $z_\nu = (x_\nu, y_\nu)$; $\nu = 1,2,3,\ldots$.

A very regular arrangement, as in Figure 5.2, represents a plane crystal.

Let Σ be the connection type indicated by Figure 5.1 and its connected subsets. Each generator g_{nm} has arity = 8 double arrows with bond values denoted by β_{nm}^{iu} where i takes the values 1,2,3,4 corresponding to the E,N,W, and S directions, and u takes the values "x" and "y" to indicate x- or y-components. The bond relation will be taken as EQUALITY.

The generator g_{nm} has the attribute $z_{nm} = (x_{nm}, y_{nm}) \in R^2$ and we shall compute the bonds by

$$\beta_{nm}^{iu} = \begin{cases} x_{nm} + \frac{1}{2}\xi_{1x} & \text{if } i = 1,\ u = \text{"x"} \\ x_{nm} + \frac{1}{2}\xi_{2x} & \text{if } i = 2,\ u = \text{"x"} \\ x_{nm} - \frac{1}{2}\xi_{1x} & \text{if } i = 3,\ u = \text{"x"} \\ x_{nm} - \frac{1}{2}\xi_{2x} & \text{if } i = 4,\ u = \text{"x"} \end{cases} \tag{5.1}$$

and similarly for u = "y".

In Chapter 4 we shall allow the ξ's to depend upon (n,m) and we will then get some fairly irregular point patterns. With the present assumption, however, we get just the usual plane crystal.

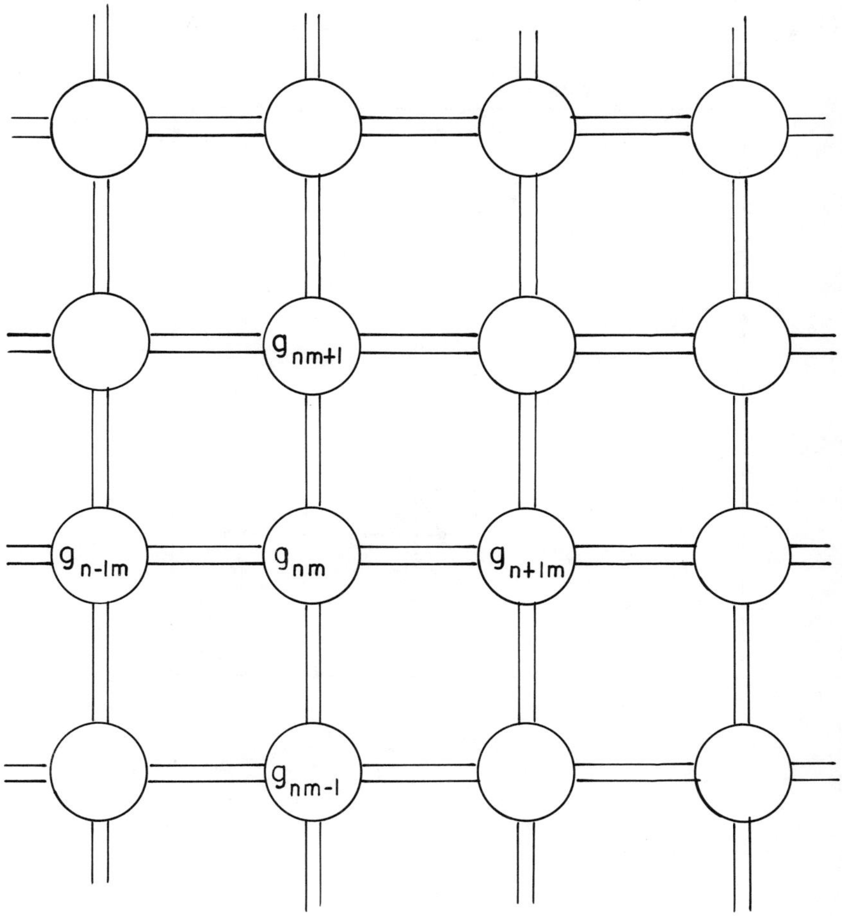

Figure 5.1

<u>Case 5.1</u>. (<u>Plane crystal</u>). Consider the image given as the lattice

$$I = \{z_{\nu\mu} = a+\nu_1\xi_1+\nu_2\xi_2\} \tag{5.2}$$

where a,ξ_1,ξ_2 are plane vectors and ν_1 and ν_2 arbitrary integers. The <u>phase vector</u> <u>a</u> is arbitrary while the <u>basis vectors</u> $\xi_1 = (\xi_{1x},\xi_{1y})$ and $\xi_2 = (\xi_{2x},\xi_{2y})$ should be linearly independent. The phase vector is usually of subordinated importance, it means just a translation, and we shall assume for the moment that it is zero.

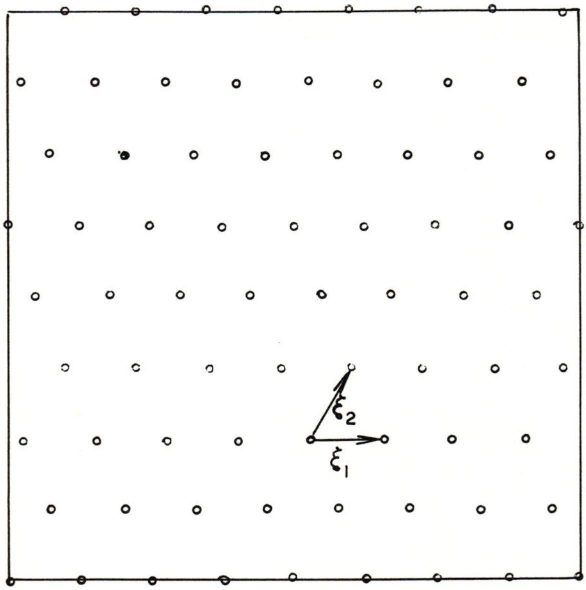

Figure 5.2

Introduce the new vectors

$$\begin{cases} \xi_1' = a_{11}\xi_1 + a_{12}\xi_2 \\ \xi_2' = a_{21}\xi_1 + a_{22}\xi_2 \end{cases} \tag{5.3}$$

and the lattice I' spanned by them. If a_{ij} are integers we have $I' \subseteq I$. In order to make $I' = I$ we must have integers μ_{ij} solving

$$\begin{cases} \xi_1 = \mu_{11}\xi_1' + \mu_{12}\xi_2' \\ \xi_2 = \mu_{21}\xi_1' + \mu_{22}\xi_2' \end{cases} \tag{5.4}$$

or, since ξ_1 is linearly independent of ξ_2

$$\begin{cases} 1 = \mu_{11}a_{11} + \mu_{12}a_{21} \\ 0 = \mu_{11}a_{12} + \mu_{12}a_{22} \end{cases} \tag{5.5}$$

and

$$\begin{cases} 1 = \mu_{21}a_{12} + \mu_{22}a_{22} \\ 0 = \mu_{21}a_{11} + \mu_{22}a_{21} \end{cases} \tag{5.6}$$

Hence

$$\begin{cases} \mu_{11} = \frac{a_{22}}{D} \\ \mu_{12} = -\frac{a_{12}}{D} \\ \mu_{21} = -\frac{a_{21}}{D} \\ \mu_{22} = \frac{a_{11}}{D} \end{cases} \tag{5.7}$$

with $D = \det(A) = a_{11}a_{22} - a_{12}a_{21}$, so that

$$\mu_{11}\mu_{22} - \mu_{12}\mu_{21} = \frac{1}{D} = \text{integer} \tag{5.8}$$

which is only possible if $D = \pm 1$. The two bases ξ_1, ξ_2 and ξ_1', ξ_2' then span the same lattice and are said to be <u>equivalent</u>.

A <u>fundamental cell</u> is a parallellogram spanned by two basis vectors. Such a cell P has the area

$$\text{area}(P) = |\xi_{1x}\xi_{2y} - \xi_{1y}\xi_{2y}| \tag{5.9}$$

for the basis ξ_1, ξ_2. For the equivalent base ξ_1', ξ_2' from (5.2) we get similarly

$$\text{area}(P') = |\xi_{1x}'\xi_{2y}' - \xi_{1y}'\xi_{2x}'| = |D| \text{ area}(P') \tag{5.10}$$

so that two equivalent bases have fundamental cells of equal area which is geometrically rather obvious.

The most important features defined on lattice patterns are the linear functions (λ, z) where $\lambda \in R^2$. It is convenient to study them in terms of the <u>reciprocal lattice</u> I' defined through bases (η_1, η_2) such that

$$\begin{cases} (\eta_1, \xi_1) = (\eta_2, \xi_2) = 1 \\ (\eta_1, \xi_2) = (\eta_2, \xi_1) = 0 \end{cases} \tag{5.11}$$

Then if

$$\begin{cases} \lambda = \lambda_1 \eta_1 + \lambda_2 \eta_2 \\ z = \mu_1 \xi_1 + \mu_2 \xi_2 \end{cases} \tag{5.12}$$

we get simply

$$(\lambda, z) = \lambda_1 \mu_1 + \lambda_2 \mu_2. \tag{5.13}$$

This will simplify the calculation of Fourier coefficients. Note that the area of a fundamental cell of I' is

$$|\eta_{1x}\eta_{2y} - \eta_{1y}\eta_{2x}| = 1/\text{area}(P) \tag{5.14}$$

which follows by direct calculation from (5.11).

<u>Case 5.2</u>. Let $Z = \{z_\nu;\ \nu = 1, 2, \ldots, n\}$ be a finite set of points in R^2, such that its covering group with respect to a given S consists only of the identity; then the pattern $\{s\,Z\}$, $s \in S$, is called a <u>rigid plane point pattern</u>.

In this case an image $I = sZ$ determines uniquely the similarity transformation s that has generated the image from the prototype Z.

If one gives up strict regularity as in Cases 5.1-2 but insists only on a statistical regularity, one is led to <u>diffuse point patterns in the plane</u>. With S as the Euclidean group this leads us to the much-studied two-dimensional and stationary stochastic point processes, among which we shall consider two special cases.

Case 5.3. The stationary Poisson process is characterized by

$$\begin{cases} 1. & \text{for a set } A \text{ of area } a,\ Z(A) = \text{Poisson-distributed with mean } m(a) \\ 2. & \text{if } A_1, A_2, \ldots, A_n \text{ are disjoint sets then } Z(A_1), Z(A_2), \ldots, Z(A_n) \text{ are stochastically independent.} \end{cases} \tag{5.15}$$

Here we have used the notation $Z(A)$ = number of point generators in A.

This is often a starting point for constructing more elaborate diffuse point patterns. The Poisson process is in a sense the exact opposite to a perfectly regular arrangement and presents a very chaotic appearance. Patterns with more structure are generated by the next case.

Case 5.4. The center satellite process of Neyman (1939) is generated in two steps. First the centers are generated from a Poisson process. In the second step one starts from a center $z_o = (x_o, y_o)$, determines the number of satellites n_z to this point from a given distribution D and, finally, generates n_z points from a frequency function $f(x-x_o, y-y_o)$.

Hence this diffuse pattern is characterized by three quantities, the intensity λ of the Poisson process, the distributions D and f. If f is concentrated and if D gives weight to large values the resulting patterns will show a clustering effect with many satellites surrounding the center. In Figure 5.3 we show three such pictures, where

(a) represents a Poisson process and (b) and (c) are center satellite processes; (c) shows more clustering than (b).

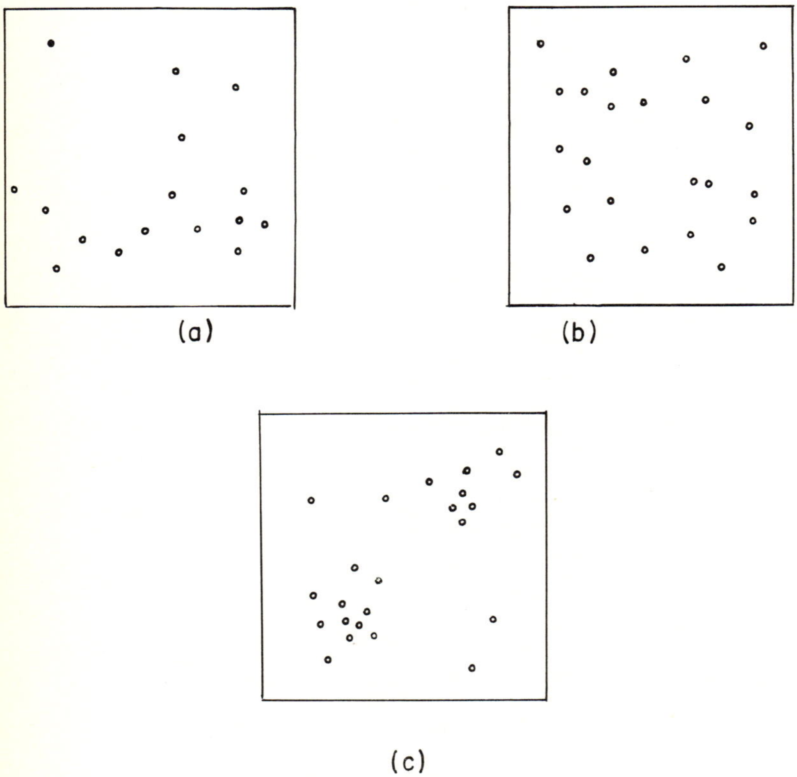

Figure 5.3

More generally one can characterize arbitrary stationary point processes by their covariance and spectral densities; the reader is referred to Bartlett (1964).

Turning to <u>line patterns</u> the generators will be chosen as plane curves. More precisely a generator will be an analytic arc. With S = Euclidean group it is natural to choose representations that are invariant with respect to S which leads us to the <u>intrinsic representation</u>

$$\frac{d\phi}{ds} = \rho(s), \quad 0 \le s \le L, \tag{5.16}$$

where ϕ is the direction of the tangent, s is arc length and L is the total length, $0 < L \le +\infty$. Then ρ means curvature and the full description of a generator g will consist of the quintuple

$$(x_o, y_o, \phi_o, L, \rho) \tag{5.17}$$

with (x_o, y_o) meaning the endpoint and ϕ_o the direction of the tangent at this point.

For linear patterns, let I be a discrete set of lines in R^2 that remains invariant with respect to a (non-trivial) discrete subgroup of the translation group in the plane. We then get very regular arrangements as for example the one shown in Figure 5.4.

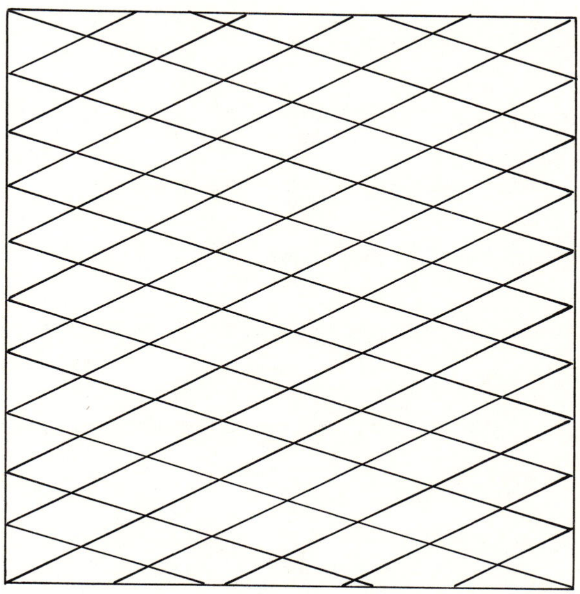

Figure 5.4

A less regular image algebra is the following one. Given a finite set of ordinary differential equations of the second degree $L_\alpha y = 0$, the generators are defined as finite directed arcs satisfying these equations and classified into generator classes G^α according to the value of α. S will now be chosen as a one-dimensional translation group meaning that external effects only act perpendicularly to these translations.

Each generator will have two endpoints e_1 and e_2. Regular configurations are those that satisfy given bond relations where endpoints coincide. These relations express how the directions of the arcs are related to each other and to the values of α for the arcs involved at the point.

As a simple illustration of this type of image algebra we use the three generator classes below

$$\begin{cases} G^1 = \text{straight lines} \\ G^2 = \text{parabolas } \dfrac{dy^2}{dx^2} = a_1 \\ G^3 = \text{parabolas } \dfrac{d^2y}{dx^2} = a_2 > a_1 \end{cases} \tag{5.18}$$

The bond relations are as follows.

$$\mathscr{R} \begin{cases} \text{the number of arcs starting at an endpoint is one,} \\ \text{two or three;} \\ \text{if one starting arc } \alpha_2 \text{ goes to the one with endpoint} \\ (\alpha_1) \text{ then } F_{\alpha_1\alpha_2}(y'_{\alpha_1}, y'_{\alpha_2}) = 0; \\ \text{if two starting arcs } \alpha_2, \alpha_3 \text{ go to the one with} \\ \text{endpoint } (\alpha_1) \text{ then } F_{\alpha_1\alpha_2\alpha_3}(y'_{\alpha_1}, y'_{\alpha_2}, y'_{\alpha_3}) = 0; \end{cases}$$

where y'_α means the derivative evaluated along the arc α

and its endpoint where it joins the other arc(s) and $F_{\alpha_1\alpha_2}$, $F_{\alpha_1\alpha_2\alpha_3}$ are given functions. A typical pattern of this type is given in Figure 5.5.

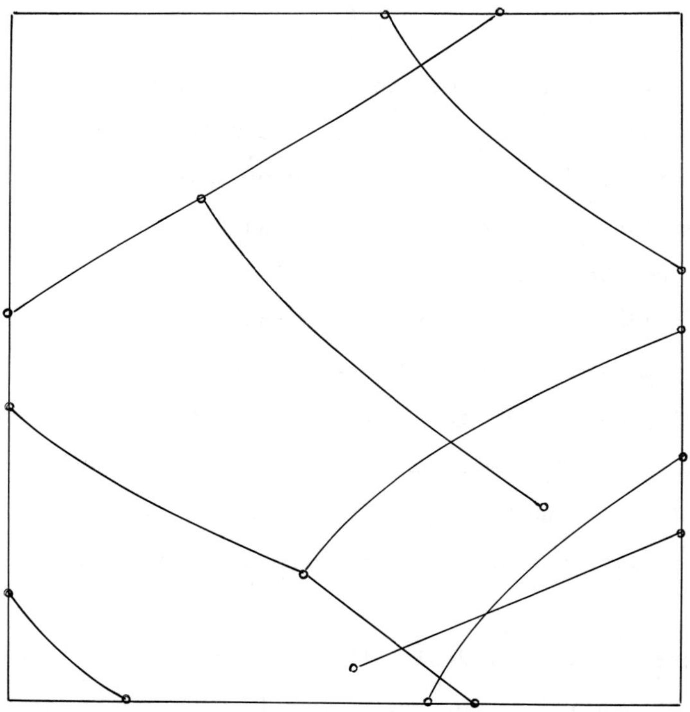

Figure 5.5

Line patterns can also appear in many diffuse forms but we shall only consider one here, the <u>diffuse network pattern</u>.

A very chaotic line pattern is the Poisson line process. Representing an oriented line by the equation

$$x \cos \phi + y \sin \phi - p = 0, \; 0 \le \phi < 2\pi, \; -\infty < p < \infty, \tag{5.19}$$

we can identify it with the truth set - a half plane - of a feature in the linear threshold logic of Section 3.9. The μ-measure used there

$$d\mu = d\phi dp \tag{5.20}$$

is EG(2) invariant. Generate a Poisson point process in the (ϕ,p)-space. This will induce the Poisson line process when (ϕ,p) are interpreted as the coordinates of a line.

Similarly clustering line processes can be defined but this will not be pursued here.

Consider n nodes with coordinates (x_i,y_i) in the plane, see Case 2.6.1. In very large networks the detailed connectivity may not be completely specified, only the coarse topology. By this we mean that a generator, in this case a segment, is only known to occur with some given probability. This probability may depend upon where the two nodes are located.

Here let us only deal with the case when S is the Euclidean group and we let the probability be invariant with respect to S so that it is the same for two congruent segments. That means that it depends only upon the length of the segment. This regulates only the presence or absence of a segment joining any two nodes and for a full specification we also need the simultaneous probabilities which here will be introduced by demanding that all segments be stochastically independent.

The above construction leads to a statistically homogeneous topology. We could also deal with heterogeneity by splitting up the entire network into blocks in such a way that each block is statistically homogeneous but the blocks differ as to their coarse topology.

Case 5.5. (Diffuse network patterns in the plane). In the plane n nodes are given and two nodes are connected with a probability that depends only upon the distance between the nodes. All segments are stochastically independent.

Of course the segments can also have attributes such as length, diameter, propagation velocity, attenuation factor. This becomes necessary when studying pattern processing networks, see Chapter 19.

In Figure 5.6 we have illustrated Case 5.5 with

$$p = \begin{cases} p_o - d_{ij}/d_o & \text{if} \quad d_{ij} \leq p_o d_o \\ 0 & \text{else} \end{cases} \tag{5.21}$$

where d_{ij} is the Euclidean distance

$$d_{ij} = [(x_i - x_j)^2 + (y_i - y_j)]^{1/2}. \tag{5.22}$$

Above we had S = full Euclidean group so that the network will have isotropic coarse topology. Now let S = translation group, so that we lose the isotropic property, but we shall still use the term statistically homogeneous topology. As an illustration of this we present Figure 5.7 with p given as in (5.21) but with the non-isotropic definition

$$d_{ij} = [2(x_i - x_j)^2 + \tfrac{1}{2}(y_i - y_j)^2]^{1/2}. \tag{5.23}$$

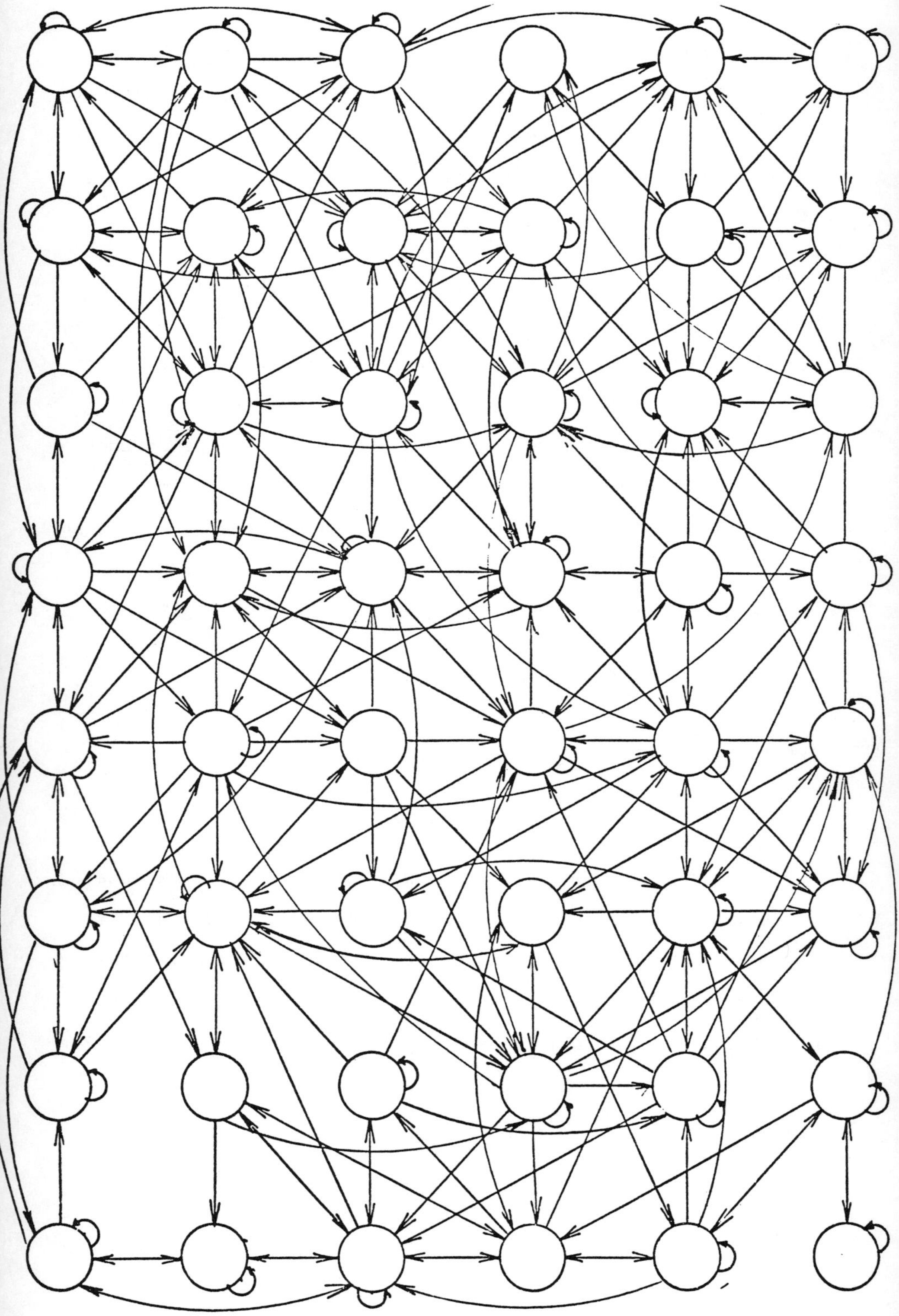

Figure 5.6

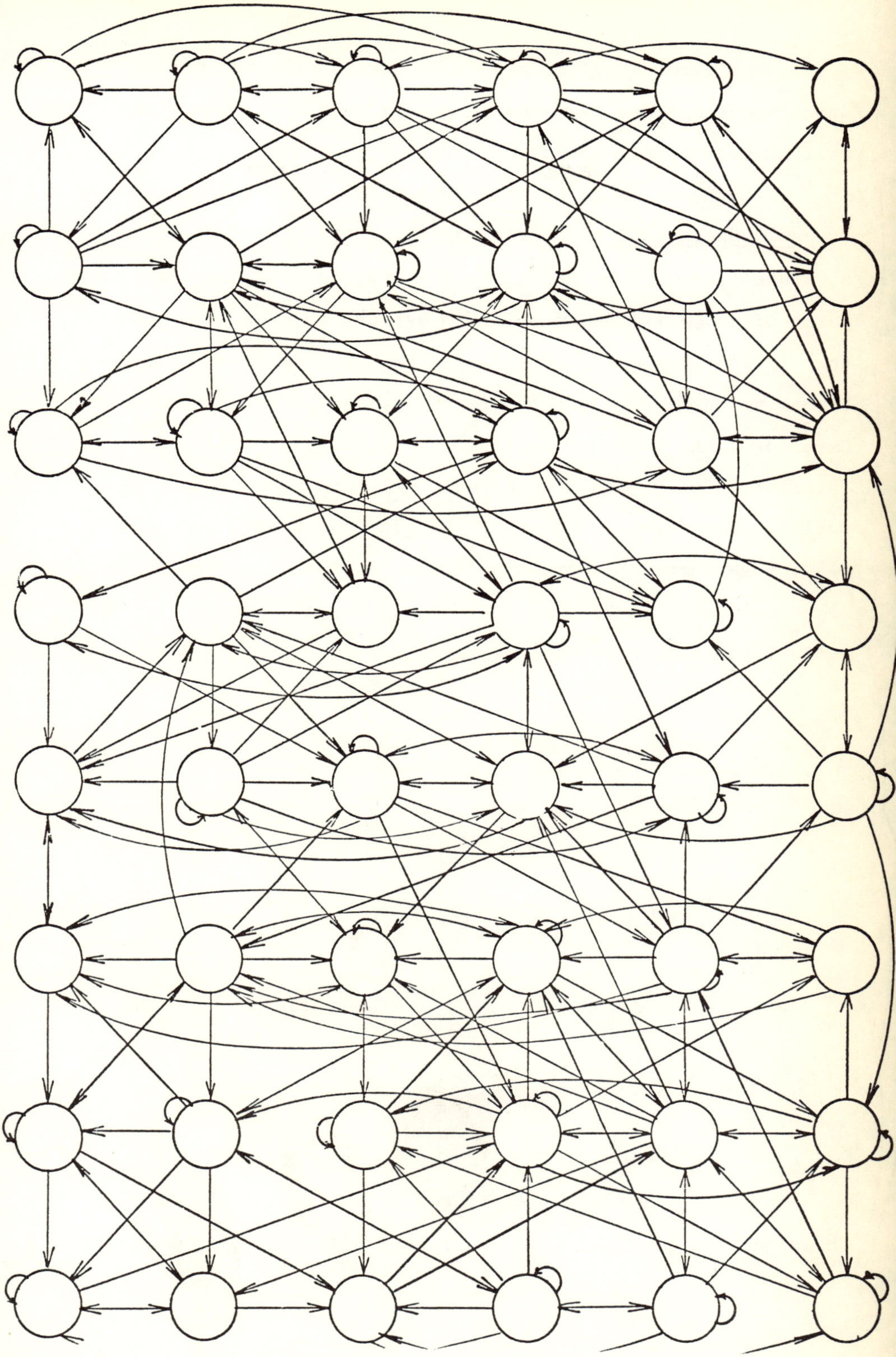

Figure 5.7

The directional effect can be seen. The properties of such images will be examined in detail in Part IV where they will be used to process patterns.

Another diffuse network pattern that has been used in geomorphological studies of drainage systems starts from n sources, uses Σ = TREE and ρ says that the starting point of a new arc falls on the previous arc. We also assume that links (segments either from a source or a junction downstream to the next junction or the outlet) join only pairwise. Exterior links begin at sources, the other links are called interior. See Figure 5.8 where $n = 9$ and 0 denotes the outlet. The figures denote the magnitudes μ of the links, i.e., the number of sources upstream, so that magnitudes add at junctions and take the value one at sources and n at the outlet, as illustrated in the figure.

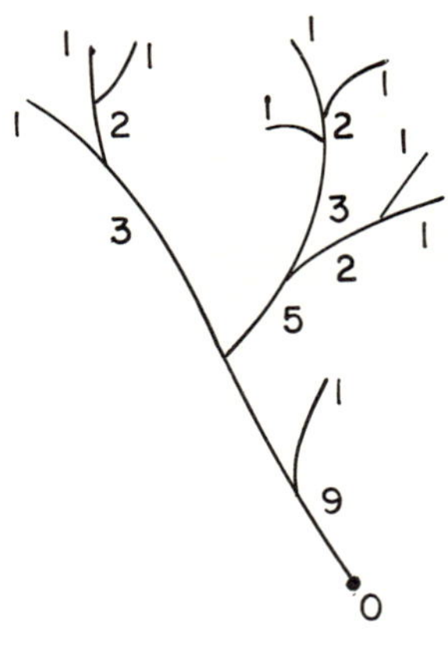

Figure 5.8

A more descriptive attribute is the stream number due to Horton and Strabler, see e.g., Horton (1945), Shreve (1966,

1967, 1969), Wernitty (1972), Smart (1972). One version, due to Strahler, defines them as follows:

$$\begin{cases} \text{(i) channels starting at sources have order } \omega = 1 \\ \text{(ii) two streams of order } \omega \text{ meeting produce a stream of order } \omega+1 \\ \text{(iii) two streams of different order meeting produce a stream of order equal to the highest of the two orders.} \end{cases} \tag{5.24}$$

An example is given in Figure 5.9.

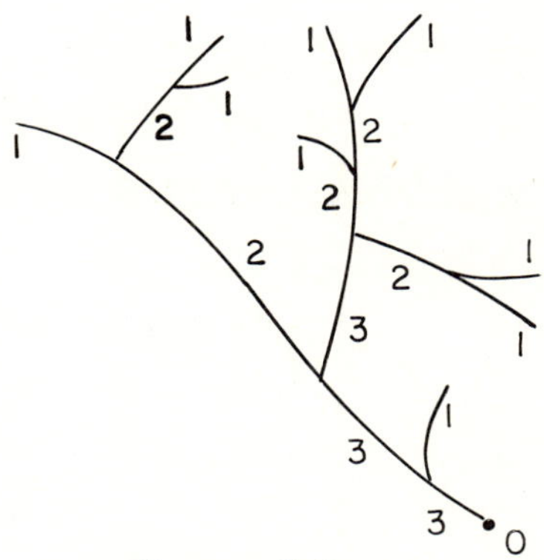

Figure 5.9

The highest order Ω involved is called the <u>order of the network</u>. A <u>stream</u> is a set of contiguous links of the same order.

Let n_ω be the number of streams of order ω, $\omega = 1,2,\ldots,\Omega$. Of course these numbers are not arbitrary since every network must contain at least two streams of order $\omega-1$ for each one of order ω,

$$n_\omega - 2n_{\omega+1} \geq 0. \tag{5.25}$$

A similar reasoning gives

$$\sum_{\omega=1}^{\Omega-1} (n_\omega - 2n_{\omega+1}) + 2^{\Omega-1} = n. \tag{5.26}$$

For the number of links and junctions we also have restrictions since adding one source to a network increases the number of links by two and junctions by one:

$$\begin{cases} \#\,(\text{links}) = 2n-1 \\ \#\,(\text{exterior links}) = n \\ \#\,(\text{interior links}) = n-1 \\ \#\,(\text{junctions}) = n-1. \end{cases} \tag{5.27}$$

Based on empirical observations (see Chapter 1), Horton announced his famous conjecture that for most natural drainage networks the bifurcation ratios are approximately constant

$$\frac{n_{\omega-1}}{n_\omega} \simeq R_\omega, \quad \omega = 2,3,\ldots,\Omega . \tag{5.28}$$

so that the sequence $\{n_\omega\}$ is approximately a geometric series. Hence R_ω would characterize the network partially. Furthermore he found that R_ω is typically fairly close to 4.

An analogous empirical law says that stream lengths ℓ_ω satisfy the approximate relation

$$\frac{\ell_{\omega-1}}{\ell_\omega} \simeq R_L, \quad \omega = 2,3,\ldots,\Omega \tag{5.29}$$

where ℓ_ω is the average of the lengths of the streams of order ω in the network.

For the moment let us only consider topological properties of the network, taking S to consist of topological transformations of the plane preserving orientations. We shall treat as different only such networks as are distinct modulo S, so that the network in Figures 5.9 is in the same pattern class as the one in Figure 5.10.

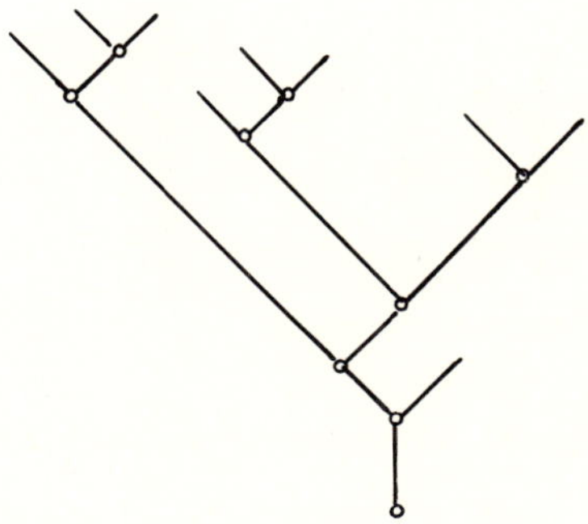

Figure 5.10

For a given number of sources we can, at least in principle, enumerate the (topologically) different networks. For $n = 5$ there are 14 possibilities shown in Figure 5.11 together with the corresponding stream numbers.

To find the number of different arrangements of stream numbers we turn to the following.

Theorem 5.1. (Shreve, 1966). *For a channel system with* n *sources the number* $N(n)$ *of topologically distinct networks is*

$$N(n) = \frac{1}{2n-1}\binom{2n-1}{n} \quad \text{for} \quad n \geq 2. \tag{5.30}$$

The number $N[n_1, n_2, \ldots, n_{\Omega-1}, 1]$ *of topologically distinct networks of order* Ω *with* $n_1, n_2, \ldots, n_{\Omega-1}, 1$ *streams of order* $1, 2, \ldots, \Omega$ *respectively is*

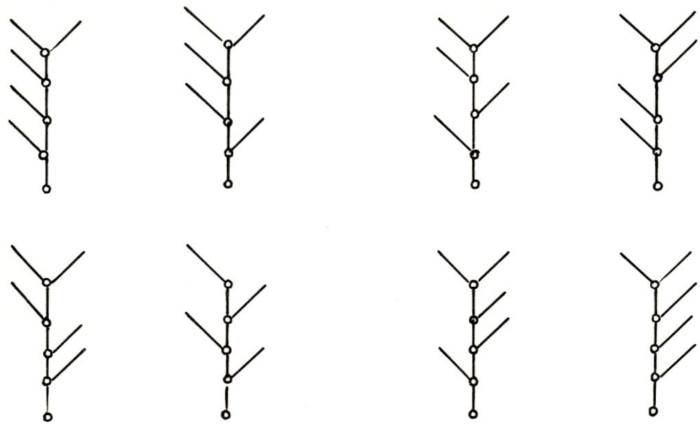

5,1

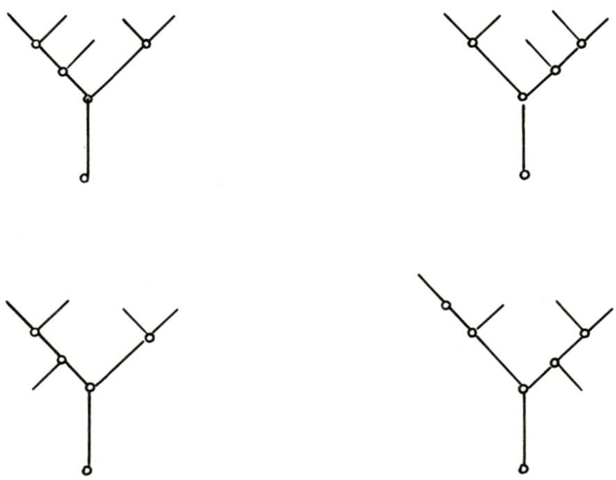

5, 2, 1

5,2,1

Figure 5.11

$$N[n_1, n_2, \ldots, n_{\Omega-1}, 1] = \prod_{\omega=1}^{\Omega-1} 2^{(n_\omega - 2n_{\omega+1})} \binom{n_\omega - 2}{n_\omega - 2n_{\omega+1}} \tag{5.31}$$

Proof: It is obvious that $N(1) = 1$. For $n \geq 2$ one can view the network as composed of two, starting from the outlet. Denote the number of sources of one of them by i, the other one must have $n-i$ sources. Forming all pairs for the possible numbers $i = 1,2,\ldots,n$ we get the recursion

$$N(n) = \sum_{i=1}^{n-1} N(i)N(n-i); \quad n \geq 2. \tag{5.32}$$

In order to get a bound for $N(n)$ when n is large we shall try the form

$$N(n) \leq \frac{a^{n-1}}{n^2} .$$

Assume that this holds for $n = 1,2,\ldots,m-1$. Then

$$N(m) = \sum_{1}^{m-1} N(i)N(m-i) \leq \frac{a^{m-2}}{m^2} s_m, \quad s_m = m^2 \sum_{i=1}^{m-1} \frac{1}{i^2} \frac{1}{(m-i)^2}$$

But

$$s_m = \sum_{i=1}^{m-1} \frac{1}{i^2} \left(\frac{m}{m-i}\right)^2 = \sum_{\substack{2i<m \\ i=1}}^{m-1} + \sum_{\substack{2i>m \\ i=1}}^{m-1} \frac{1}{i^2} \left(\frac{m}{m-i}\right)^2 \leq 4 \sum_{i=1}^{\infty} \frac{1}{i^2}$$

$$+ 4 \sum_{1}^{\infty} \frac{1}{i^2} = \frac{8\pi^2}{6}$$

so that the bound holds for $n = m$ if $a \geq (8\pi^2)/6$. However it holds for $n = 1$ and therefore for all values of n. The generator function expansion converges for $|z| < 1/a$ so that $g(z)$ is analytic inside this circle.

Introduce the generating function

$$g(z) = \sum_{n=1}^{\infty} N(n)z^n \tag{5.33}$$

which converges for small values of z

$$g(z)-z = \sum_{n=2}^{\infty} N(n)z^n = \sum_{n=2}^{\infty} \sum_{i=1}^{n-1} N(i)z^i N(n-i)z^{n-i}$$
$$= \sum_{i=1}^{\infty} N(i)z^i \sum_{n=i+1}^{\infty} N(n-i)z^{n-i} = g^2(z). \tag{5.34}$$

Hence

$$g(z) = \frac{1}{2}[1-(1-4z)^{1/2}] \tag{5.35}$$

taking the branch of the square root function that makes $g(0) = 0$. The coefficient of z^n in the Taylor expansion of g at $z = 0$ is then, for $n \geq 2$,

$$N(n) = -\frac{1}{2}(-4)^n \binom{1/2}{n} = \frac{1}{2n-1}\binom{2n-1}{n} \tag{5.36}$$

as in (5.30).

To prove (5.31) let us consider n_ω streams of order ω. $2n_{\omega+1}$ of them must join in pairs to form the $n_{\omega+1}$ streams of the next higher order. This will leave $n_\omega-2n_{\omega+1}$ streams of order ω that go into the $2n_{\omega+1}-1$ links of order at least equal to $\omega+1$ before lower order streams join. But each of the remaining $n_\omega-2n_{\omega+1}$ streams can join the links from right or left accounting for the factor $2^{n_\omega-2n_{\omega+1}}$. The number of distinguishable distributions of r objects in s cells is (see Feller (1957), I, p. 36)

$$\binom{r+s-1}{r}. \tag{5.37}$$

Here $r = n_\omega-2n_{\omega+1}$ and $s = 2n_{\omega+1}-1$ giving us the expression in (5.31).

The theorem gives directly the probability measure with the discrete probabilities

$$p = \frac{N(n_1,n_2,\ldots,n_{\omega-1},1)}{N(n)} \tag{5.38}$$

of finding a network with stream numbers $n_1, n_2, \ldots, n_{\omega-1}, 1$ if n is the number of sources. For drainage networks the following pattern has been suggested in the absence of geologic controls.

<u>Case 5.7</u>. (<u>Topologically random channel networks</u>). Among all channel networks with n sources we define a probability distribution over the pattern classes by assigning to them probabilities given by (5.38).

As an example consider Table 5.1 for $n_1 = 60$ where the ten first rows indicate the stream numbers and bifurcation ratios for the most probable networks. The fifth column gives the cumulated probabilities. The ten first are of order 4 and have bifurcation ratios R_ω quite close to 4. The ten last rows represent less probable networks and the bifurcation ratios vary a good deal more.

Since natural channel networks are parts of larger ones it is reasonable to express this via the idealization that the whole network is infinite and that we observe samples from it. This leads to a diffuse pattern with infinite networks.

<u>Theorem 5.2</u>. (Shreve (1967)). <u>The probability</u> $u(\omega)$ <u>of drawing a link or order</u> ω <u>is</u>

$$u(\omega) = 2^{-\omega}; \quad \omega = 1,2,\ldots \tag{5.39}$$

<u>and the probability</u> $v(\mu)$ <u>of drawing a link of magnitude</u> μ <u>is</u>

$$v(\mu) = \frac{2^{-(2\mu-1)}}{2\mu-1}\binom{2\mu-1}{\mu}; \quad \mu = 1,2,\ldots \tag{5.40}$$

<u>The probability</u> $s(\omega)$ <u>of drawing a stream of order</u> ω <u>is</u>

$$s(\omega) = 3.4^{-\omega}, \quad \omega = 1,2,\ldots$$

Stream Number Probabilities for $N_1 = 60$

N_2	N_3	N_4	p	Σ_p	R_s
15	4	1	0.0675	0.0675	3.91
14	3	1	0.0594	0.1269	3.79
16	4	1	0.0591	0.1860	3.95
15	3	1	0.0563	0.2423	3.83
14	4	1	0.0554	0.2978	3.87
13	3	1	0.0459	0.3436	3.75
16	3	1	0.0394	0.3830	3.87
17	4	1	0.0375	0.4206	3.98
13	4	1	0.0321	0.4527	3.83
16	5	1	0.0296	0.4822	4.01
19	4	2	0.0015	0.9679	2.60
20	4	1	0.0015	0.9694	4.07
12	5	1	0.0014	0.9707	3.85
20	5	2	0.0013	0.9720	2.65
20	6	2	0.0013	0.9733	2.69
18	7	2	0.0012	0.9745	2.69
13	1	0	0.0011	0.9756	8.59
17	7	2	0.0011	0.9767	2.67
18	2	1	0.001	0.9777	3.82
12	5	2	0.001	0.9788	2.52

Table 5.1

Proof: Introduce the simultaneous probability $\gamma(\mu,\omega)$ of getting a link of magnitude μ and order ω. Then $\gamma(1,1) = 1/2$; this corresponds to an exterior link and we know that exterior and interior links should occur equally often, see (5.37). Also $(\mu,1)$ is impossible if the magnitude μ is greater than one so that $\gamma(\omega,1) = 0$. Similarly $\gamma(\mu,1) = 0$ for $\omega > 1$.

If μ and $\omega > 1$ then to get a (μ,ω) link we must first get an interior link, which happens with probability 1/2. Going upstream the junction can be an $\omega-1$ link joined with another $\omega-1$ link. Remembering that magnitudes add this gives us the probability

$$\sum_{\alpha=1}^{\mu-1} \gamma(\alpha,\omega-1)\gamma(\mu-\alpha,\omega-1). \tag{5.41}$$

We can also have a junction of an ω link with a β link, $\beta < \omega$. This happens with probability

$$\gamma(\alpha,\omega) = \sum_{\beta=1}^{\omega-1} \gamma(\mu-\alpha,\beta) \tag{5.42}$$

for fixed α, and of course the two links mentioned can be exchanged for each other so that (5.42) should be multiplied with 2.

Combining (5.41) and (5.42) we get the recurrence relation

$$\gamma(\mu,\omega) = \frac{1}{2}\sum_{\alpha=1}^{\mu-1} [\gamma(\alpha,\omega-1)\gamma(\mu-\alpha,\mu-1)$$

$$+ 2\gamma(\alpha,\omega)\sum_{\beta=1}^{\mu-1} \gamma(\mu-\alpha,\beta)]; \quad \mu,\omega > 1;$$

with the stated boundary conditions. Summing (5.33) over ω one gets for the marginal probabilities $v(\mu)$

$$\begin{cases} v(\mu) = \frac{1}{2}\sum_{\alpha=1}^{\mu-1} v(\alpha)v(\mu-\alpha), & \mu > 1 \\ v(1) = \frac{1}{2} \end{cases} \tag{5.44}$$

This equation is quite similar to what we had before and can be solved in the same way leading to the expression in (5.40).

On the other hand, if we sum (5.43) over μ we get instead

$$\begin{cases} u(\omega) = \frac{1}{2}[u^2(\omega-1)+2u(\omega)\sum_{\beta=1}^{\omega-1} u(\beta)]; & \omega > 1; \\ u(1) = \frac{1}{2} \end{cases} \tag{5.45}$$

But we have

$$[\sum_{\omega=2}^{p-1} u(\omega)]^2 = \sum_{\omega=2}^{p-1} [u^2(\omega)+2u(\omega)\sum_{\beta=1}^{\omega-1} u(\beta)] \tag{5.46}$$

so that we get

$$[\sum_{\omega=1}^{p-1} u(\omega)]^2 = u^2(1) + \sum_{\omega=2}^{p-1} [u^2(\omega)-u^2(\omega-1)]$$

$$+ \sum_{\omega=2}^{p-1} [u^2(\omega-1)+2u(\omega) \sum_{\beta=1}^{\omega-1} u(\beta)]$$

$$= u^2(p-1) + 2 \sum_{\omega=2}^{p-1} u(\omega)$$

$$= u^2(p-1) + 2 \sum_{\omega=1}^{p-1} u(\omega) - 1.$$

Solving this quadratic equation in $\Sigma u(\omega)$ we get

$$\sum_{\omega=1}^{p-1} u(\omega) = 1 - u(p-1) \tag{5.48}$$

which can be substituted in (5.47) to give

$$u(\omega) = \frac{1}{2} u(\omega-1), \quad \omega > 1 \tag{5.49}$$

which gives

$$u(\omega) = 2^{-\omega}, \quad \omega \geq 1 \tag{5.50}$$

as stated.

Finally, to get the probability of streams of order ω, it is enough to note that they correspond exactly to links having junctions of type $\omega-1$ with $\omega-1$. Using (5.50) this gives us

$$\frac{s(\omega+1)}{s(\omega)} = \frac{\frac{1}{2} u^2(\omega)}{\frac{1}{2} u^2(\omega-1)} = \frac{1}{4}, \quad \omega > 1 \tag{5.51}$$

so that the proof of the theorem is complete.

Case 5.8. (Infinite topologically random channel networks). Infinite channel networks where the probability of a link of magnitude μ and order ω satisfies the recurrence relation and boundary conditions in (5.43-44).

Note that in the infinite pattern the expression for $s(\omega)$ corresponds exactly to Horton's law for stream numbers, and that these networks patterns have implicitly in them a statistical homogeneity assumption.

The reader interested in non-topological properties of channel networks, i.e. in metric properties such as stream lengths and basin area should consult Shreve (1969), Smart (1972), Wermitty (1972).

Leaving the line patterns, we now turn our attention to plane set patterns, beginning with the very simple ones consisting of closed and bounded convex sets. They will reappear later as generators in more structured image algebras.

A convex set can be written as an intersection of its supporting closed half planes

$$c = \bigcap_{\alpha \in A} H_\alpha \tag{5.52}$$

where the power of the index set A can be finite, denumerable, or continuous. It is natural to consider the generators as half planes and identification R simply as identification of the resulting sets. We shall see below how this point of view is fruitful in pattern analysis of deformed convex sets. Here and in the following cases we employ Σ = FULL and with the bond relation NOT IMPLY.

Case 5.9. Consider the image algebra $\mathscr{K}$ consisting of closed, bounded, convex sets in the plane with three operations: intersections, scalar multiplication for $c > 0$

$$cK = \{x \mid x = cu \quad \text{with} \quad u \in K\}, \tag{5.53}$$

and Minkowski addition

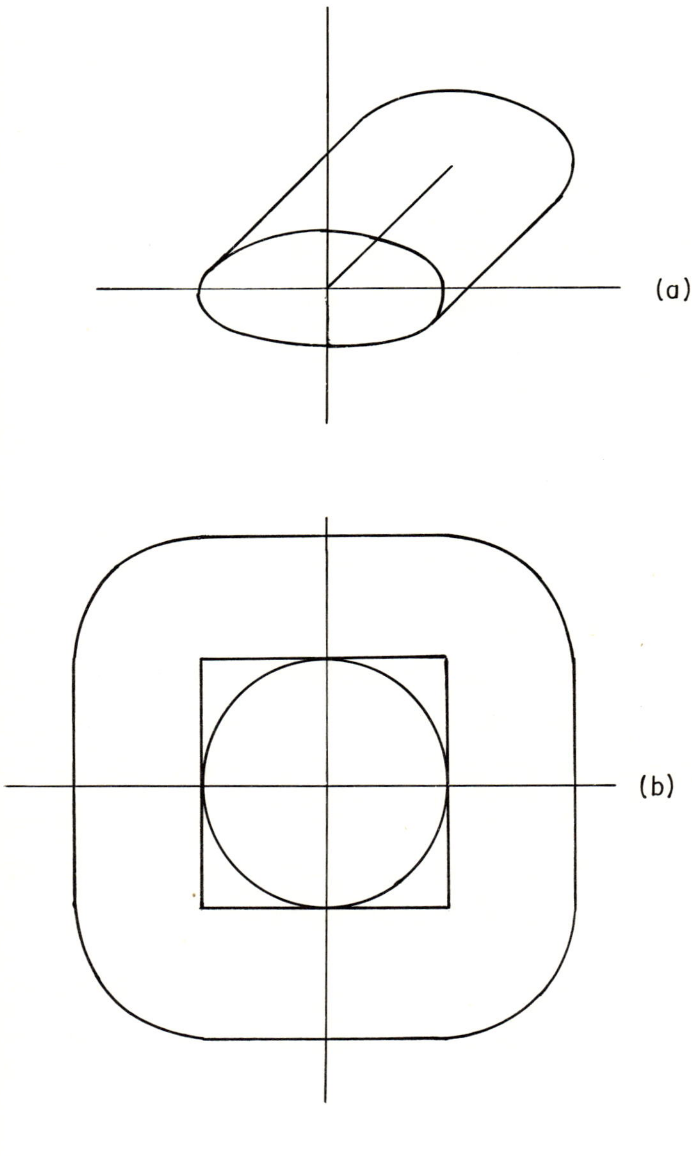

Figure 5.12

$$K_1+K_2 = \{x|x = x_1+x_2 \text{ with } x_1 \in K_1, x_2 \in K_2\}. \quad (5.54)$$

In particular, if K_o consists of the only point (x_o,y_o) then adding K_o to K means translating K by the vector (x_o,y_o). If K_o is a line segment, as in Figure 5.12(a), then adding K_o means stretching in the direction of the line segment: a line segment has been added to an ellipse. To add cD, where D is the unit disk $x^2+y^2 \leq 1$, means to form the parallel set at the distance c outwards from ∂K, see Figure 5.12(b) where a disk has been added to a square.

$\mathcal{K}$ will be topologized by the Hausdorff metric

$$d(K_1,K_2) = \inf(c|K_1 \subseteq K_2 + cD, K_2 \subseteq K_1 + cD) \quad (5.55)$$

The three operations mentioned above are continuous in this topology.

A convenient way of characterizing convex plane sets is through the <u>support function</u> $\underline{s}$

$$s(\phi) = \max_{(x,y)\in K} \{x \cos \phi + y \sin \phi\} . \quad (5.56)$$

The geometric significance is shown in Figure 5.13.

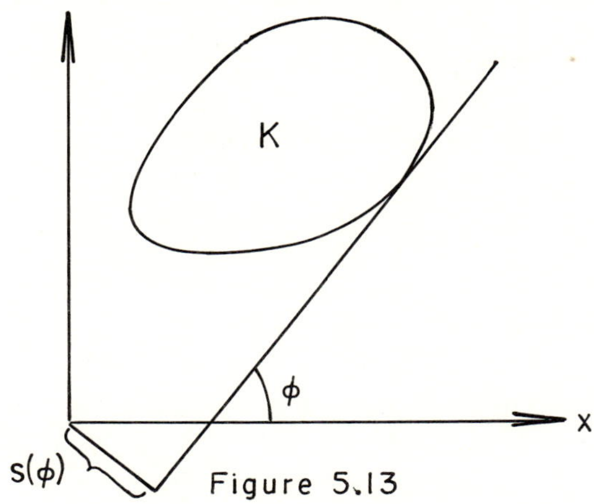

Figure 5.13

Then $s(\phi)$ is continuous and periodic 2π.

It follows directly from the definitions above that

$$\begin{cases} cK & \text{has support function } cs \\ K_1+K_2 & \text{has support function } s_1+s_2 \\ K_1 \subseteq K_2 & \text{if and only if } s_1 \le s_2 \\ d(K_1,K_2) = \sup_{\phi} |s_1(\phi)-s_2(\phi)| = ||s_1-s_2||_\infty \end{cases} \tag{5.57}$$

where the last statement says that the resulting topology is equivalent to the topology induced by the sup-norm for the support functions.

The following result gives a useful representation of support functions.

<u>Theorem 5.3.</u> (R. Vitale, (1974)) <u>In order that</u> $s(\phi)$ <u>be a support function of a set in</u> $\mathscr{K}$ <u>it is necessary and sufficient that it can be written as</u>

$$s(\phi) = a \cos \phi + b \sin \phi + \int_0^\phi \sin(\phi-\psi)R(d\psi) \tag{5.58}$$

<u>where</u> a <u>and</u> b <u>are real constants and</u> R <u>is a bounded measure on</u> $(0,2\pi)$ <u>satisfying the orthogonality relations</u>

$$\int_0^{2\pi} \cos \phi \; R(d\phi) = \int_0^{2\pi} \sin \phi \; R(d\phi) = 0. \tag{5.59}$$

<u>Proof</u>: For given s as in (5.58) define

$$\begin{cases} x(\phi) = a - \int_0^\phi \sin \psi \; R(d\psi) \\ y(\phi) = b + \int_0^\phi \cos \psi \; R(d\psi) \end{cases} \tag{5.60}$$

and put $K' = \text{conv}\{(x(\phi),y(\phi)),\ 0 \le \phi \le 2\pi\}$. We find that

$$x(\phi)\cos \phi + y(\phi)\sin \phi = s(\phi). \tag{5.61}$$

We have that for any convex combination

$$(x_o,y_o) = (\Sigma\alpha_\nu x(\phi_\nu),\ \Sigma\alpha_\nu y(\phi_\nu)),\ \Sigma\alpha_\nu = 1,\ \alpha_\nu \geq 0, \tag{5.62}$$

we get

$$x_o \cos\phi + y_o \sin\phi = a\cos\phi + b\sin\phi + \Sigma\alpha_\nu \int_o^{\phi_\nu} \sin(\phi-\psi)R(d\psi) . \tag{5.63}$$

We have however the integral inequality

$$\int_0^{\beta} \sin(\phi-\psi)R(d\psi) \leq \int_o^{\phi} \sin(\phi-\psi)R(d\psi) \tag{5.64}$$

with equality for $\beta = \phi$. Indeed

$$\int_o^{\phi} \sin(\phi-\psi)R(d\psi) - \int_0^{\beta} \sin(\phi-\psi)R(d\psi) = \int_{\beta+}^{\phi} \sin(\phi-\psi)R(d\psi). \tag{5.65}$$

The integrand $\sin(\phi-\psi)$ is non-negative if $\beta \in [\phi-\pi,\phi]$ so that the integral is non-negative. In the opposite case we use the orthogonality relation in (5.65) rewriting the integral as

$$-\int_{\phi}^{\beta+2\pi} \sin(\phi-\psi)R(d\psi). \tag{5.66}$$

This implies that for any $(x_o,y_o) \in K'$ the maximum of the left hand side of (5.64) is at most equal to $s(\phi)$ with equality reached when $(x_o,y_o) \in \partial K'$. Hence s is the support function of a compact convex set K'.

On the other hand if K is an arbitrary convex compact polygon we can always approximate it in the Hausdorff metric. For such a polygon K_n one can write the support function as

$$\begin{aligned} s_n(\phi) &= a\cos\phi + b\sin\phi + \sum_{\phi_k \leq \phi} \ell_k \sin(\phi-\phi_k) \\ &= a\cos\phi + b\sin\phi + \int_0^{\phi} \sin(\phi-\psi)R_n(d\psi), \end{aligned} \tag{5.67}$$

where ℓ_ν are the lengths of the sides and ϕ_k are the angles. This is of the form (5.58) and defines measures R_n of variation equal to the perimeter of K_n, hence uniformly bounded. We can then select a weakly convergent subsequence converging to some R and get, in the limit, the relation (5.58). This proves the assertion.

If K has a continuous radius of curvature r then s is twice differentiable and

$$Ls = \ddot{s} + s = r \tag{5.68}$$

so that in this case the radius of curvature is simply Ls. Also, integrating (5.68) we get

$$\int_0^{2\pi} r\, d\phi = \int_{\partial K} ds = \text{perimeter} = \int_0^{2\pi} s(\phi)d\phi. \tag{5.69}$$

The fact that the homogeneous equation $Ls = 0$ has the solutions $\cos\phi$, $\sin\phi$ is the basis for the possibility of representing a support function trigonometrically in (5.58).

Before leaving the convex sets let us point out that some interesting image algebras can be obtained similarly to (5.52) by replacing the half planes H_α by other sets.

Let the prototype I_o consist of a closed convex set with smooth boundary without any straight sections so that the curvature $\rho = d\phi/ds$ is a continuous positive function of <u>arc length</u> s. Since ϕ and s (assuming a fixed position for $s = 0$) are in one-to-one correspondence, we can express curvature as a function of ϕ: $\rho = \rho(\phi)$.

<u>Case 5.10</u>. With S as the set of translations we get the image algebra consisting of all images of the form

$$I = \bigcap_{s \in S_o} (sI_o) \tag{5.70}$$

where S_o is an arbitrary subset of S. Then I will also be a convex set and can be characterized through its boundary $b = \partial I$, which can now have corners but not straight sections.

It is possible to describe this image algebra $\mathscr{T}$ in terms of the boundary.

<u>Theorem 5.4</u>. <u>Let</u> E <u>be a closed convex set without straight sections and with the curvature</u> $\rho = \psi(\phi)$ <u>defined a.e. Then</u> $E \in \mathscr{T}$ <u>if and only if the curvature is uniformly large in the sense of the inequality</u>

$$\psi(\phi) \geq \rho(\phi), \quad \text{a.e.} \quad \phi \in [0,2\pi]. \tag{5.71}$$

<u>Proof</u>: Let $E \in \mathscr{T}$ so that it can be represented as in (5.70). If S_o is a finite set then b consists of a finite number of arcs from ∂I_o and the inequality (5.71) holds with equality everywhere except at the corners of b. On the other hand, if S_o is infinite we can approximate E by a set $E' \supseteq E$, where E' has a finite representation and is arbitrarily close to E. Let ϕ_1, ϕ_2 be two angles such that the corresponding points P_1 and P_2 on b are not corners. In the same way, denote by P_1' and P_2' the points on E' corresponding to the same angles (see Figure 5.14) where crosses denote the corners. Consider the arc joining two corners of $\partial E'$ and the arc of E corresponding to the same values of ϕ: ϕ_i, ϕ_{i+1}. The first arc is at least as long as the second one. Adding, we get

$$\ell(P_1,P_2) \leq \ell(P_1',P_2') \leq \sum_i \int_{\phi_i}^{\phi_{i+1}} \frac{1}{\rho(\phi)}\, d\phi = \int \frac{d\phi}{\rho(\phi)} \tag{5.72}$$

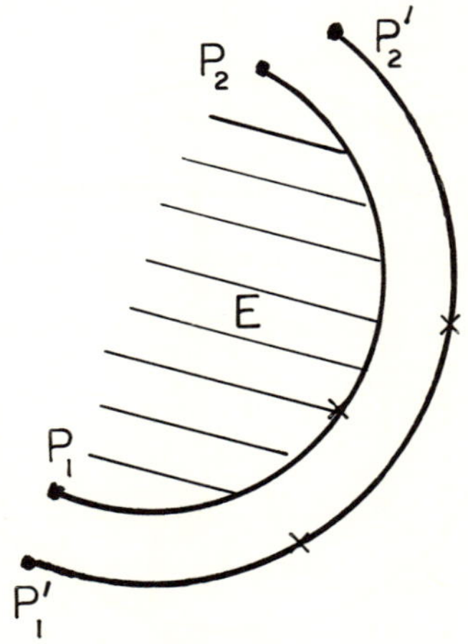

Figure 5.14

where $\ell(a,b)$ stands for the length of the arc joining a and b. The second inequality is due to the fact that ϕ has saltuses at the corners ϕ_i. Taking the limit $P_2 \to P_1$ in (5.72), we get a.e.

$$\frac{ds}{d\phi} \leq \frac{1}{\rho(\phi)} \tag{5.73}$$

which implies (5.71).

Conversely, if the inequality (5.71) is satisfied for a closed, convex set E, consider a point on E which is not a corner (see Figure 5.15). The radius vector $\vec{PQ}(\phi)$, $0 \leq \phi \leq 2\pi$, cuts $\partial E'$ at Q. Here E' stands for the translate sI_o which brings the point of I_o with angle ϕ to the point P.

Denoting by $\vec{u}(\alpha)$ a unit vector in the direction α, we have (with ds = arc element of ∂E)

$$\overrightarrow{PQ(\phi)} = \int_{\phi_o}^{\phi} \vec{u}(\alpha)ds = \int_{\phi_o}^{\phi} \vec{u}(\alpha)\frac{ds}{d\alpha}\,d\alpha = \int_{\phi_o}^{\phi} \vec{u}(\alpha)\frac{d\alpha}{\psi(\alpha)} \tag{5.74}$$

and similarly

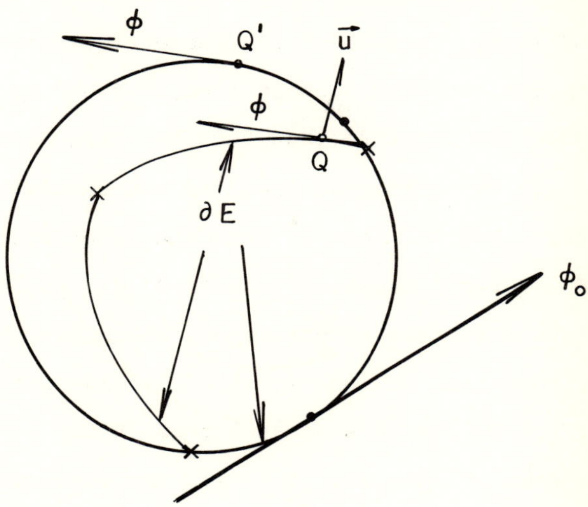

Figure 5.15

$$\overrightarrow{PQ'(\phi)} = \int_{\phi_o}^{\phi} \vec{u}(\alpha) \frac{d\alpha}{\rho(\alpha)} . \tag{5.75}$$

Hence

$$P\overrightarrow{Q(\phi)} = \int_{\phi_o}^{\phi} \frac{\rho(\alpha)}{\psi(\alpha)} d\overrightarrow{PQ'}(\alpha). \tag{5.76}$$

Let $\vec{u}$ be a unit vector at Q perpendicular to the support line ϕ (see the figure), and form the inner product

$$\begin{aligned} x = (\vec{u},\overrightarrow{PQ(\phi)}) &= \int_{\phi_o}^{\phi} (u,u(\alpha)) \frac{d\alpha}{\psi(\alpha)} \\ &= \int_{\phi_o}^{\phi} \cos(\phi - \tfrac{\pi}{2} - \alpha) \frac{d\alpha}{\psi(\alpha)} = \int_{\phi_o}^{\phi} \sin(\phi-\alpha)\frac{d\alpha}{\psi(\alpha)} \end{aligned} \tag{5.77}$$

and similarly

$$x' = (\vec{u},\overrightarrow{PQ'(\phi)}) = \int_{\phi_o}^{\phi} \sin(\phi-\alpha) \frac{d\alpha}{\rho(\alpha)} . \tag{5.78}$$

Then, from the inequality (5.71), we get

$$x'-x = \int_{\phi_o}^{\phi} \sin(\phi-\alpha) \left[\frac{1}{\rho(\alpha)} - \frac{1}{\psi(\alpha)}\right] d\alpha \geq 0 \tag{5.79}$$

if the <u>sin</u> in the integrand is non-negative, hence for $\phi_o \leq \phi \leq \phi_o+\pi$. Because of the geometric significance of $x'-x$ this implies that the support line in the direction ϕ for E' is further out than that for E. This circumstance, together with the analogous ones for the remaining range of ϕ, means that $E \subseteq E'$. Now we form the intersection of all sets like E' for all starting points P on the boundary of E. It follows that this intersection, which has the form of (5.70), contains E. But it must coincide with E since any point $P_o \notin E$ can be separated from E by a "support curve" through some P on ∂E. Noticing the construction (see Figure 5.15), it follows that P_o is excluded from E', which completes the proof of the theorem.

Note that with an image of this type we can associate a <u>minimal set</u> S_o <u>of the translations generating it</u>: S_o can be looked at as the dual of I. The operation $I_1 \cap I_2$ in $\mathcal{T}$ corresponds to the operation $S_1 \cup S_2$ in S. Also, it is clear that S_o is obtained from all the non-corner points of ∂I.

One could ask for similar results when S is a different similarity group, e.g. the set of all rigid motions of the plane.

<u>Case 5.11</u>. Let us form another image algebra, starting from a finite subset $\mathcal{K}_o \subseteq \mathcal{K}$, with some indexing $\mathcal{K}_o = \{K_i\}$, and form

$$\mathcal{T} = \text{intersection and translation closure of } \mathcal{K}_o \text{ subsets.}$$

In other words we allow all translations of members of $\mathcal{K}_o$ and translations thereof, as well as limits of sequences of such sets. We can characterize a set in $\mathscr{K}_0$ by its radius of curvature $R(\phi)$ as a function of the angle, and to avoid a trivial complication we shall assume that R be continuous.

Introduce the function class Φ by the conditions that its members satisfy

$$\begin{cases} \text{a) any } f \in \Phi \text{ is defined on } (-\frac{\pi}{2}, \frac{\pi}{2}) \text{ and is non-negative and } L_1 \\ \text{b) } \int_{-\pi}^{\pi} \cos\psi \, f(\psi)d\psi = 0 \\ \text{c) } \int_0^{\phi} \cos(\phi-\psi)f(\psi)d\psi \geq 0 \text{ for all } \phi \in (-\frac{\pi}{2}, \frac{\pi}{2}). \end{cases} \tag{5.80}$$

The set Φ is a convex cone in $L_1(-\frac{\pi}{2}, \frac{\pi}{2})$.

Given two sets $K_1, K_2 \in \mathscr{T}$ with radius of curvature functions R_1 and R_2 we shall say that R_1 is <u>cosine-dominated</u> by R_2, written $R_1 < R_2$, if $R_2-R_1 \in \Phi$. This introduces a partial ordering among the R's and we shall briefly look at some simple properties of it.

First, if $R_1(\psi) \geq R_2(\psi)$ for all $\psi \in (-\frac{\pi}{2}, \frac{\pi}{2})$, then $R_1 > R_2$. If both functions are continuous at $\psi = 0$, then $R_1 > R_2$ implies $R_1(0) \geq R_2(0)$. This follows from

$$R_1(0)-R_2(0) = \lim_{h\to 0} \frac{1}{h} \int_0^h \cos(h-\psi)[R_1(\psi)-R_2(\psi)]d\psi \geq 0. \tag{5.81}$$

Third, if $R < R_1$ and $R < R_2$, then $R < \frac{1}{2}(R_1+R_2)$. Again we just use the non-negativity of the cosine factor in (5.80).

<u>Theorem 5.5. In order that the set</u> $E \in \mathscr{K}$ <u>with continuous radius of curvature</u> R <u>belong to the image algebra</u> $\mathscr{T}$ <u>defined</u>

<u>in</u> Case 5.11 <u>it is necessary and sufficient that for each</u> $\alpha \in (0,2\pi)$ <u>there is at least one index</u> i <u>such that</u> $R_i(\phi+\alpha)$ <u>is cosine dominated by</u> $R(\phi+\alpha)$.

<u>Proof</u>: In order to prove sufficiency let α be an arbitrary angle and associate it with one index i and the set K_i translated to the point of ∂E corresponding to α. Using the same proof as in Theorem 5.4 it follows that this translate of K_i contains E. Therefore the intersection of them is just E so that $E \in \mathcal{T}$.

On the other hand if $E \in \mathcal{T}$, then for any point of ∂E corresponding to an angle α there is a translate of some K_i containing S and tangential to ∂E at the point. But this can be expressed in a statement relating $R(\alpha+\phi)$ to $R_i(\alpha+\phi)$ by the partial ordering and the result follows.

<u>Case 5.12</u>. (<u>Euclidean set patterns</u>). The generators are subsets of the plane arranged into a finite set of generator classes with the operations union and intersection and S = Euclidean group. Identification is by equality of the resulting sets.

Here a generator class could consist of all (or a subset of all) lines, another one by all (or a subset of all) circles and so on.

If we relax the condition imposed on the prototype(s) we can let I_o be, say, the complement of a 90° wedge as in Figure 5.16(a) and S = all Euclidean motions, and we get the polygonal images in (b) and (c). Note that the prototype can be written as

$$I_o = \{(x,y) \mid x \leq 0 \quad \text{or} \quad y \leq 0\} = g_x \vee g_y, \tag{5.82}$$

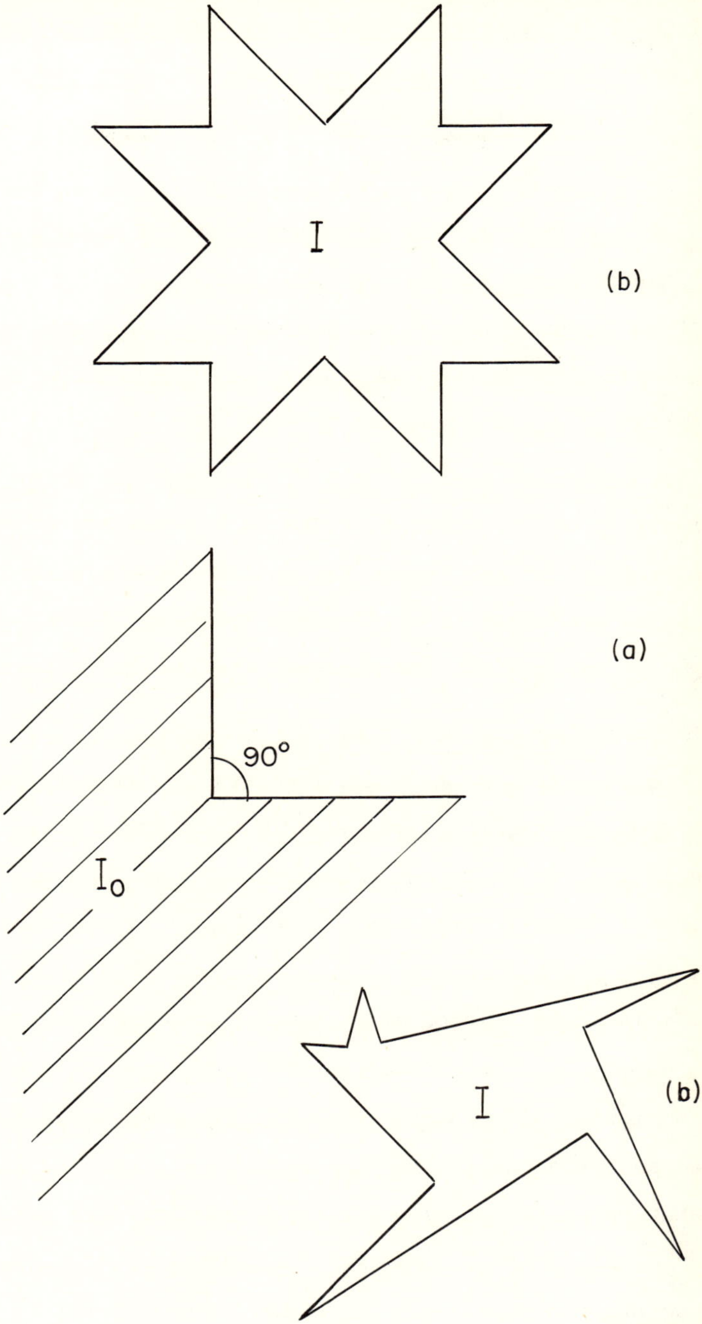

Figure 5.16

in terms of the simple linear threshold logic features (see Section 5.9)

$$\left.\begin{aligned} g_x &= \{(x,y)\,|\,x \le 0\} \\ g_y &= \{(x,y)\,|\,y \le 0\} \end{aligned}\right\} \tag{5.83}$$

so that we now operate with a feature logic of order two with the halfplane as the simple feature.

Such images can also be characterized in terms of their boundary. To look more closely into this question we shall study a case that will reappear in Part III.

<u>Case 5.13</u>. Form an image algebra consisting of all images of the form

$$I = \bigwedge_{(s_1,s_2)\in S} ((s_1 I_o) \vee (s_2 I_o)) \tag{5.84}$$

with I_o as in (5.82) and S consisting of all translations and rotations that are multiples of $\pi/2$.

The conjunction in (5.84) can be understood to say that certain open quadrants $Q_i(z)$ should be excluded where z is some point (x,y) in the plane and $Q_i(z)$ is the quadrant with z as the corner and $i = 1,2,3,4$ with the usual numbering.

To indicate some of the complications that can occur in this seemingly simple image algebra, an image has been given in Figure 5.17 and the reader is recommended to identify the quadrants that have been eliminated.

Notice that tentacles can be formed, at most four (one in the figure), that I need not be connected, and that corner points can separate a connected component into two.

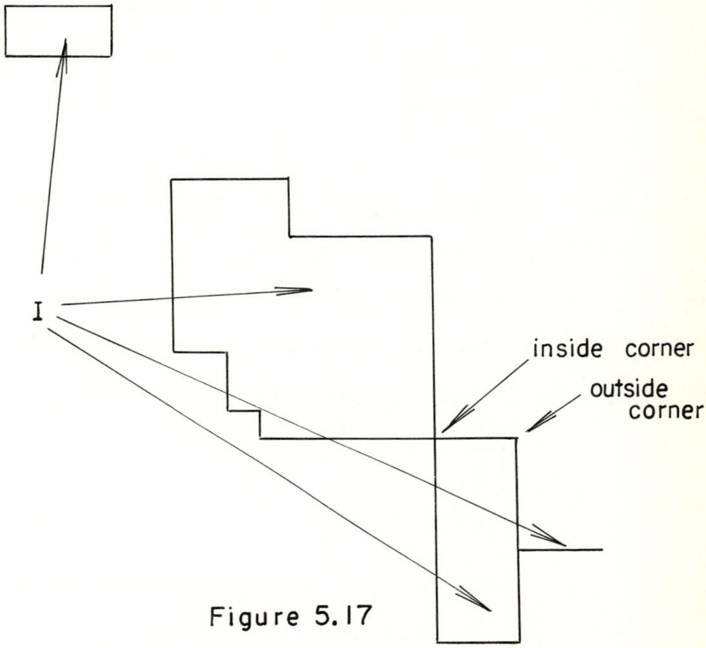

Figure 5.17

Lemma 5.1. Consider an arbitrary image in $\mathscr{T}$ and cut it with a horizontal line $y = y_o$. The resulting set

$$H(y_o) = \{x \mid (x,y_o) \in I\} \tag{5.85}$$

is a closed interval including the possibilities empty, single-point, half-infinite, infinite. Similarly for vertical cuts $V(x_o)$.

Proof: The proof is easy. Using the representation (5.84) we get

$$H(y_o) = \cap\{x \mid (x,y_o) \in s_1 I_o \text{ or } s_2 I_o\} = \cap H(s_1,s_2). \tag{5.86}$$

But each $H(s_1,s_2)$ is either empty, a single point, closed interval, half-infinite or infinite line, so that the same will hold for the intersection of them.

A consequence of the lemma is that if $\underline{I}$ is connected then it is also singly connected, a fact that will be used without being stated explicitly each time.

Let us consider an $\underline{I}$ of finite numerical complexity (it has structural complexity 2 by definition) and let us deal only with bounded, connected ones. The boundary consists of vertical and horizontal line segments, forming inside and outside corners. The role of the tentacles is easy to grasp and it means no essential restriction to assume that they do not occur; more about this below. We can write

$$I = E_1 \cap E_2 \cap E_3 \cap E_4 \tag{5.87}$$

where E_i is the conjunction of all the double features occurring in (5.84) belonging to excluded quadrants numbered i. Of course an E_i can very well be the entire plane if no i-quadrants are eliminated. The boundary of any E_i has inner and outer corners alternating and if we run through it in the positive direction its segments can be characterized by the inequalities:

$$\begin{cases} \partial E_1: & \Delta x \leq 0, \quad \Delta y \geq 0 \\ \partial E_2: & \Delta x \leq 0, \quad \Delta y \leq 0 \\ \partial E_3: & \Delta x \geq 0, \quad \Delta y \leq 0 \\ \partial E_4: & \Delta x \geq 0, \quad \Delta y \geq 0 \end{cases} \tag{5.88}$$

When we run through ∂I in the positive direction and go from one E_i to the next, the two boundaries are joined at an outer corner. This is a complete characterization under the given conditions and gives us

<u>Theorem 5.6</u>. <u>Let</u> I <u>be a tentacle-free bounded and connected</u>

image of finite numerical complexity. Then ∂I *consists of at most four arcs with each arc consisting of alternating vertical and horizontal line segments, as in* (5.82), *and inside corners alternating with outside. Arcs are joined at outside corners. Conversely, if* I *is a closed, connected, bounded set with* ∂I *as described above, it can be represented as in* (5.87) *with a finite feature set.*

Proof: It only remains to prove the converse part. Form the E_i corresponding to the four (at most) arcs making up ∂I. It is then seen that $I' = E_1 \cap E_2 \cap E_3 \cap E_4$ has the boundary ∂I, since no cancellation is possible, and hence coincides with I.

To deal with the tentacles it is enough to note that a tentacle-free image in this algebra can be given tentacles anywhere on the (at most) four extreme intervals, where x or y achieve their minima and maxima respectively. Conversely, any tentacle must maximize or minimize x or y.

The situation when the numerical complexity is infinite can also be dealt with. Say that I is bounded, connected and has a boundary with a continuous tangent. Corresponding to the construction of the ∂E_i we now get four arcs with tangents behaving similarly to the inequalities in (5.88).

Indeed, consider ∂E_1 run through in the positive direction. Its tangent must then be in the direction of the second quadrant, the tangent of ∂E_2 in the third quadrant, and so on, and we can formulate

Theorem 5.7. A bounded, connected image I *of structural complexity* 2 *and with a continuous tangent has a boundary*

<u>that can be decomposed into four arcs</u> a_1, a_2, a_3, a_4 <u>such that the direction of the tangent along</u> a_i <u>is in the</u> (i+1)st <u>quadrant</u> (<u>addition modulo</u> 4). <u>Conversely, a closed</u>, <u>bounded</u>, <u>connected set</u> I <u>with a continuous tangent satisfying this direction condition is of structural complexity</u> 2 <u>in the image algebra</u>.

<u>Proof</u>: Only the converse part remains to be proven. Along the arc a_i we associate a set

$$E_i = \bigcap_{z \in a_i} ({\sim}Q_i(z)). \tag{5.89}$$

None of the excluded quadrants $Q_i(z)$ in (5.89) cuts into I because of the constraints on the direction of the tangents along a_i. Hence $I' = E_1 \cap E_2 \cap E_3 \cap E_4 \supseteq I$. On the other hand, $\partial I' \subset \partial I$ and the proof is easily completed.

Image algebras with logical features as generators and of given structural complexity will be studied in a more general context, not necessarily planar patterns, in Section 5.9.

<u>Starshaped images</u> are related to convex sets: the generators are no longer half planes but directed line segments. The generator index α is the direction of the segment, $0 \leq \alpha \leq 2\pi$, the length $\ell(\alpha)$ is another attribute. We have $\omega_{in} = \omega_{out} = 1$ with the common bond value $b = (x_o, y_o, \alpha)$ where (x_o, y_o) is the starting end point of the directed segment.

With the bond relation

$$b\rho b' \iff x_o = x_o', \quad y = y_o', \quad \alpha \neq \alpha' \tag{5.90}$$

and with the connection type Σ = FULL, we shall identify two configurations if the resulting unions of generators coincide

$$c \equiv c' (\text{mod } R) \iff \bigcup_{\alpha} g_{\alpha} = \bigcup_{\alpha} g'_{\alpha}. \tag{5.91}$$

This represents a starshaped image centered at the common starting point (x_o, y_o) and described in polar coordinates $r = \ell(\alpha)$, where r is the radius from (x_o, y_o) to ∂I.

Similarly if we use disks $D(x_o, y_o, r)$ as generators and start from a line image as given through the equation

$$\phi(x,y) = c, \tag{5.92}$$

the bond values are now $b = (x_o, y_o)$ with the bond relation

$$b \rho b' \iff \phi(x_o, y_o) = \phi(x'_o, y'_o) \text{ and } (x_o, y_o) \neq (x'_o, y'_o). \tag{5.93}$$

If, for example, $\phi(x,y) = x \sin \psi + y \cos \psi$ and $r = r(x_o, y_o)$ is a given function of $-x \cos \psi + y \sin \psi$ we get <u>medial axis</u> images as in Figure 5.18, see Blum (1967). The line image $\phi(x,y) = c$ is called the <u>skeleton</u> of the figure and $r(x,y)$ the <u>quench</u> function.

Using other generators and bond relations we will arrive at other image algebras, but this is left to the reader.

To study the metric aspect of feature images we shall consider halfplanes picked at random and satisfying the bond relation "NOT IMPLY" as before. With some loss of generality we shall assume that the halfplane generated are

$$x \cos \phi_k + y \sin \phi_k \leq \ell_k, \quad k = 1,2,\ldots,n \tag{5.94}$$

where $\phi_k = \frac{2\pi k}{n}$ and ℓ_k are non-negative i.i.d. variables with a frequency function f on the positive real line. What happens when n becomes large: do we get any interesting random set measures in the limit?

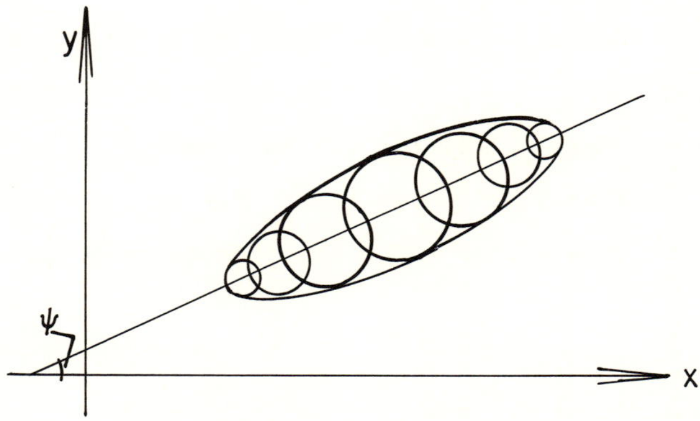

Figure 5.18

It is obvious that the answer is negative with the way the question was posed above. Indeed, if ℓ_{inf} is the infimum of the support of f, it is clear that the feature generated image will be a circle with radius ℓ_{inf}. In particular, if $\ell_{inf} = 0$, we get just a single point.

To get a non-trivial limit it is necessary to let f depend upon n, or, which is equivalent, to normalize the set by scaling it appropriately. Say that $f_{inf} = 0$ and that

$$f(\ell) = A\ell^p(1+o(1)), \quad \ell \downarrow 0, \quad A > 0.$$

Then we can calculate the joint "distribution function"

$$G_n(t_1,t_2,\ldots,t_m) = P\{r_1 > t_1, r_2 > t_2,\ldots,r_m > t_m\} \quad (5.96)$$

when we have written the boundary in polar form $r = r(\phi)$ and put $r = r(\psi_\nu)$, where the ψ_ν are fixed angles ordered in increasing order $0 \leq \psi_1 < \psi_2 \ldots < \psi_m < 2\pi$.

By direct inspection of the geometry we find that G_n is the probability of the compound event

$$\left.\begin{array}{lll} \dfrac{\ell_k}{\cos(\phi_k-\psi_1)} > t_1 & \text{for} & |\phi_k-\psi_1| < \frac{\pi}{2} \\ \dfrac{\ell_k}{\cos(\phi_k-\psi_2)} > t_2 & \text{for} & |\phi_k-\psi_2| < \frac{\pi}{2} \\ \cdot\ \cdot\ \cdot\ \cdot\ \cdot & & \\ \dfrac{\ell_k}{\cos(\phi_k-\psi_m)} > t_m & \text{for} & |\phi_k-\psi_m| < \frac{\pi}{2} \end{array}\right\} \tag{5.97}$$

It can be seen that this can be expressed as an <u>infimal convolution</u>: given two sequences $\{a_k\}$ and $\{b_k\}$ the infimal convolution is the new sequence whose general element is $c_k = \inf_\nu [a_\nu b_{k-\nu}]$. In the present case the b sequence consists of $1/\cos \phi_k$.

Introduce the functions

$$i_\nu(x) = \text{indicator function of } (\psi_\nu - \tfrac{\pi}{2}, \psi_\nu + \tfrac{\pi}{2}) \tag{5.98}$$

and put

$$i(x;t_1,t_2,\ldots,t_m) = \max_\nu \{t_\nu i_\nu(x)\cos(x-\psi_\nu)\} \tag{5.99}$$

Then the argument used in extreme value theory gives us

$$G_n = \prod_{k=1}^{n} \{1 - F[i(\tfrac{2\pi k}{n})]\}. \tag{5.100}$$

But $F(x) \sim Ax^{p+1}(p+1)^{-1}$ as $x \downarrow 0$ so that with the normalization $t_\nu = n^\alpha s_\nu$ and $\alpha = -(p+1)^{-1}$ we get

$$\begin{aligned} &\lim_{n\to\infty} G_n(n^\alpha t_1, n^\alpha t_2, \ldots, n^\alpha t_m) \\ &\quad = \exp - \{\frac{A}{p+1}\int_0^{2\pi} i^{p+1}(x;t_1,t_2,\ldots,t_m)\}. \end{aligned} \tag{5.101}$$

In other words we have

<u>Theorem 5.8.</u> <u>If the image is normalized by dividing by</u> n^α,

<u>and if</u> (5.90) <u>holds</u>, $\alpha = -(p+1)^{-1}$, <u>the limiting measure is given by</u>

$$G(t_1,t_2,\dots,t_m) = \exp - \{\frac{A}{p+1}\int_0^{2\pi} i^{p+1}(x;t_1,t_2,\dots,t_m)dx\}. \tag{5.102}$$

<u>In particular the one-dimensional marginal is given by</u>

$$G(t) = \exp - \{\frac{A}{p+1}\sqrt{\pi}\frac{\Gamma(\frac{p+2}{2})}{\Gamma(\frac{p+3}{2})} t^{p+1}\} \tag{5.103}$$

<u>and the two-dimensional marginal for</u> $\delta = \psi_2-\psi_1$, $0 \le \delta \le \pi$

$$G(t_1,t_2) = \exp - \frac{A}{p+1}\{t_1^{p+1}\int_{-\pi/2}^{x_o} \cos^{p+1}x dx + t_2^{p+1}\int_{x_o}^{\pi/2+\delta} \cos^{p+1}(x-\delta)dx\} \tag{5.104}$$

<u>where</u> x_o <u>satisfies</u> $t_1 \cos x_o = t_2 \cos(x_o-\delta)$, $-\frac{\pi}{2}+\delta \le x_o \le \frac{\pi}{2}+\delta$.

Let us look at the case $p = 0$. Then (5.103) reduces to an exponential distribution with mean $1/(2A)$. The covariance function is obtained from (5.104) after integrating by parts, say for $A = 1$,

$$\begin{aligned} r(\delta) &= \int_o^\infty\int_o^\infty G(t_1,t_2)dt_1dt_2 \\ &= \int_o^\infty\int_o^\infty \exp - \{t_1 \sin x_o+t_1+t_2-t_2\sin(x_o-\delta)\}dt_1dt_2 \end{aligned} \tag{5.105}$$

where it should be noted that x_o depends upon t_1,t_2, and δ.

The covariance function in (5.105) is continuous in δ which is easiest seen by going back to the representation in (5.102). Also the integrand in (5.105) is differentiable with a derivative bounded by an integrable function, so that $r(\delta)$

is differentiable with $r'(0) = 0$. It is not twice differentiable, however. The boundary ∂I of the image then has the property that when written in polar coordinates it is continuous in the mean but not differentiable in the mean.

To illustrate the appearance of such images we have simulated this feature image algebra and display five cases in Figure 5.19. The parameter values of p are $-8/9, 0,1,9,19$ in the figures a,b,c,d and e. The number of generators is 100.

As p increases the boundary becomes more circular in shape which follows from the form of the limiting distribution above when one calculates $\mathrm{Var}(r)/E^2(r)$.

The pictures also seem to support the conjecture that the reason why $r(\phi)$ is not differentiable in the mean is the clustering of the corners, especially close to $r = 0$. This is barely visible in the reproductions of the original graphs.

Plane contrast images represent a function - here to be taken to be real or complex valued - on some subset of the plane. Note however that it may also have external bonds.

Writing $I = I(x_1,x_2) = I(z)$, where $z = (x_1,x_2)$, the function $I(z)$ is the result of combining contrast generators with each other.

To start with an extremely regular case let the generators be trigonometric functions, say in complex form

$$g(x_1,x_2) = A \exp i\,\{\lambda_1 x_1+\lambda_2 x_2-\phi\} \tag{5.106}$$

or in real form

$$g(x_1,x_2) = A \cos\,\{\lambda_1 x_1+\lambda_2 x_2-\phi\} \tag{5.107}$$

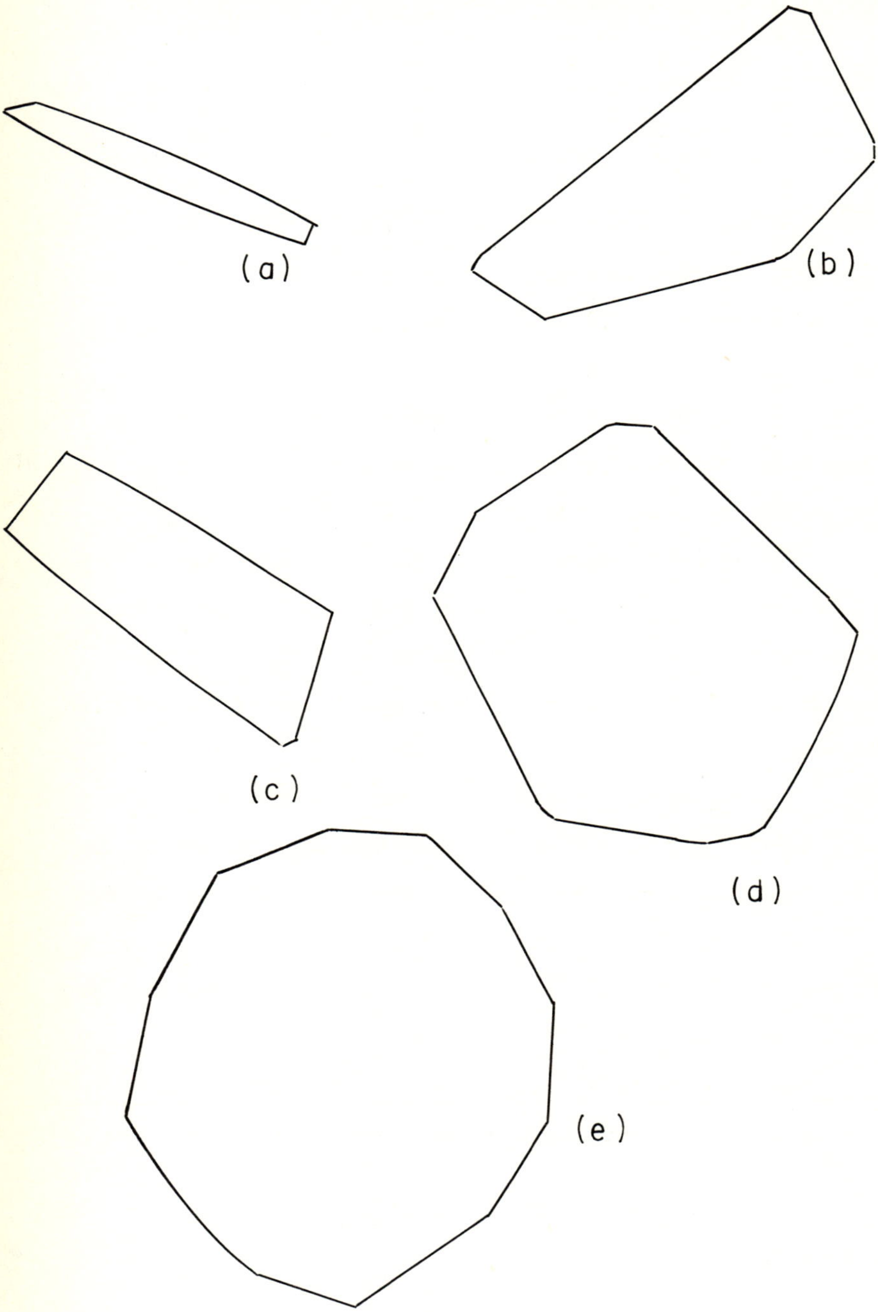

Figure 5.19

where A is the amplitude, ϕ is the phase angle, and (λ_1,λ_2) is the frequency vector or wave number.

Such a generator alone gives rise to the image in Figure 5.20 where the value of I(z) has been indicated by

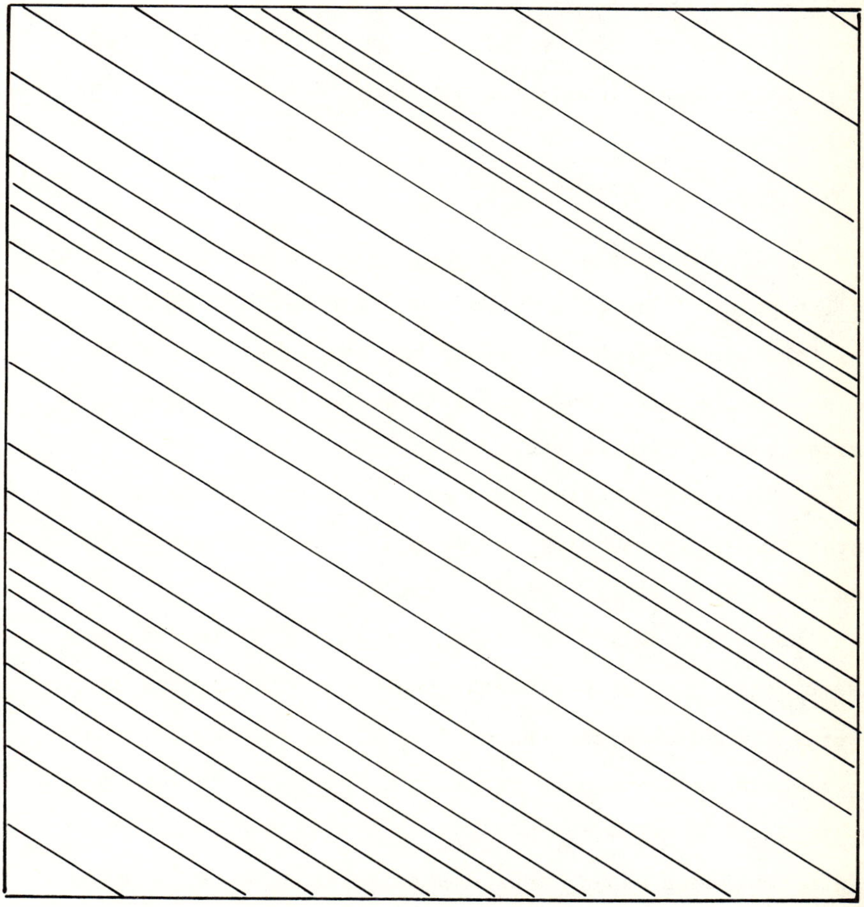

Figure 5.20

the density of the lines plotted. In general the iso-curves $I(x_1,x_2) = \xi_1$ will give an idea of the behavior of the contrast image. We will also use gradient curves: a family of curves orthogonal to the iso-curves. At a point z_o we may use a local coordinate system (ξ_1,ξ_2) where ξ_1 is a parametrisation of iso-curves and ξ_2 one of gradient curves.

This construction will not work if the gradient vanishes

$$\text{grad } I = \left(\frac{\partial I}{\partial x}, \frac{\partial I}{\partial y}\right) = 0 \tag{5.108}$$

and z_o is then called a singular point, otherwise it is a regular point. The behavior of I in a neighborhood of z_o will then depend upon the values of the higher derivatives, and we get the well known classification into types such as maxima, saddle points, etc.

The local behavior at a regular point can be partially characterized by the slope $||\text{grad } I||$ and the aspect arg grad I as well as by their derivatives.

In 6 we shall discuss the notions of hills, valleys, passes, etc. that give a topological characterization of the contrast image when the value of $I(z)$ means height. Another characterization is in terms of texture, for example roughness which measures deviation from linearity. The absolute roughness of I over a set E is given by the expression

$$R_{abs}(I) = \frac{1}{m(E)} \iint_E \left(\left(\frac{\partial^2 I}{\partial x^2}\right)^2 + \left(\frac{\partial I^2}{\partial y^2}\right)^2\right) dxdy \tag{5.109}$$

and the relative roughness by

$$R_{rel}(I) = R_{abs}(I) / \frac{||I||^2}{m(E)} \tag{5.110}$$

where $||I||$ denotes the L_2-norm of the image.

Case 5.14. (Plane Fourier images). Let the generators be as in (5.106) with the generator index $\alpha = (\lambda_1, \lambda_2)$ and with S = translations. Put Σ = FULL. We assume infinite arity with the bond values equal to α and with the bond relation ρ = DIFFERENT. Two regular configurations $c = (g_1, g_2, \ldots, g_m)$ and $c' = (g_1', g_2', \ldots, g_m')$ will be identified if

$$\sum_{i=1}^{n} g_i(z) = \sum_{i=1}^{m} g_i'(z) \quad \text{for all} \quad z. \tag{5.111}$$

If we allow infinite cardinality we also need a criterion to ensure convergence in the appropriate topology.

Let the cardinality be finite for simplicity. It is not true that given the image the generators can be identified uniquely. However, the configuration can be chosen in a canonical form by collapsing all generators belonging to the same generator class G^α. Then, since trigonometric functions with different frequency vectors are linearly independent it follows that the generators are uniquely determined in this canonical form.

A metric study of these Fourier images can be carried out as follows applying the general procedure in 4.10. To the different α's associate weights in the following manner. Let there be given an absolutely continuous bounded measure F on the two-dimensional α-space. Pick a natural number N and consider frequency vectors of the form $(\frac{k}{N}, \frac{\ell}{N}) = \lambda$ where $-N \leq k, \ell < N$. To each such λ we associate a phase angle $\phi_\lambda = \phi_{k,\ell}$ uniformly distributed over $(0, 2\pi)$ and form the configuration

$$c^{(N)} = (A_k \exp i[(\lambda_{k\ell}, z) + \phi_{k\ell}];\ k, \ell). \tag{5.112}$$

The weight A_k should be the square root of the F-measure of the rectangle with sides $(\frac{k}{N}, \frac{k+1}{N})$ and $(\frac{\ell}{N}, \frac{\ell+1}{N})$ respectively. To make $c^{(N)}$ regular we must let the frequencies be different and we let k and ℓ range over all values from $-N, N-1$. This defines a P^N-measure over $\mathscr{C}(\mathscr{R})$.

It is interesting to note that when $N \to \infty$ the probability measure P^N converges to that of a stationary Gaussian stochastic process with mean zero and covariance function of the form

$$\int_{-\infty}^{\infty}\int_{-\infty}^{\infty} \exp[i(\lambda,z)]F(d\lambda) \tag{5.113}$$

so that F is simply the spectral measure of the process.

This representation can be used to calculate various parameters for the image. For example the expected absolute roughness, see (5.109), turns out to be

$$\int_{-\infty}^{\infty}\int_{-\infty}^{\infty} (\lambda_1^4+\lambda_2^4)F(d\lambda_1,d\lambda_2) \tag{5.114}$$

and other parameters can be derived similarly. The topological ones are more difficult, however.

Let us now consider a more structured case. The previous one has been quite efficient in describing practically observed plane images, e.g. in forestry (see Matern 1960) and in oceanography (see Pierson (1955)). It is quite general and may need further specification in cases when more is known of the image structure.

To present such a case let us start from a finite set of real valued functions $\phi_\alpha(z)$; $\alpha = 1,2,\ldots,m$ is the generator index. Let S be the direct product $S_1 \times S_2 \times S_3$ where S_1 is the translation group $z \to z+z_o$, S_2 is the linear group $z \to Az$, A non-singular 2×2 matrix, and S_3 consists of multiplication with positive numbers.

Form the set G^α of all $s\phi_\alpha$, when s is an arbitrary element in S and the union G of all the G^α. Obviously each G^α is closed under similarity transformation. We shall also assume that g^α's with different α are linearly independent.

Case 5.15. (Direct topographic images). With the given G, Σ = FULL, and ρ = DIFFERENT we get a configuration space where we shall identify configurations just as in (5.106).

The resulting image algebra again has the property that if we collapse generators with the same index into one generator then this canonical form is uniquely determined from the image.

The prototypes $\phi^\alpha(x_1,x_2)$ will determine the qualitative behavior of the images. Writing $r = \sqrt{x_1^2+x_2^2}$ we mention the following examples of topographic generators.

Index	Generator	
hill	$\phi(x_1,x_2) = \psi(r),\psi$	as in Figure 5.21a
pit	$\phi(x_1,x_2) = -\psi(r),\psi$	"
valley	$\phi(x_1,x_2) = -\psi(x_1),\psi$	"
ridge	$\phi(x_1,x_2) = \psi(\lvert x_1\rvert),\psi$	"
plateau	$\phi(x_1,x_2) = \psi(\lvert x_1\rvert),\psi$	as in Figure 5.21b
crater	$\phi(x_1,x_2) = \psi(r),\psi$	as in Figure 5.21c

Table 5.2

The metric treatment is like the one for Fourier images. For a given natural number N consider a finite set

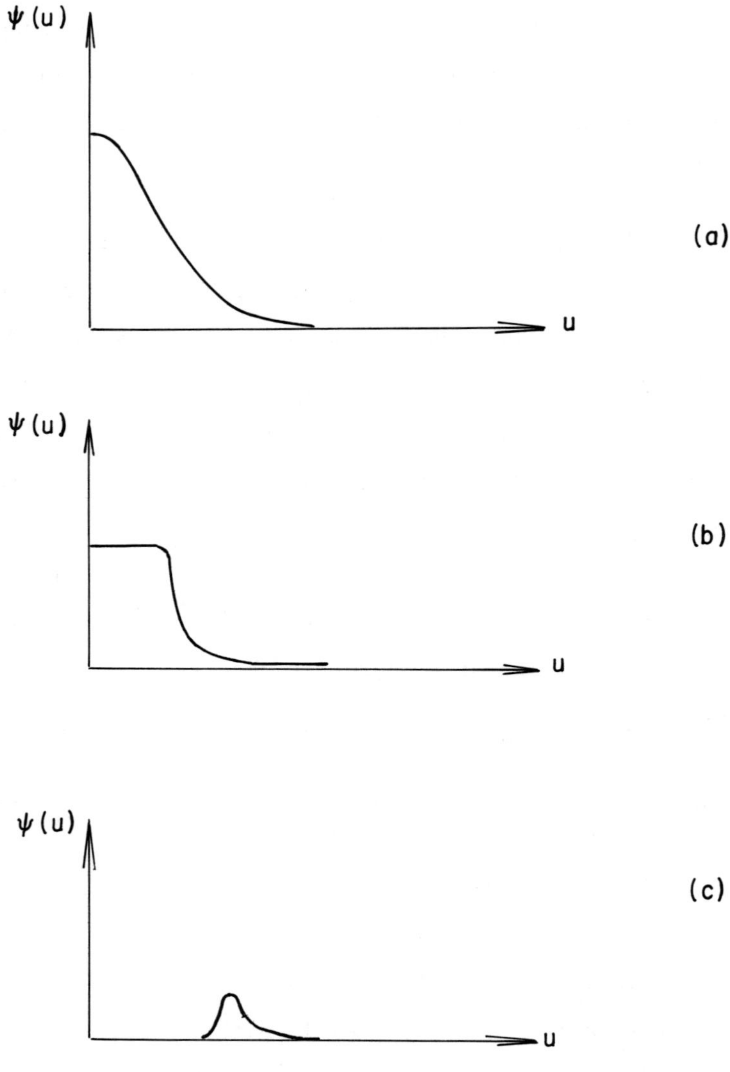

Figure 5.21

$E_n \subset E = A \times S_1 \times S_2$, where A is the generator index set, such that E_N becomes asymptotically dense in E as $N \to \infty$. Define the random image

$$I = \sum_{e \in E_N} a_e \, \phi^{\alpha(e)}(s_1(e)z + s_2(e)) \qquad (5.115)$$

where $e = (\alpha(e), s_1(e), s_2(e))$ with $\alpha(e) \in A$, $s_1(e) \in S_1$, and $s_2(e) \in S_2$. The a_e are independent stochastic variables. Their distribution will determine the qualitative topography in particular for large values of N which will be the main case of interest. We will return to this image algebra in Chapter 17.

Let us return now to curve patterns. Consider two images F and G given by

$$\begin{cases} F\colon f(x,y) = k \\ G\colon g(x,y) = \ell \end{cases} \qquad (5.116)$$

where f and g are two smooth functions and k and ℓ take integers as values.

It has been known for a very long time that if two such images are superimposed under suitable conditions a derived curve image will appear. The new curves will be formed by the shortest diagonals of the (approximately) parallelogram subsets. In Figure 5.22 this would be the dotted curves, in other words the curve pattern H

$$H\colon f(x,y) - g(x,y) = m \qquad (5.117)$$

which is called the subtractive Moiré fringes. With other orientation the visible derived pattern will be based on the relation obtained from (5.117) by replacing the minus sign by the plus sign: additive Moiré fringes. Actually both additive and subtractive fringes can appear in the same picture, and they are then separated by a commutation boundary.

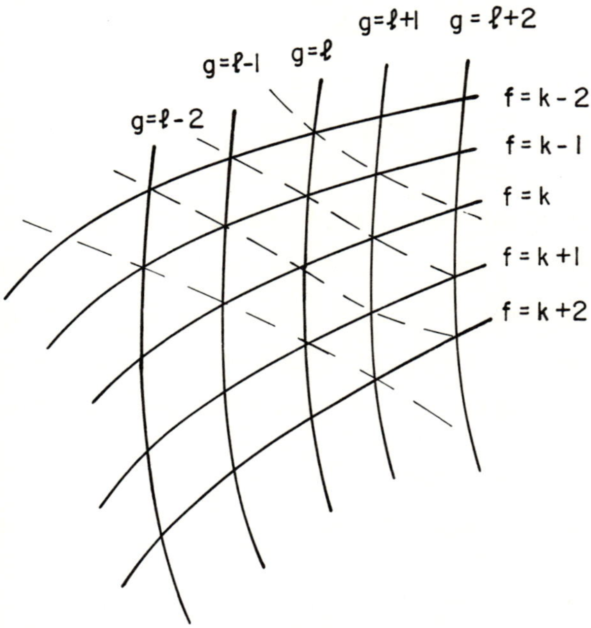

Figure 5.22

To be able to see the derived line pattern clearly the diagonals should be of quite different lengths and the F and G curves should be closely spaced.

Let us say that g is a (small) translation of f, $g(x,y) = f(x-\varepsilon_1, y-\varepsilon_2)$. Then we get approximately for H

$$f'_x \varepsilon_1 + f'_y \varepsilon_2 = \text{grad} f \cdot \varepsilon = m \qquad (5.118)$$

with $\varepsilon = (\varepsilon_1, \varepsilon_2)$ so that the derivation of H is practically equivalent to calculating the gradient of the level curves in the F-system and then forming the dot product with the vector ε.

Let us say that F is the Archimedean spiral

$$\sqrt{x^2+y^2} = a \arctan \frac{y}{x}, \tag{5.119}$$

or to bring out the fact that we need all the branches of the arctan function

$$f(x,y) = \frac{1}{a}\sqrt{x^2+y^2} - \arctan \frac{y}{x} = 2\pi k. \tag{5.120}$$

In a small area of the (x,y)-plane we use (5.120) with one particular branch of the arctan function. In order that the Moiré fringes be clearly visible, the value of $\underline{a}$ should be fairly small.

Let $\varepsilon = (\varepsilon_1, 0)$; we would expect to get the derived H system from

$$\frac{\partial f}{\partial x} = \frac{x}{a\sqrt{x^2+y^2}} + \frac{y}{x^2+y^2} = m\cdot c \quad . \tag{5.121}$$

If $\underline{a}$ is small this will give approximately the system

$$\frac{x}{\sqrt{x^2+y^2}} = m.c_1 \tag{5.122}$$

which is a system of straight lines through the origin and the system is symmetric around the y-axis. This is illustrated in Figure 5.23.

Figure 5.23

6. Spatial Patterns.

Much of the content of the last two sections carries over with only minor changes to the three dimensional case and we can therefore be rather brief when discussing spatial patterns. This does not mean that the extension from two to three dimensions is always easy. In the study of crystals, for example, their group properties become much more involved and especially the analysis of crystal patterns leads to some extremely difficult questions.

In the last part of the previous section we discussed contrast images $I(x,y)$ with the plane as background space and taking scalar values. If the value is interpreted as height it is more natural to talk of a set image in space, describing the topography of a geographical region.

In Case 5.15 we introduced patterns that describe landforms directly: hills, valleys, etc. However we can instead start from a height pattern $z = I(x,y)$, assumed to have been generated one way or another, and then <u>derive patterns</u> from it, still using the terminology of peaks, valleys, etc., but with another interpretation. We shall not attempt a full analysis of how this can be done, and present only an informal discussion of "typical" situations.

We start again with the singular points of the height image where the gradient of the height field vanishes. All other points are called regular. Through a regular point passes exactly one <u>level curve</u> $I(x,y) = \text{const.}$, and, at right angle to it, a <u>slope curve</u> with a direction, so that we can speak of the downward or upward slope.

Let $P = (x_o,y_o)$ be a critical point and consider the

matrix of second derivatives evaluated at P

$$\left\{ \begin{matrix} \dfrac{\partial^2 I}{\partial x^2} & \dfrac{\partial^2 I}{\partial x \partial y} \\ \\ \dfrac{\partial^2 I}{\partial y \partial x} & \dfrac{\partial^2 I}{\partial y^2} \end{matrix} \right\} \tag{6.1}$$

and let λ_1 and λ_2 be its eigenvalues. If $\lambda_1, \lambda_2 > 0$, P is classified as a pit, if $\lambda_1, \lambda_2 < 0$ as a peak, and if the eigenvalues have different signs as a saddle-point. For the moment we disregard the possibility that λ_i can vanish.

Consider two peaks as in Figure 6.1 and the family of slope lines originating from one of them. The line of indifference joining the two hills is called a ridge line and leads through a pass.

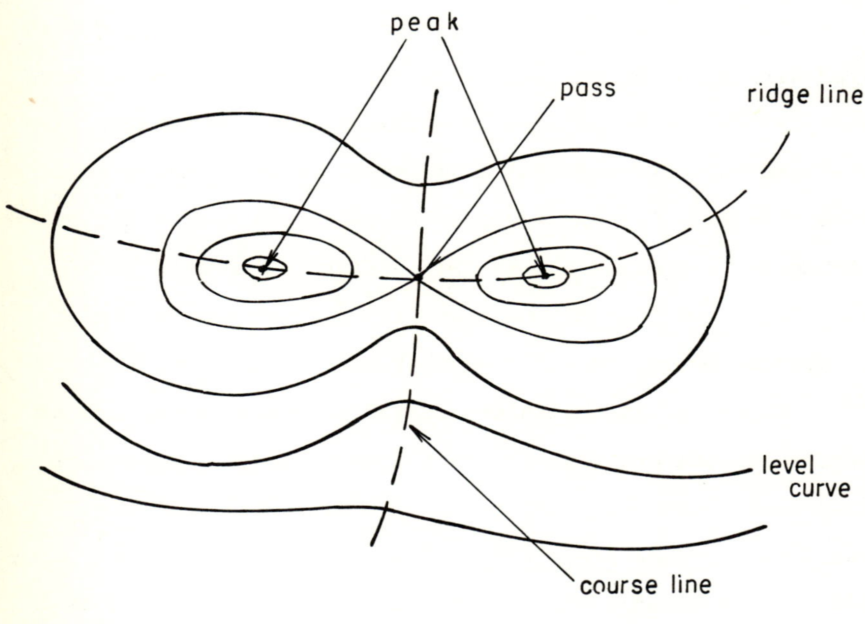

Figure 6.1

Similarly for two pits one is led to course lines and pales. Course lines delimit hills while ridges delimit dales.

Case 6.1. (Derived topographic patterns). For a given peak consider the set of points from which the upward slope lines lead to this peak. This set is called a hill and the partition generated in this way is called the hill image. Similarly we construct the dale images starting with pits and slope lines upwards from them.

Take one of these derived images for a large region. At the corners of the partition a number of lines (edges) meet, and it has been argued that for stability reasons that number is most often three, see Woldenberg (1972). But if three edges meet at a corner and since each edge has two corners we get

$$3n_c = 2n_e \tag{6.2}$$

if $n_c = \#(\text{corners})$, $n_e = \#(\text{edges})$. We also introduce $n_a = \#(\text{areas})$, the number of subsets generated by the partition. Because of Euler's law we have

$$n_a - n_e + n_c = 2. \tag{6.3}$$

If a certain area has n edges, it will correspond to $n/2$ edges in the partition except for the boundary edges, and it follows that, if we neglect the boundary effect

$$n_e \simeq \frac{1}{2} \sum_n n\, N_n \tag{6.4}$$

with $N_n = \#(\text{areas with } n \text{ edges})$. Using (6.2) and (6.3) this gives the approximate relation, neglecting the 2 on the right side of (6.3),

$$\sum_n N_n - \frac{1}{2}\sum_n n\,N_n + \frac{1}{3}\sum_n n\,N_n \tilde{=} 0 \qquad (6.5)$$

or

$$\Sigma(6-n)N_n \tilde{=} 0 \qquad (6.6)$$

so that the <u>average number of edges</u> in the image

$$\frac{\Sigma n\ N_n}{\Sigma\ N_n} \tilde{=} 6. \qquad (6.7)$$

If we also include the terms neglected in the above argument it follows that the average number of edges is at most equal to 6.

In Section 5.5 we have seen an attempt to explain Horton's laws by probabilistic-topological reasoning. It is also possible to derive them from arguments as above, but we shall not discuss this possibility here.

The approximate relation (6.7) will be of use later on when studying segmentation of contrast images in Chapter 13.

The reader will notice that the above discussion is not rigorous. We have not gone into the detailed differential topology, accounting for multiple passes or pales, non-isolated extremum points and so on. Nevertheless it should give some idea of the "typical" form of derived topographic images.

This case starts from a three-dimensional image and leads to a partition of the planes. In many situations, for example in metallurgy and plant morphology, three-dimensional space itself is partitioned into cells by a family of surfaces.

In Case 6.2-3 the generators will be surfaces and points in R^3 respectively, the configurations will be free and of denumerably infinite cardinality. Hence two arbitrary configurations can be joined together. Identifying configurations that lead to the same partition we get a commutative binary operation in $\mathcal{T}$ in each case and we shall write it in the form I_1+I_2. If the surfaces, which play the role of generators here, are just planes one speaks of a <u>tesselation pattern</u>. Let us parametrize an arbitrary plane P as $P = (\theta,\phi,p)$ where the unit vector $u = (\sin\theta\cos\phi, \sin\theta \sin\phi, \cos\phi)$ represents a normal to P and $0 \le \theta < \pi$, $0 \le \phi < \pi$. The value p is the length of the perpendicular from the origin to the foot point pu of P.

<u>Case 6.2</u>. (<u>Homogeneous Poisson plane process</u>). Form the inhomogeneous Poisson point process with intensity

$$\lambda = \frac{1}{2\pi} \sin\theta \tag{6.8}$$

in (θ,ϕ,p)-space and form the resulting partition from the random set of planes.

The cells of such partitions are a.c. convex polyhedra with each vertex belonging to exactly three faces of the polyhedron. The characterization of the random polyhedra in terms of volume, surface area, edge lengths etc., leads to difficult distribution problems, but much is known about their moments of low order, see Miles (1972).

Another tesselation pattern that has attracted much attention is the <u>Voronoi pattern</u>.

<u>Case 6.3</u>. (<u>Voronoi tesselations</u>). Introduce a homogeneous Poisson point process with intensity λ in R^3. The points will be called the nuclei of the tesselation. To each nucleus we associate the set of points closer to this nucleus

than to any other nucleus. The resulting cells will again be convex polyhedra.

Just as for the homogeneous Poisson plane process the general distribution problems are difficult but lower order moments are known, see Miles (1972).

For two given images I_1 and I_2 form a third one I_1+I_2 as described previously. Now, if we have two stochastically independent diffuse patterns from Case 6.3 (6.2 is treated in the same way), so that two marginal probability measures are given over $\mathcal{T}$, the binary operation leads to another diffuse pattern. Since "adding" Poisson point process by the union of their points leads to a new Poisson process whose intensity is the sum of the original intensities it follows that the two diffuse tesselation patterns combine into a new one of the same type but with a new intensity. In this sense the diffuse patterns of the last two cases are closed under their binary operations.

A related, but more complicated, spatial pattern is when the region of space under consideration is made up of two phases, a parent phase which dominates to begin with when nuclei of the new phase are formed. They grow into separate particles which later on establish contact with each other and grow into complex shapes due to their interaction. The connectivity of each of the phases expresses an important property of such patterns. In spite of the great theoretical interest of such topological pattern analysis, it will not be discussed here, and the reader is referred to De Hoff in De Hoff-Rhines (1968) and Stelle (1972).

The three dimensional configurations of excluded volume

can of course be extended from generators of spherical shape, Case 4.9.2, to more general shapes. If the bond relations do not only prevent overlap of generators but also imply contiguity we get the following sort of spatial patterns.

Let each generator be given as a smooth closed and bounded subset of R^3 parametrized by the coordinates of certain points of its boundary to which the bonds will be affixed. To each bond we shall associate three types of bond values:

a) the generator index

b) the coordinates of the boundary point

c) the direction cosines of the normal at the point.

In its simplest form the bond relation will demand that

a) the two generator indices are appropriately related

b) equality of the two connected boundary points

c) the angle between the two normals is bounded by a constant that may depend upon the two generator indices.

The condition in c) may have to be replaced by one involving not only the angle between the normals but two angles, azimuth and elevation.

Case 6.4. (Anatomical patterns). Generators are given with bonds and bond values as above and images are identified as unions of the sets involved in the configurations.

As an example see the configuration in Figure 6.2 representing in highly schematic form the bone structure of the human hand. To make the figure simple the symbols for the bond values and the arrows have been left out.

Such anatomical patterns may be used to analyze motion

patterns for animal locomotion.

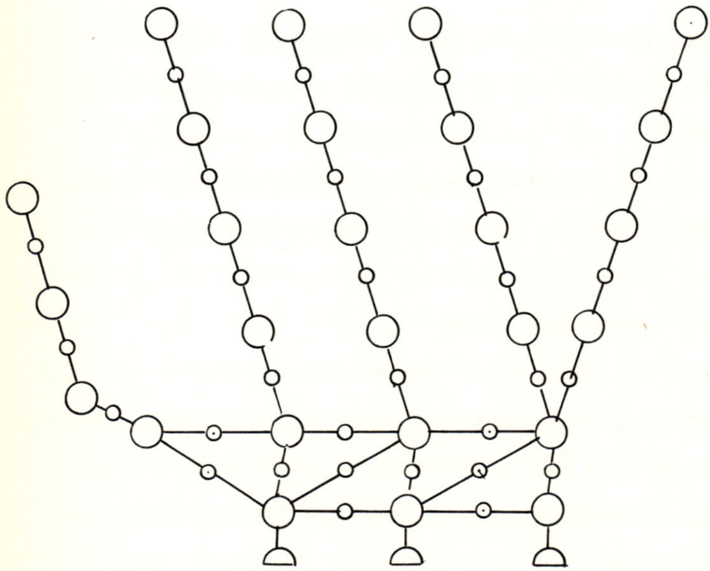

Figure 6.2

7. Space-time Patterns: Motion and Behavior.

Many of the image algebras studied in the preceding sections can be generalized to more than three dimensions in a natural manner. There is, however, one class of important image algebras in four dimensions that deserves special treatment, the space-time patterns.

The background space is then $X = R^3 \times R^1$, where R^1 stands for time. These patterns owe their particular role among the higher-dimensional ones to the fact that time carries an arrow - it is directed - which will influence the choice of bond relations below.

Among space-time patterns we shall study those describing motions in this section. We shall also mention briefly behavior patterns, which can be viewed as motions in a space more general than R^3.

Another important group of space-time patterns occurs in the study of growth and decay. It will be discussed in Section 3.8.

The generators used to build up motion configurations will have the following properties. Both the in-arity and out-arity are unbounded, and all the values of the inbonds of a particular generator equal a real number t_{in}. Similarly, all the values of the outbonds equal some real number t_{out} with $t_{out} \geq t_{in}$. The role of the generator index α is to separate the motions into different types and we shall speak of G as <u>the repertoire of motions</u>. If two configuration spaces are constructed in the same way except that one starts from a generator set G and the other from G', with $G \subset G'$, we shall say that the second one is <u>more versatile</u> then the first. Then the structural complexity will be greater for the second configuration space, see Section 2.1.

The similarity transformations shall include time translations, $t \to t+h$. The effect on the bond values of a generator g will be to shift them to the values $t_{in}+h$, $t_{out}+h$. Occasionally certain space transformations will also be included, but they will have no effect on the bond values. As usual the generator classes G^{α} must be S-invariant.

When we combine elementary motions together we will have to be careful so that they are performed in the right order. This leads us to the connection type Σ = PARTIAL ORDER and all arrows in a σ will be unidirectional.

For the same reason we decide that the bond relation $\beta_{out}\rho\beta_{in}$ will be true if and only if $t_{out} \leq t_{in}$; the arrow goes from β_{out} to β_{in}: one task has to be completed before the next one is started. Note that this ρ is S-invariant as is always required, see Section 4.2.

This determines $\mathscr{R} = \langle\Sigma,\rho\rangle$ and together with G and S the set $\mathscr{L}(\mathscr{R})$ of regular configurations is specified.

To get the image algebra we must choose an identification rule R and here we have more latitude. Three rules will be considered.

If c and c' are two regular space-time configurations each one determines a <u>total motion</u>: a medium represented in R^3 is carried from one arrangement to another. We shall say that $c \equiv c' (\text{mod } R_1)$ if c and c' have the same external bonds and they induce the same total motion of the medium. This does not mean that the two motions are identical in space-time, only that their total effects are the same.

On the other hand if c and c' have the same external bonds and bond values and if they represent the <u>same motion throughout</u> we shall say that $c \equiv c' (\text{mod } R_2)$. Then the motions are equal in every detail.

Finally, if $c = c'$ we write $c \equiv c' (\text{mod } R_3)$ so that R_3 is the trivial rule of identification by equality of configurations, see Section 5.1.

These rules satisfy Definition 1.1 and determine three image algebras $\mathscr{T}_k = \mathscr{L}(\mathscr{R})/R_k$, $k = 1,2,3$. Obviously $R_1 > R_2 > R_3$ with the resulting homomorphisms in Theorem 1.2(ii).

To study more complex space-time configurations it will be convenient to introduce macro-generators as certain often occurring configurations, see Section 4.2.

Before looking at some concrete realizations of the above, let us point out that the important problem will usually be how do space-time images lead to space images, when the initial space image is given? In other words we look at motion images as pattern processors of space images. The study of general pattern processors will not be undertaken until we get to Part IV (see Appendix); here we shall only be concerned with a very special case. In particular we can ask the following question. If the final space image is given as well as the initial one, what motion patterns induce the change from the latter to the first one. Typically the answer is not unique and one would then like to find the most "economical" one of the possible motion images in the class.

To sum up, we start with a certain repertoire of motions, combine them together, and find how the medium responds. This will depend upon the properties of the medium and what sort of tools are applied to it. The definition of the generators will implicitly contain information about the tools to which the motions are applied. They could be simply the hands of the operator or sophisticated mechanical devices.

Case 7.1. (Motion patterns). With the generators given as motions in space form configurations with Σ = PARTIAL ORDER and ρ as "$\leq$". The regular configurations are then identified modulo R_i, i = 1,2,3, which defines the image algebras.

We are now ready to look at a concrete case: fundamental hand motions which will be the generators now. If there is a single human operator employing both hands, the generators will be the hand motions with an index indicating left or right, L or R.

The pioneer in motion studies, F. B. Gilbreth, defined a set of such hand motions, called <u>therbligs</u>, and the definition is still in use after some modifications.

A list of the symbols used together with the definitions is given in Table 7.1. They can be used in a double row, one row for each hand, to describe the simultaneous motion of the hands including the time values. This is just

SYMBOL	NAME	DESCRIPTION
	SEARCH	Locate an article
	FIND	Mental reaction at end of search
→	SELECT	Selection from a number
∩	GRASP	Taking hold
	HOLD	Prolonged grasp
	TRANSPORT LOADED	Moving an article
9	POSITION	Placing in a definite location
#	ASSEMBLE	Putting parts together
U	USE	Causing a device to perform its function
	DISASSEMBLE	Separating parts
0	INSPECT	Examine or test
	PRE POSITION	Placing an article ready for use
	RELEASE LOAD	Release an article
	TRANSPORT EMPTY	Movement of a body member
	REST	Pause to overcome fatigue
	UNAVOIDABLE DELAY	Idle—outside person's control
	AVOIDABLE DELAY	Idle—within person's control
	PLAN	Mental process before action

Table 7.1

the sort of motion configuration we have mentioned above. With our formalism a small motion chart could look like Figure 7.1 where SYN stands for a special generator with $\omega_{in} = \omega_{out} = 2$, all the bond values equal. The role of it is to synchronize events at certain times. All the generators have bonds as indicated but the values have not been shown in the figure to avoid overcrowding it with symbols.

Obviously the characteristic property of such configurations is not the fact that they happen to describe hand motions. What is important is that they are combinations of certain basic activities which are linked together in a determined order with time constraints imposed. One meets other similar configurations in the <u>analysis of industrial processes</u>. Many systems have been suggested for the basic activities. One of them employs five symbols as in Table 7.2.

○	operation
▭	inspection
Ɖ	delay
⇛	transport
▽	storages

Table 7.2

The generators are therefore grouped into the five classes indicated. They may have more attributes, identifiers, measures, etc.

As an example, consider Figure 7.2 which describes the process of reconditioning brake shoes. Except that the time

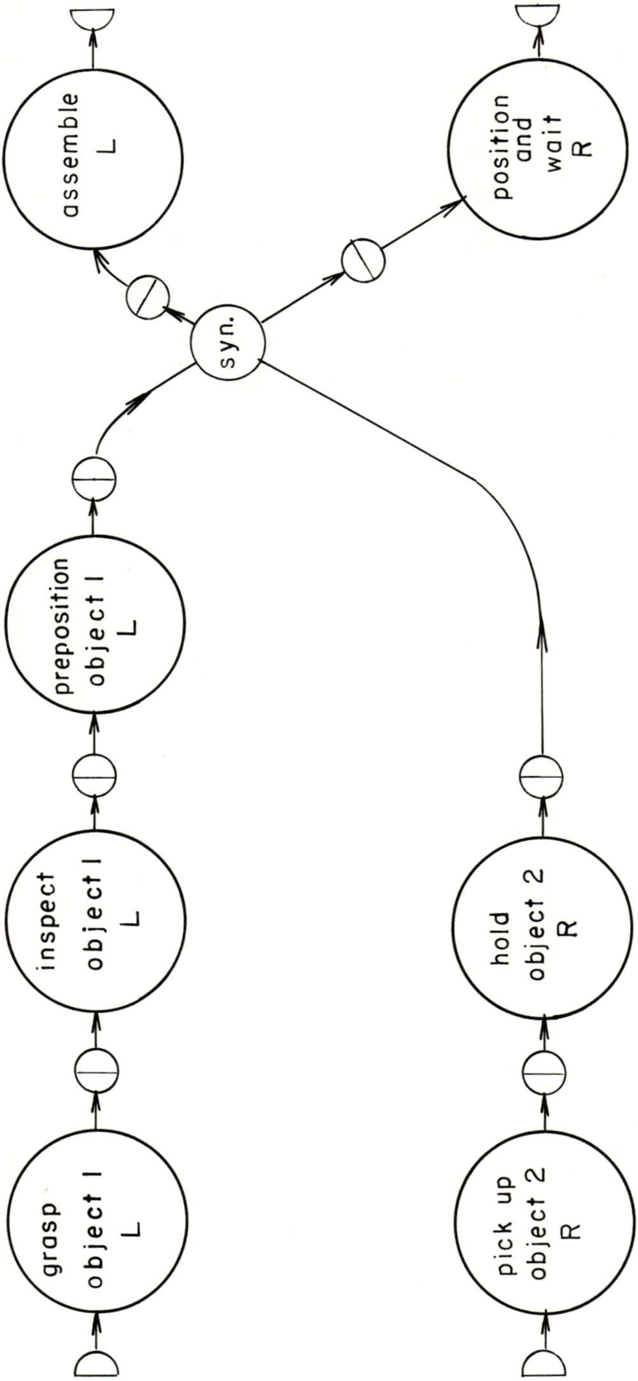
grasp object 1 L
inspect object 1 L
preposition object 1 L
assemble L
syn.
pick up object 2 R
hold object 2 R
position and wait R

Figure 7.1

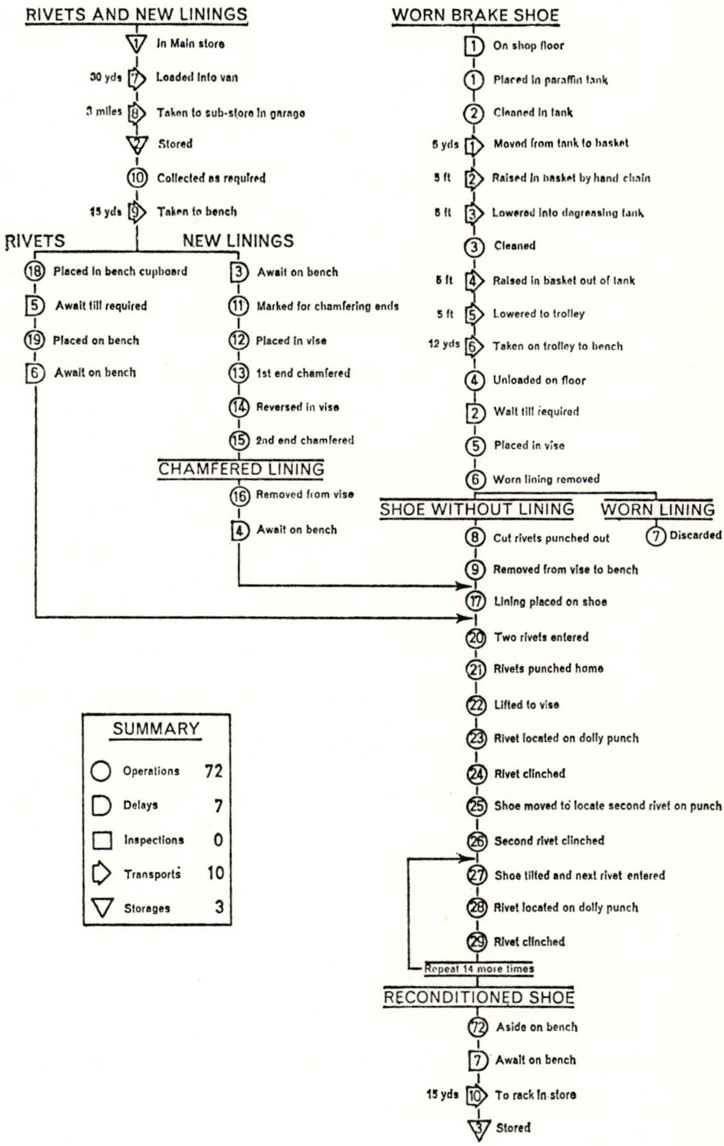

Figure 7.2

(from Larkin (1969))

values are not shown explicitly its structure coincides with the one we have discussed earlier. Consider, however, the activities 27-28-29, which are repeated 14 more times. We have not allowed such cycles in the presently used connection type and would instead have introduced a special macro-generator for the cycle. This is only a matter of convenience of course.

The PERT-diagrams employed in Planning, Evaluation, Review Technique are also of this form but the space in which activities are described can be more general than R^3. A small PERT-diagram is shown in Figure 7.3 where the attributes h_i denote the duration of activity A_i so that the bond values satisfy the equation $t_i = s_i+h_i$. The value of h_i can be a fixed number or given as a stochastic variable. Due to the partial order connection type and the role of ρ an activity can only be started when all the preceding ones have been completed.

A well known problem is to determine from the PERT diagram the minimum total time when we "pack" the activities as closely as possible, eliminating slack as far as is possible. In the figure the total time can be expressed simply as

$$T = \max\{h_1+h_3+\max(h_5,h_6),h_2+h_4\} + h_7 \tag{7.1}$$

but in practically occurring cases one encounters much more complicated expressions involving max and addition. If the h_i are stochastic variables the analysis of such expressions is not straightforward, and we shall return to this topic in Chapter 14 of Part III, see Appendix.

The effect of the motions or more general activities will depend upon what tools are used and upon the properties

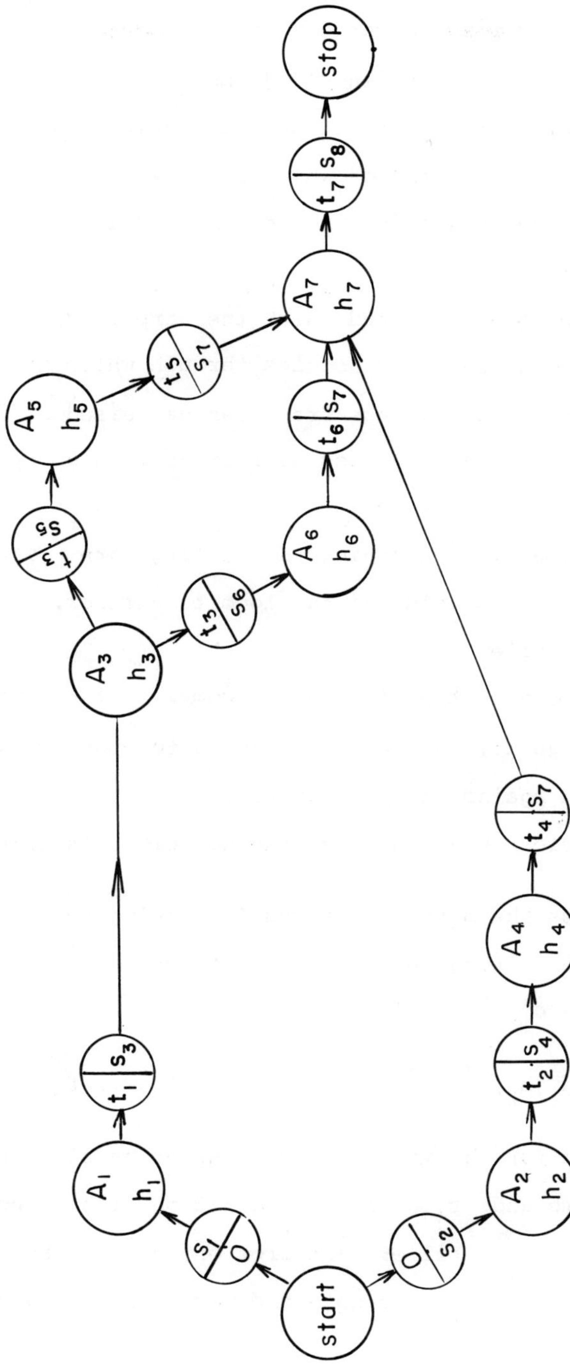

Figure 7.3

of the materials to which the tools are applied. In the previous cases the results are rather obvious, but if the mechanisms or materials are more involved one will need a more detailed discussion of the resulting patterns. We shall look at two such cases, the first of which is weaving.

The most common weaving process employs a loom with five main parts:

1. The warp beam around which the warp yarns are wound.

2. The harnesses with heddles through which the warp yarn has to pass. Each harness frame can be raised or lowered to let the filling yarn pass over or under a group of warp yarns.

3. The shuttle containing the filling yarn. The shuttle passes from one side of the loom to another. The filling is also called the weft.

4. The reed looking like a big comb. All the warp yarns pass through it, and its function is to push or batten the filling yarn against the already formed fabric.

5. The cloth beam on which the new cloth is taken up.

To express the motion patterns in weaving formally we shall let the generators be the basic motions divided into two classes $G = G^{(1)} \cup G^{(2)}$. Here

$$G^{(1)} = \{h_1, h_2, \ldots, h_r\}, \qquad G^{(2)} = \{f_1, f_2, \ldots, f_s\} \qquad (7.2)$$

where h_i stand for the motion consisting of raising the ith harness frame and f_j denotes the motion of passing the shuttle with the jth filler yarn from one side to another.

Let $W = \{1, 2, \ldots v\}$ denote the warp yarns and consider the partition

$$W = h_1 \cup h_2 \cup \ldots \cup h_r \tag{7.3}$$

into disjoint subsets. Configurations of h's and f's will then produce weave patterns and we shall briefly examine some of the most common ones. The figures will be in black and white but one could very well associate colors and varying size and texture to the different yarns in the warp and fillers. These would then be attributes for the generators. Let the kth warp yarn be of type $w(k)$.

Motion patterns produce plane weave patterns. To identify these we shall use an identification rule that is based only on the value (sort of yarn) visible at each of the discrete crossings of warp yarn with the filler. This neglects some aspects of the resulting weave but contains all the crucial combinatory information.

With two harnesses only, $r = 2$, and one filler, $s = 1$, one can produce the simple weave pattern by repeating the motion configuration in Figure 7.4 periodically. The warp yarns pass through harness 1 and 2 in an alternating manner. These motion patterns produce the plane pattern shown in the figure.

A simple variation of this is the basket weave of which an example is shown in Figure 7.5. The filler consists of two yarns, a double pick, and the plane pattern produced by the motion pattern has a characteristic checkerboard appearance.

The twill weaves that need at least three harnesses (shafts) are illustrated by Figure 7.6. It is a three-shaft twill and the produced plane pattern has the diagonals typical for twills. More complicated shapes can be produced by varying the diagonal through changes in the warp threading.

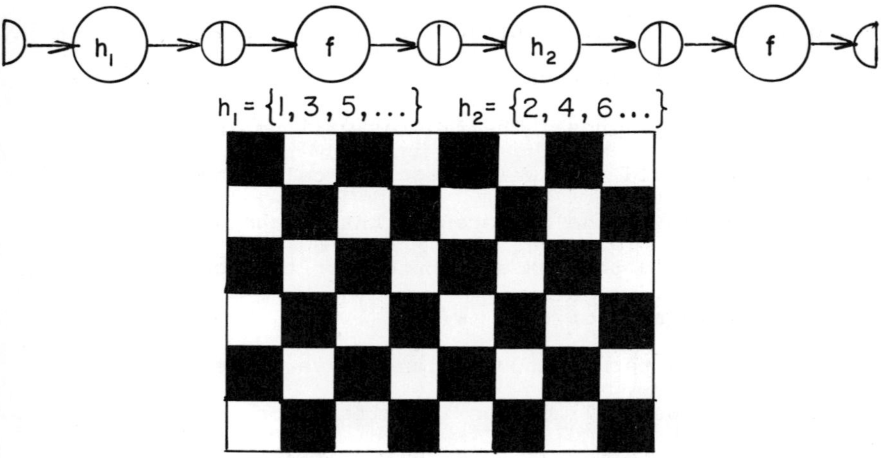

Figure 7.4

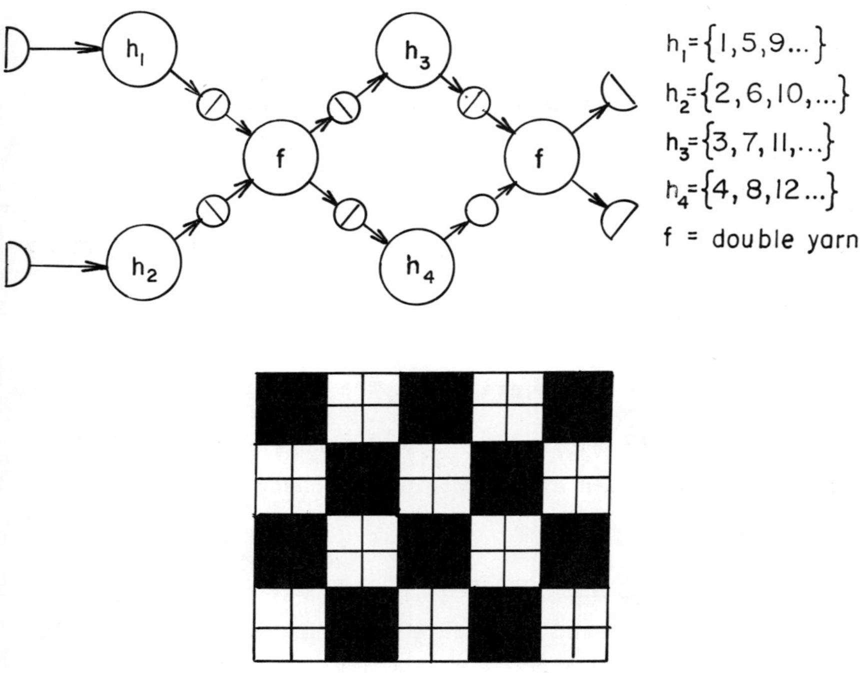

Figure 7.5

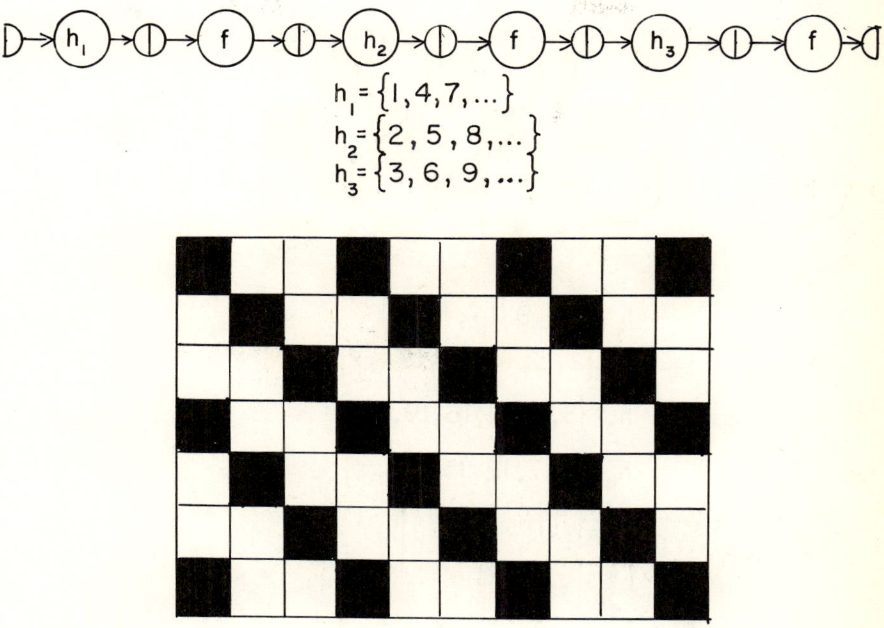

Figure 7.6

One such variation is shown in Figure 7.7. It employs four harnesses in which the threading order is given by the difference scheme ... 5,3,5,3,5,3,..., so that starting with warp yarn no. 1 we get 1,6,9,14,17... . This gives us the first harness, h_1, and so on for the others. The space-time configuration has a cycle as shown in the figure with the harnesses changed by adding 1 to its number modulo 4. These motions produce the <u>herringbone patterns</u> in the figure.

For <u>satin weaves</u>, finally, one needs at least five harnesses, a smaller number would only give twill patterns. An

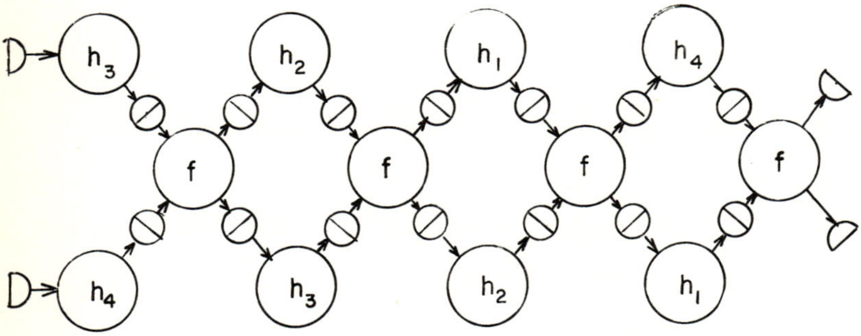

$$h_1 = \{1, 6, 9, 14, 17, \ldots\}$$
$$h_2 = \{2, 5, 10, 13, 18 \ldots\}$$
$$h_3 = \{3, 8, 11, 16, 19, \ldots\}$$
$$h_4 = \{4, 7, 12, 15, 20, \ldots\}$$

Figure 7.7

example of a five shaft satin is shown in Figure 7.8 where the warp threading is given by the difference scheme ...5,5,5,..., so that starting with warp yarn no. 1 we get the first harness 1,6,11,16,... and so on. The time pattern for successive harnesses is given by adding 3 modulo 5 to their number. The result will depend upon whether the surface is dominated by the warp - warp satin, or by the weft - weft satin.

The plane pattern produced is a five-shaft weft satin. Note the long floats typical of satins which tend to reflect light and give the surface its luster.

More formally, if we look upon motions as images in the background of space-time, we can write

$$\begin{cases} f = f(t) \\ h = h(t) \end{cases} \tag{7.4}$$

at a time t when harness $h(t)$ has been raised, followed by using the shuttle with filler $f(t)$. The values of f are natural numbers, but those of h are subsets of W. Let x be a coordinate across the warp and y another at right angle to it. Integral values of x and y, which are the only values used here, correspond to crossings between warp and weft. Identifying y with t, the space-time image given by (7.4) produces a space (or plane, rather) image I. At a point (x,y) of the weave the value is the type of yarn shown at the point. If the x-yarn in the warp, with value $w(x)$, shows it means that it has been raised by the harness through which it is threaded. Otherwise the filler $f = f(y)$ is shown.

Hence

$$h_1 = \{1, 6, 11, \ldots\}$$
$$h_2 = \{2, 7, 12, \ldots\}$$
$$h_3 = \{3, 8, 13, \ldots\}$$
$$h_4 = \{4, 9, 14, \ldots\}$$
$$h_5 = \{5, 10, 15, \ldots\}$$

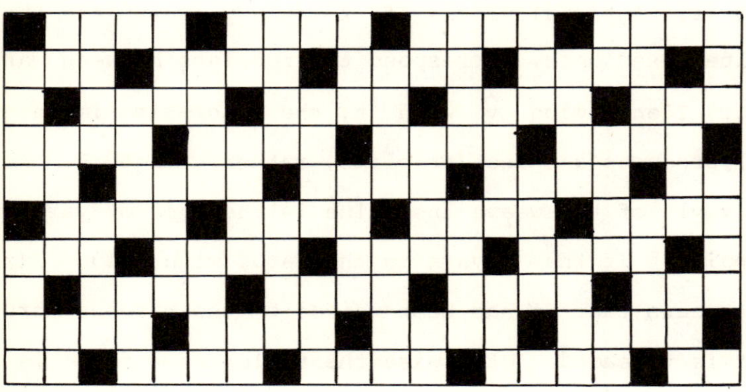

Figure 7.8

$$I(x,y) = \begin{cases} w(x) & \text{if } x \in h(y) \\ f(y) & \text{if } x \notin h(y) \end{cases} \tag{7.5}$$

or

$$I(x,y) = f(y) + [w(x)-f(y)] \{x \in h(y)\} \tag{7.6}$$

where the curly bracket stands for the indicator function of the expression inside.

It is the relation (7.6) that transforms the motion image $\{f(t),h(t)\}$ into the plane image I: <u>the morphogenetic equation</u>. If we negate the curly bracket and change left and write, $x \to -x$, we get the plane image on the reverse side of the weave. For example a warp satin goes over into a weft satin.

In the case of weaving the morphogenetic equation is quite simple but when we consider the analogous equation for other combinations of tools and material we will meet more difficult equations regulating the pattern synthesis. A case of considerable interest is the creation of shape using a rotating volume of deformable material, say in pottery or glass blowing.

Let the x-axis be along the axis of rotation and consider rotationally symmetric bodies initially forming a circular cylinder of radius R and height 1, say $0 \leq x \leq 1$, see Figure 7.9. As time increases from $t = 0$ the shape is allowed to change preserving the rotational symmetry. Let the radius at height x and time t be $R(x,t) = R+y(x,t)$.

To simplify the derivatives we shall treat the range of x-values as T^1, so that the $\mathcal{T}$-functions will be assumed to be periodic with period 1.

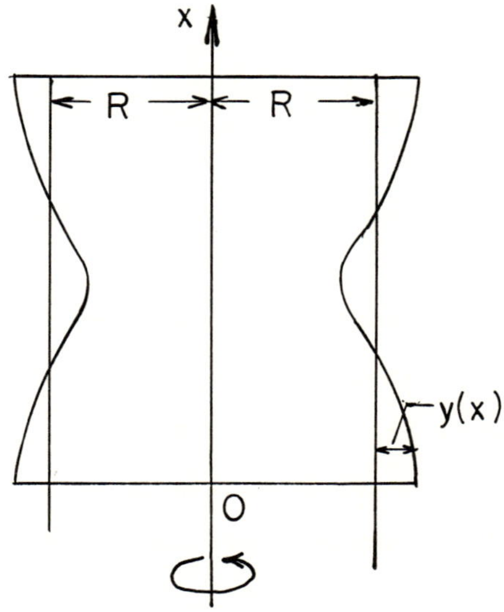

Figure 7.9

The shape $R(\cdot,t)$ depends upon the tools, repertoire of basic motions, and the properties of the material. The latter will be assumed to be elastic or quasi-elastic with internal friction, and almost, but not quite, incompressible. We shall assume that the morphogenetic equation is of the form

$$\frac{\partial y(x,t)}{\partial t} = s\,\frac{\partial^2 y(x,t)}{\partial x^2} + F(x,t) - c\int_0^1 y(x,t)\,dx \tag{7.7}$$

where the left side corresponds to internal friction. On the right the first term is due to the elastic behavior with a proportionality constant s that we shall put equal to one. This amounts to a choice of unit of length. The second term $F(x,t)$ is the force density applied at x for the time t. The third term tries to express the almost incompressible nature of the material, c is a large positive number.

We do not claim that (7.7) is based on sound physical principles. What follows should be viewed as a preliminary attempt to understand how pattern synthesis can be understood, at least in principle, in the present context. Of course one should try to replace (7.7) by more realistic equations.

It may be noted that inertial and centrifugal forces have been left out from the left side in (7.7). This does not seem to be a serious omission. More questionable is the form of the elastic force, and even more so, the incompressibility term which is quite artificial.

Anyway, the equation is not unreasonable and gives a starting point for the pattern synthesis: from motion to shape. The generators which describe how the tool is applied first in one position, then moved to another and so on, will be assumed to express the force density as

$$F = g = Ae^{2\pi i\alpha x} \tag{7.8}$$

where A is a complex constant and α an integer. The set of α's is bounded, $|\alpha| \leq a$, so that the larger a is the more versatile is the operation. To a given generator belong two time values, the inbond s and the outbond t, $t \geq s$. The average square

$$\int_0^1 |g(x)|^2 dx = |A|^2 \tag{7.9}$$

expresses how much effort was applied per time unit.

The generators are applied in sequence following Σ = LINEAR and same ρ as before. Note that they do not span the subspace $L_n = [e^{2\pi i\alpha x};\ |\alpha| < a]$ since we do not add the generators, only concatenate them. However each $G^{(\alpha)}$ is closed under time translations and multiplications with complex scalars.

We have examined what happens when uniformly moving force fields are applied, $g(x-vt)$, v = velocity. It turned out, however, that the versatility did not increase and we shall only give the results for stationary force fields, $v = 0$.

Only the above generators have been examined. It would be of interest to study morphogenesis for other generators, for example of the type $g(x) = B$ in (x_1, x_2) and zero outside, or fields with linear dependence upon x, and so on.

Consider the configuration in Figure 7.10, where A_i is the complex amplitude of the ith generator and α_i is the generator index.

Introducing the fundamental solution

$$K(x,\xi;\tau) = \frac{1}{\sigma\sqrt{2\pi\tau}} \sum_{k=-\infty}^{\infty} \exp\left\{-\frac{(x-\xi-k)^2}{2\sigma^2 t}\right\} \tag{7.10}$$

of the heat equation on T^1

$$\frac{\partial K}{\partial \tau} = \frac{\sigma^2}{2}\frac{\partial^2 K}{\partial x^2}, \quad \tau > 0 \tag{7.11}$$

with $K \to \delta(x-\xi)$ for $\tau \downarrow 0$, we can write

$$\begin{aligned} y(x,t) &= \int_0^1 y(\xi,0)L(x,\xi;t)d\xi \\ &\quad + \int_{\xi=0}^1 \int_{s=0}^t F(\xi,s)L(x,\xi;t-s)dsd\xi \end{aligned} \tag{7.12}$$

with

$$L(x,\xi;\tau) = K(x,\xi;\tau) + e^{-\frac{c\tau}{\mu}} - 1 . \tag{7.13}$$

The constant σ in (7.11) should be calculated from

$\sigma^2/2 = 1/\mu$.

The resulting expression for y is indeed the solution. First, it is obvious that as $t \downarrow 0$ in (7.12) we get $y(x,t) \to y(x,0)$. Second, direct calculation of y from (7.12), using the properties of K shows that (7.7) is satisfied.

Consider a motion configuration with bond values as indicated in Figure 7.10. Then $y(\cdot,t_n)$ can be written as

$$y = L_1 g_n + L_2(t_n,t_{n-1})L_1 g_{n-1} + L_2(t_n,t_{n-2})L_1 g_{n-2} + \\ + \ldots L_2(t_n,t_1)L_1 g_1. \tag{7.14}$$

Here we have assumed that the ith generator in the configuration has the attribute A_i for the amplitude and α_i is its generator index. Its inbond value is s_i and the outbond value is t_i. The bond relation ρ says that $t_i \leq s_{i+1}$.

The two linear operators L_1 and L_2 appearing in (7.14) come from equation (7.12) and are

$$(L_2(t+h,t)y(\xi))(x) = \int_0^1 y(\xi)L(x,\xi;h)d\xi, \quad h > 0. \tag{7.15}$$

It transforms space-functions. The other one is defined as

$$(L_1(t,s)y(\xi,s))(x) = \int_{\xi=0}^1 \int_{u=s}^1 y(\xi,u)L(x,\xi;t-u)dud\xi \tag{7.16}$$

and transforms space-time functions.

We can carry out (7.15), (7.16) in closed form for n any generator and get

$$(L_2(t,s)e^{2\pi ik\xi})(x) = \\ = e^{2\pi ikx}\left[e^{-4\pi^2k^2\sigma^2(t-s)} + \left[e^{-\frac{c(t-s)}{\mu}} - 1\right]\delta_{ko}\right] \tag{7.17}$$

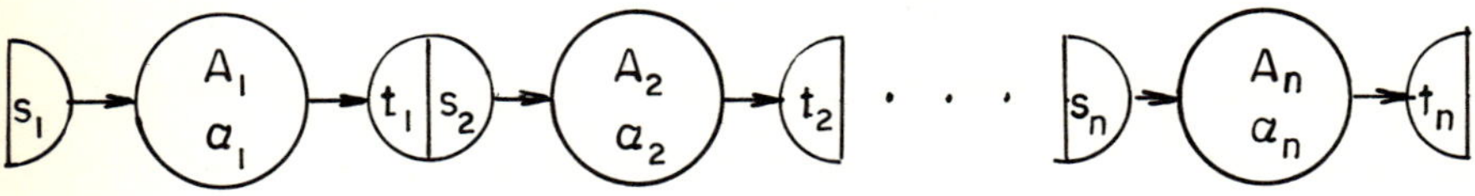

Figure 7.10

Similarly we get for L_1

$$(L_1(t+h;t)e^{2\pi ik\xi})(x) = \frac{1-e^{-4\pi^2k^2\sigma^2h}}{4\pi^2k^2\sigma^2}\, e^{2\pi ikx} \tag{7.18}$$

for $k \neq 0$, and for $k = 0$ the alternative expression

$$= \frac{\mu}{c}\,(1 - e^{-\frac{ch}{\mu}}). \tag{7.19}$$

Combining these expressions the total contribution of the general term in (7.14) for $h = 0$ is

$$g(x)\psi_k(-t) \tag{7.20}$$

where we have introduced the continuous and decreasing function

$$\psi_k(u) = e^{-4\pi^2k^2\sigma^2u}, \quad k \neq 0 \tag{7.21}$$

with $\psi_k(0) = 1$ and $\psi_k(+\infty) = 0$. For the special case $k = 0$ we get instead

$$g(x)\psi_o(-t) \tag{7.22}$$

with

$$\psi_o(u) = e^{-\frac{cu}{\mu}}. \tag{7.23}$$

We can now get a better idea of how the space-time patterns of applying forces at different places and different times produce space patterns. If $\Delta t = t-s$ is small for a particular generator in the space-time configuration, then $\Delta\psi_k$ will be small: the material has not been given enough time to yield. If Δt is moderately large on the other hand, the response will be noticeable but inversely proportional to k^2. In other words, generators with considerable spatial variations lead to smaller response than for slow spatial variations.

How difficult is it to shape a certain spatial pattern $y = y(\xi)$ starting from $y \equiv 0$ and extending over some time interval ending at $t-T$, which value we shall choose as 0? To make the question precise, let us measure the difficulty by the L_2 norm

$$||F||^2 = \int_{\mu=0}^{1} \int_{-\infty}^{0} |F(\xi,s)|^2 d\xi ds. \tag{7.24}$$

Among all space-time patterns F leading to y, what is the minimum value of $||F||^2$?

With our atomistic set up, at any given t we can only have one space harmonic $e^{2\pi ikx}$ present, and they change abruptly at the times occurring at the bond values of the finite number of generators. The minimum can be shown not to be attained over $\mathscr{L}(\mathscr{R})$ so we ask instead for $\inf ||F||^2$.

Let us denote by $A_k(t)$ the amplitude of the kth space harmonic present at time t in the F-pattern. Then

$$||F||^2 = \sum_k \int_{-\infty}^{0} |A_k(t)|^2 \, dt \tag{7.25}$$

and the kth space harmonic in the resulting space pattern will have the coefficient

$$\eta_k = \int_{-\infty}^{0} A_k(t)\psi_k(-t)dt \tag{7.26}$$

(see equation (7.20)) for $k \neq 0$, and similarly

$$\eta_o = \int_{-\infty}^{0} A_o(t)\psi_o(-t)dt. \tag{7.27}$$

In other words, we look for the infimum of $||F||^2$ in (7.25) when η_ν, $\nu = 0,\pm 1,\pm 2,\ldots m$ have prescribed values as the Fourier coefficients of y.

Denoting the derivative of ψ_k by ϕ_k, and applying Schwartz' inequality we get the bound

$$\int_{-\infty}^{0} |A_k(t)|^2 dt \geq \frac{1}{J_k} \left| \int_{-\infty}^{0} A_k(t)\psi_k(t)dt \right|^2 = \frac{1}{J_k}|\eta_k|^2 \tag{7.28}$$

where

$$J_k = \int_{-\infty}^{0} \psi_k^2(t)dt = \frac{1}{8\pi^2 k^2 \sigma^2}, \quad k \neq 0 \tag{7.29}$$

and similarly

$$\int_{-\infty}^{0} |A_o(t)|^2 dt \geq \frac{2c}{\mu}\eta_o^2 . \tag{7.30}$$

Hence

$$\begin{aligned} ||F||^2 &\geq \sum_{k\neq o} \frac{k^4|\eta_k|^2}{J_k} + \frac{2c}{\mu}|\eta_o|^2 \\ &= 8\pi^2\sigma^2 \sum_k k^2|\eta_k|^2 + \frac{2c}{\mu}|\eta_o|^2 \end{aligned} \tag{7.31}$$

so that

$$||F||^2 \equiv 2\sigma^2 \int_0^1 |y'(x)|^2 dx + \frac{2c}{\mu}\left|\int_0^1 y(x)dx\right|^2 . \tag{7.32}$$

This completes the proof of the following

Theorem 7.1. *The infimum value of* $||F||$ *to form the final space pattern* $y = y(x)$ *is given by* (7.32) *where the first term expresses the irregularity of the space patterns and the second the total change of size.*

As an illustration we show the space-time patterns in Figure 7.11 generated as above except that the type of genera-

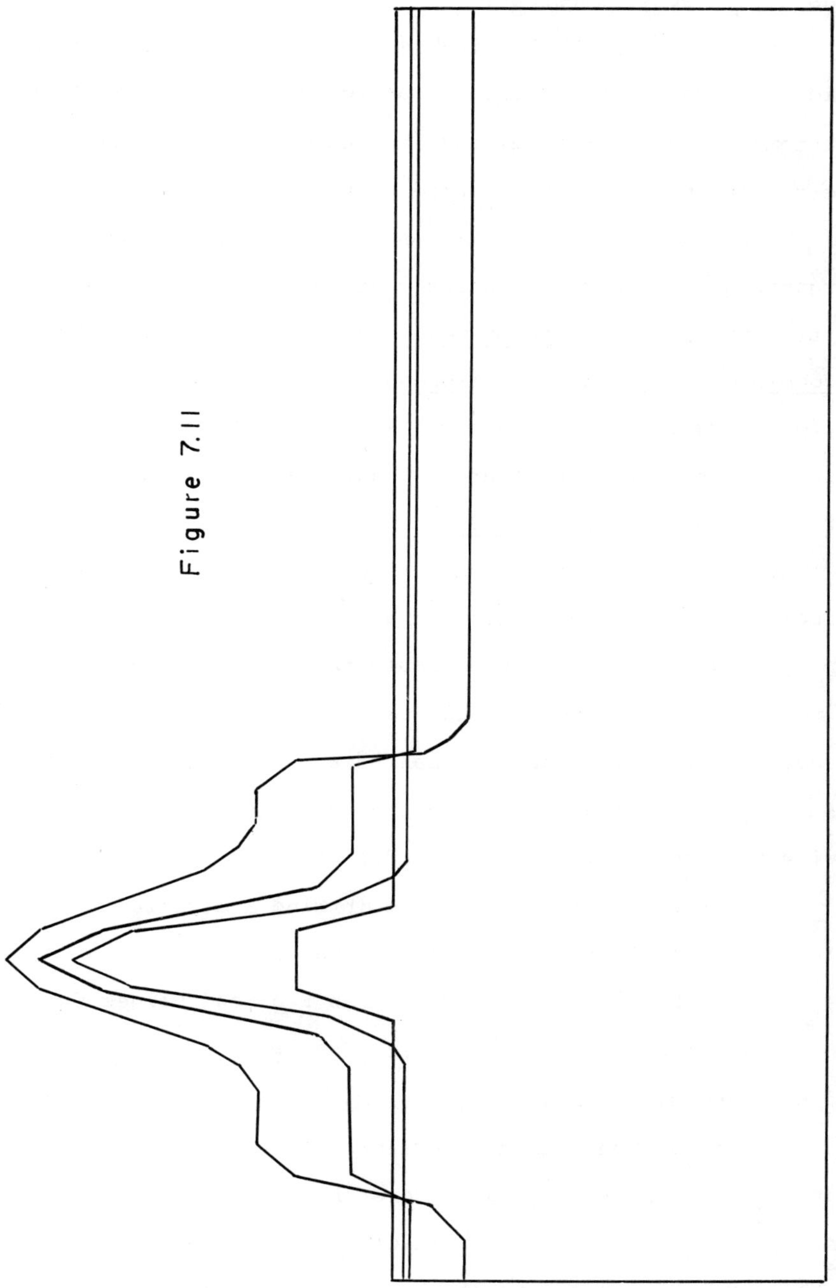

Figure 7.11

tor is different. One can see how the material successively yields to the applied forces.

Lathes, cutting and grinding machines could also be studied from the point of view of pattern synthesis. We shall not attempt to do that here, and instead move on to other space-time patterns in which space is not necessarily R^3.

To characterize the behavior of a person, a group, or an organization, one has to start with the set of actions that are available. <u>Behavior means the way actions are chosen and combined in a certain environment</u>. Behavior patterns therefore have the same structure as the earlier ones in this section, but the generators are now actions and we can consider G to be the <u>action space</u>.

As an example let us consider economic behavior, more precisely that of an insurance company. We have chosen this example because insurance is probably the business that has been most exposed to logical and quantitative study. This does not mean that insurance companies are run more rationally and scientifically than other businesses; only that their activity has been subjected to careful conceptual analysis.

As an example of what we have in mind a stylized and highly incomplete list of possible generators is given in Table 7.3. As before each generator has two bond values for the beginning and end of the activity. Many more generators could be added to the list.

Combining such action generators as above we get behavior configurations. A real understanding of the behavior is only possible if the configuration is correlated with the changes in economic environment that influenced the company executive to take his actions. Even this, however, does not really "explain" the behavior, it is only a formal description of it.

Generator	Qualitative attribute	Numerical attribute
reduce premium	policy group	amount
increase premium	policy group	amount
publicity	medium	amount
agent drive	target group	-
invest	type	amount
liquidate assets	type	amount
re-insure	contract, company	amount
introduce new insurance	contract type	premium scale
drop old insurance	contract type	premium scale

Table 7.3

A more ambitious attempt would be to try to express the choice of actions as the result of applying a few guiding principles. In other words, one would consider configurations in <u>strategy space</u>, not necessarily fixed in time, but hopefully of simpler appearance than action configurations.

In a recent paper by Bohman (1971) a step has been taken in this direction. He suggests three principles: equity, profitability, and solvency.

An equity group is a set of policies with fairly homogeneous risk structure. <u>Equity control</u> arises in producing at least reasonable relations for the premiums, adminstration costs, and claims between the different equity groups.

Profitability could be expressed by asking that the risk reserve should increase during a given year as a certain proportion of the premium income and another proportion of the initial value of the risk reserve during the same year.

Solvency could be formalized by saying that the amount of capital available to the company (in some precise sense) should be enough to meet the claim requirements during the year with a given probability. This amounts to a risk theoretic calculation starting from assumptions about the claim size distribution and claim probabilities.

It is an open question whether it is possible to relate action configurations in the real world to strategy configurations. Indeed, the complexities, inconsistencies, and subjective elements in a real economic environment could make such an analysis practically impossible or meaningless. Nevertheless it presents a challenging task for the applied mathematician.

8. Space-time Patterns - Growth and Decay.

Growth patterns share one important property with motion patterns - they take place against a background of spacetime. They differ in that the motion patterns we have considered in the last section involve more or less conscious decisions and choice of alternative motions in order to attain some goal. This does not apply to growth patterns, where the development is given (in a deterministic or probabilistic way) by the internal organization and its interaction with the environment. It may be that we study biological growth, or growth of social or economic units, but in each case we try to find laws of growth, expressed in terms of the basic concepts of pattern theory.

The image algebras will of course depend upon what generators are used, but before we discuss this, let us look at two general types of growth. In both cases the configurations change with time and give rise to developing spatial images. Such an image is called a snapshot of the growth image. The combination of the spatial images constitutes a growth image.

If the configurations involve the same generators except for changes in the attributes, excluding the generator indices, and with constant bond structure, then we shall speak of generator growth.

On the other hand if the generator indices are allowed to vary, or the bond structure changes, or new generators are added, then it will be called configuration growth.

Note that in the latter case the transitions, say in the generator indices, must be discontinuous. The time of these sudden transitions will divide the temporal image into sub-images, the growth regimes. We shall encounter some cases below in which the different growth regimes are radically different from each other and give the growth its characteristic form of sudden transitions separated by time intervals with more gradual development.

Growth is usually accompanied by decay, but in what follows we shall pay little or no attention to the latter.

At this point we must warn the reader that the growth patterns described below have little or no empirical support as far as their applicability to biological or social organisms is concerned. Instead they should be viewed only as mathematical constructs whose relevance for biological morphogenesis has not yet been established. The paper by Rosen (1972) on biological morphogenesis, which should be studied by any interested reader, expresses this by saying that one investigates the <u>capability of models</u>. It is this, rather than full fledged, realistic models, that will be attempted in this section.

It is clear that the mechanism behind a growth pattern should be translation invariant both in space and time: only relative time and relative space operate on it. It could also be argued that rotation invariance should hold, at least for rotations around the vertical axis; gravitational forces single out the vertical direction for the organism.

If growth rates differ in different organs of the organism - <u>allometric growth</u> - a crucial question is how information is carried to the organs to regulate the rates. If the cells "do not know where they are", how could growth rates be regulated as they obviously are? To avoid the assumption that each subset is completely programmed to grow at certain rates and in certain directions by mitosis, cell growth, and differentiation, some hypothetical regulator is needed and we shall return to this later on.

At this point we would like to make a remark about symmetry in growth patterns. There is nothing surprising in the fact that regularities are found in artifacts - they have been created intentionally by their designers. But why do we perceive patterns in nature?

In our pattern formalism, the role of the observer has been stressed via the identification rule. What patterns are perceived depends upon the environment <u>and</u> upon the observational set up.

But even so it would be futile to deny the objective existence of striking patterns and regularities in the physical world and it raises the question whether it is possible to explain some of them by invoking teleological ideas. Put in mathematical form, can the existence of certain patterns be made clear by appealing to extremum principles? An example of such a principle would be minimum potential energy.

Common sense seems to indicate that in simple surroundings, extremem principles ought, indeed, to result in regular arrangements. When one starts to look more carefully into this question, however, it is not apparent why this is so. Actually, as we shall see later, there are cases when counterintuitive situations occur as well as others where maximum regularity is the rule. Only a detailed mathematical analysis can give us the answers required, and we shall return to this in the latter part of this section.

Let us start with two simple types of generator growth: <u>convex growth</u> and <u>star-shaped growth</u>. In the first case the generators are half planes, see Section 5, and the generator index α is the angle associated with a certain half plane. A generator $g^{(\alpha)}$ also has the numerical attribute $s(\alpha)$, the value of the support function.

Assuming the convex set to be smooth as $R \in C_2(0,2\pi)$, where R is the radius of curvature, we know that s is periodic with period 2π, that $R = s''+s$ is non-negative, and that the orthogonality relation holds so that the functional

$$c(s) = \int_0^{2\pi} e^{i\alpha}\, s(\alpha)d\alpha = 0. \tag{8.1}$$

These conditions are necessary and sufficient.

Starting at time $t = 0$ with some non-interesting convex shape, say circular or almost circular, $R(\alpha,\theta) \cong R_o$, centered at the origin, we shall assume that the numerical attribute $s(\alpha)$ of the αth generator changes in a way that depends only upon the local properties at α. To fix ideas let

$$\frac{\partial s(\alpha,t)}{\partial t} = f(R(\alpha,t)). \tag{8.2}$$

If R is small, so that ∂I is very curved at α, then growth will be assumed to be slow, and conversely for large values of R. One could imagine a mechanism that depends upon the size of the intersection between a small circle of give radius located at this point of ∂I with ∂I. This means that f should be an increasing function of R.

Using the expression of R in terms of the support function we get

$$\frac{\partial R}{\partial t} = f''(R)(R'_\alpha)^2 + f'(R)R''_{\alpha\alpha} + f(R). \tag{8.3}$$

It will be illuminating to examine the simplest case of (8.3) when f is linear, $f(R) = a+bR$, where a and b are positive constants, so that the equation reduces to

$$\frac{\partial R}{\partial t} = bR''_{\alpha\alpha} + a+bR. \tag{8.4}$$

First, it is clear that <u>this does not carry us outside of the present image algebra</u>. Indeed, multiplying (8.2) by $e^{i\alpha}$ and integrating over α we get after integrating by parts twice

$$\frac{\partial}{\partial t}[c(s)] = b\int_0^{2\pi} R(\alpha,t)e^{i\alpha}d\alpha = b\int_0^{2\pi} (s+s''_{\alpha\alpha})e^{i\alpha}d\alpha = 0 \quad (8.5)$$

so that $c(s)$ remains 0 as t increases. Second, R is a smooth function of α and t and does not become negative. If it became negative, let t_o be the first time it vanished, say for $\alpha = \alpha_o$. But then $R''_{\alpha\alpha} \geq 0$ so that (8.4) shows that $\partial R/\partial t > 0$ and R takes positive values again. Thus the support function will retain the properties needed and $I = I(t)$ will be convex for all positive t's.

What happens with the shape of $I(t)$ as t becomes large? Introduce the Fourier coefficients

$$\sigma_k(t) = \frac{1}{2\pi}\int_0^{2\pi} e^{ik\alpha}R(\alpha,t)d\alpha \quad (8.6)$$

so that, using (8.4) again,

$$\begin{cases} \dfrac{d\sigma_o}{dt} = a+b\sigma_o \\ \dfrac{d\sigma_k}{dt} = b(1-k^2)\sigma_k, \quad k \neq 0 \end{cases} \quad (8.7)$$

Hence

$$\begin{cases} \sigma_o(t) = [\sigma_o(0) + \frac{a}{b}]\exp(bt) - \frac{a}{b} \\ \sigma_k(t) = \sigma_k(0)\exp(b(1-k^2)t), \quad k \neq 0. \end{cases} \quad (8.8)$$

Note that the σ_k have been separated in (8.7). These <u>modes of growth</u> do not interact with each other, a fact to which we shall return later. The coefficient $\sigma_o(t)$ grows exponentially, while the others either are zero ($k = \pm 1$) or tend to zero ($|k| > 1$). Hence the <u>normalized image</u> $e^{-bt}I(t)$ <u>tends to a disk with radius</u> $\sigma_o(0) + \frac{a}{b}$ <u>irrespective of what the initial image was</u>.

Recalling that the present image algebra is the one of Case 5.9 we can give another directly geometric interpretation to the partial differential equation in (8.2). With the notation $J(s)$ for the convex set with support function s, we can write (8.2) as a <u>set differential equation</u>

$$\frac{dI(t)}{dt} = J(f(R)) \tag{8.9}$$

which is interpreted as the limit of the image relation when $\varepsilon \downarrow 0$

$$I(t+\varepsilon) = I(t) + \varepsilon \cdot J(f(R)) \tag{8.10}$$

where the plus and multiplication signs of the image algebra denote Minkowsky addition and multiplication by scalars respectively.

Our second growth pattern will operate on <u>star-shaped regions</u>, see Case 2.9.8. Let the image at time t and $I(t)$ be formed from the generators $g^{(\alpha)}$, $g^{(\alpha)}$ = ray from the origin of length $r(\alpha)$ in the direction α. In the subconfiguration made up of the generators in the sector $(\alpha,\alpha+d\alpha)$ we assume that energy is received at ∂I and consumed in the inner part of the sector and that the local growth rate will be proportional to the surplus energy. The two terms should then be linear and quadratic respectively, and it is not unreasonable to try the assumption that the growth rate is given by

$$\frac{\partial r}{\partial t} = ar - br^2 = br(r_c-r), \quad r_c = \frac{a}{b}, \tag{8.11}$$

where a and b are positive constants expressing how fast energy is received and consumed.

An alternative to (8.11) is to assume that the rate in

area growth $rdrd\phi$ is proportional to the surplus of energy. This means that we should write

$$\frac{\partial r}{\partial t} = a-br = b(r_c-r). \tag{8.12}$$

Both equations lead to similar qualitative results and it is enough to discuss one of them.

Starting again with an initial image of approximately circular shape of radius $r_o << r_c$ we can solve (8.11) directly and get

$$\frac{1}{r_c} \ln \frac{r(\alpha,t)[r_c-r(\alpha,0)]}{[r_c-r(\alpha,t)]r(\alpha,0)} = bt \tag{8.13}$$

or

$$r(\alpha,t) = r_c \frac{r(\alpha,0)e^{at}}{r_c-r(\alpha,0)+r(\alpha,0)e^{at}}. \tag{8.14}$$

As t increases the value of $r(\alpha,t)$ grows from $r(\alpha,0)$ to r_c, so that <u>the limiting shape is a circle irrespective of the initial shape</u>. This is just as in the convex growth case and it is again closely linked to the fact that the modes of growth - here the growth of the individual generators - are not coupled. In this way we will not be able to model growth patterns to more interesting shapes than the circle, since we are not willing to modify the models by direction-dependent coefficients.

We return now to star-shaped growth but with the important modification that <u>the generators are interacting with each other</u>. In order not to obscure the discussion by inessential complications, let the configuration now be finite $c = (g_1,g_2,\ldots,g_n)$ where g_k has the generator index (here angle) $2\pi k/n$. The growth rate of g_k, i.e. $\dot{r}_k$, is as in equation (8.12) except that <u>the growth is also inhibited by the neighboring generators</u>. Since the mechanism shall have rotational symmetry the inhibition caused by g_ℓ upon g_k must only depend upon the index differences $k-\ell$. With a quadratic inhibition effect we get the set of ordinary differential equations

$$\frac{dr_k(t)}{dt} = a - br_k(t) - \sum_{\ell=1}^{n} c_{k-\ell} r_\ell^2(t) \tag{8.15}$$

where the non-negative inhibition constants c_ν are defined periodically with period n and even, $c_{-\nu} = c_\nu$.

When growth begins with a small image the constant term in (8.15) will dominate, so that the time derivatives are positive. Later on the linear and quadratic terms will come into play and it is less easy to see what will happen. Are there limiting shapes and if so, how do they look? It is clear that circular equilibrium shape with radius r is possible if r solves the equation

$$r^2 + \frac{b}{c} r = \frac{a}{c}, \quad c = \sum_{\nu=1}^{n} c_\nu. \tag{8.16}$$

Only the positive root is physically meaningful so that

$$r_c = -\frac{b}{2c} + \sqrt{\frac{b^2}{4c^2} + \frac{a}{c}}\ . \tag{8.17}$$

This is not the only singular point of the system of differential equations in (8.15), however, as we shall see in a moment. Since biological systems are inherently noisy the question arises which of them are stable. Let us answer this first for the circular shape. The matrix in the linearized version of (8.15) at this singular point is given by

$$A = -bI-2\begin{bmatrix} c_o r_1 & c_1 r_2 & \cdots & c_{n-1} r_n \\ c_{-1} r_1 & c_o r_2 & \cdots & c_{n-2} r_n \\ \cdots & & & \\ c_{-n+1} r_1 & c_{-n+2} r_2 & \cdots & c_o r_n \end{bmatrix} \tag{8.18}$$

where in this case $r_k \equiv r_c$. But $\{c_{\nu-\mu}\}$ is a circulant matrix so that the only possible values of its eigenvalues are $\gamma(\lambda)$, where

$$\gamma(\lambda) = \sum_{\nu=1}^{n} c_\nu e^{2\pi i\nu\lambda} \tag{8.19}$$

and λ takes some values of the form multiple of $1/n$, see Notes.

For asymptotic stability we want the eigenvalues of A to have negative real parts (they are actually real since in this special case A is symmetric, see e.g. Hale (1969)). Call the minimum value γ. But

$$\max(\text{eigenvalues of } A) = -b-2r\gamma$$

$$= -b[1 - \frac{\gamma}{c} + \frac{\gamma}{c}\sqrt{1 + \frac{4ac}{b^2}}\,] \tag{8.20}$$

so that we need

$$\gamma > -\frac{c}{\sqrt{1 + \frac{4ac}{b^2}} - 1}\,. \tag{8.21}$$

The inequality (8.21) is not always satisfied. To give an example, say that we have only <u>nearest neighbor inhibition</u> so that $c = 0$ except for $c_{-1} = c_1 = c/2$. Then

$$\gamma(\lambda) = c \cos \lambda \tag{8.22}$$

and it is known that for large n the eigenvalues correspond to equally spaced λ-values in $(0,2\pi)$. Hence $\gamma \tilde{=} -c$ and the stability criterion becomes

$$c < \frac{3b^2}{4a} . \tag{8.23}$$

For given values of a and b asymptotic stability holds for small c values but not for large ones. Hence circular patterns are possible growth limits for large c but not for small ones.

Consider the quite different star-shaped configuration

$$\begin{cases} r_1 = r_3 = r_5 = \cdots \\ r_2 = r_4 = r_6 = \cdots \end{cases} \tag{8.24}$$

for even n with tentacles alternating with indentations. To make it an equilibrium pattern it should be a critical point of (8.15) so that

$$\begin{cases} r_1^2 + \frac{b}{c} r_2 = \frac{a}{c} \\ r_2^2 + \frac{b}{c} r_1 = \frac{a}{c} \end{cases} \tag{8.25}$$

One solution is $r_1 = r_2 = r$ as before. The other two possibilities are

$$\begin{cases} r_1 = \frac{b}{2c} + \sqrt{\frac{a}{c} - \frac{3b^2}{4c^2}} \\ r_2 = \frac{b}{2c} - \sqrt{\frac{a}{c} - \frac{3b^2}{4c^2}} \end{cases} \tag{8.26}$$

and another obtained by permuting the indices. For certain choices of a, b, and c both roots are real and positive and hence this is a star-shaped equilibrium pattern.

As far as asymptotic stability is concerned, we need

the minimum of the eigenvalues of the matrix

$$bI + c\begin{Bmatrix} 0 & r_2 & 0 & 0 & 0 & \cdots \\ r_1 & 0 & r_1 & 0 & 0 & \cdots \\ 0 & r_2 & 0 & r_2 & 0 & \cdots \\ 0 & 0 & r_1 & 0 & r_1 & \cdots \end{Bmatrix} \tag{8.27}$$

But the smallest eigenvalue of this matrix is greater than or equal to (if we normalize by $c = 1$)

$$b - 2\sqrt{r_1 r_2} = b - 2\sqrt{b^2 - a} \tag{8.28}$$

(see Notes). If we choose

$$\frac{3}{4}b^2 < a < b^2 \tag{8.29}$$

then r_1 and r_2 are real and positive. In this range (8.28) is positive so that the linearized matrix has negative eigenvalues and this star-shaped pattern alternating large and small rays is asymptotically stable.

To get some feeling for how these effects will appear quantitatively, a computational experiment was carried out in which the equations (8.15) were solved and plotted for different values of the parameters a, b, c_k. Of course the shape of $I(0)$ must not be perfectly circular since it would then remain circular unless noise was introduced in the computations.

Two cases are shown. In Figure 8.1a is seen a situation where $r_k \equiv$ constant is an asymptotically stable equilibrium and a limiting circle. Figure 8.1b on the other hand, illustrates the situation in (8.29) and the star-shaped regions tend to have more pronounced alternating long and short

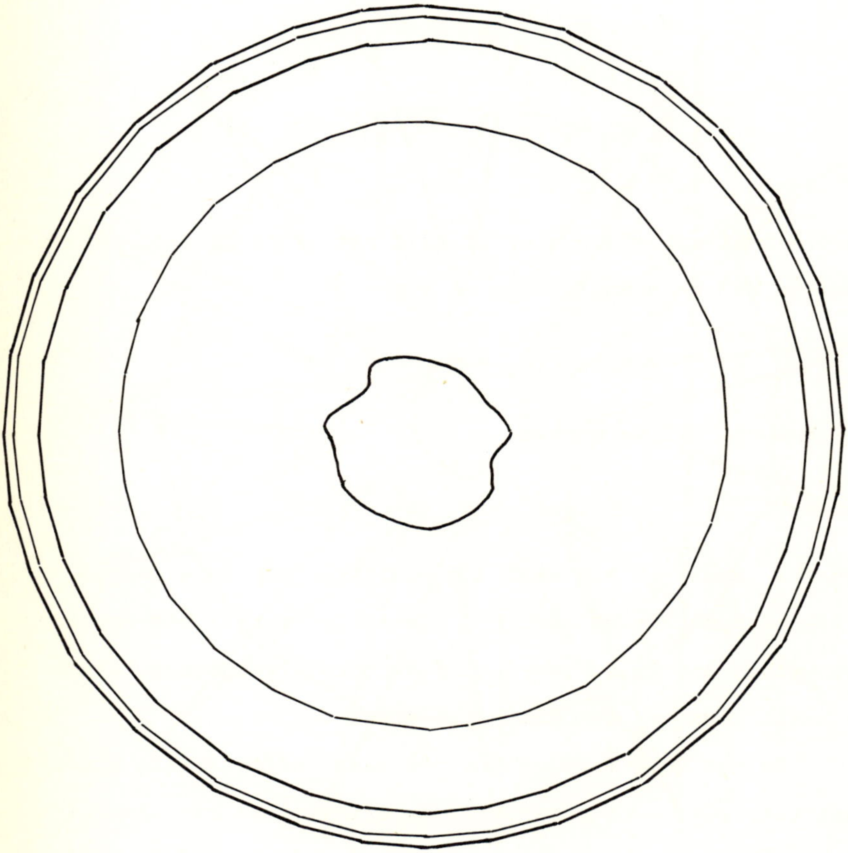

Figure 8.1a

radii.

It was interesting to observe how, in the second case, growth started in a very regular manner and the original shape grew fairly fast to circular shape. Then growth seemed to stop for a long while, but suddenly it started again and the image became more irregular and rapidly approached the

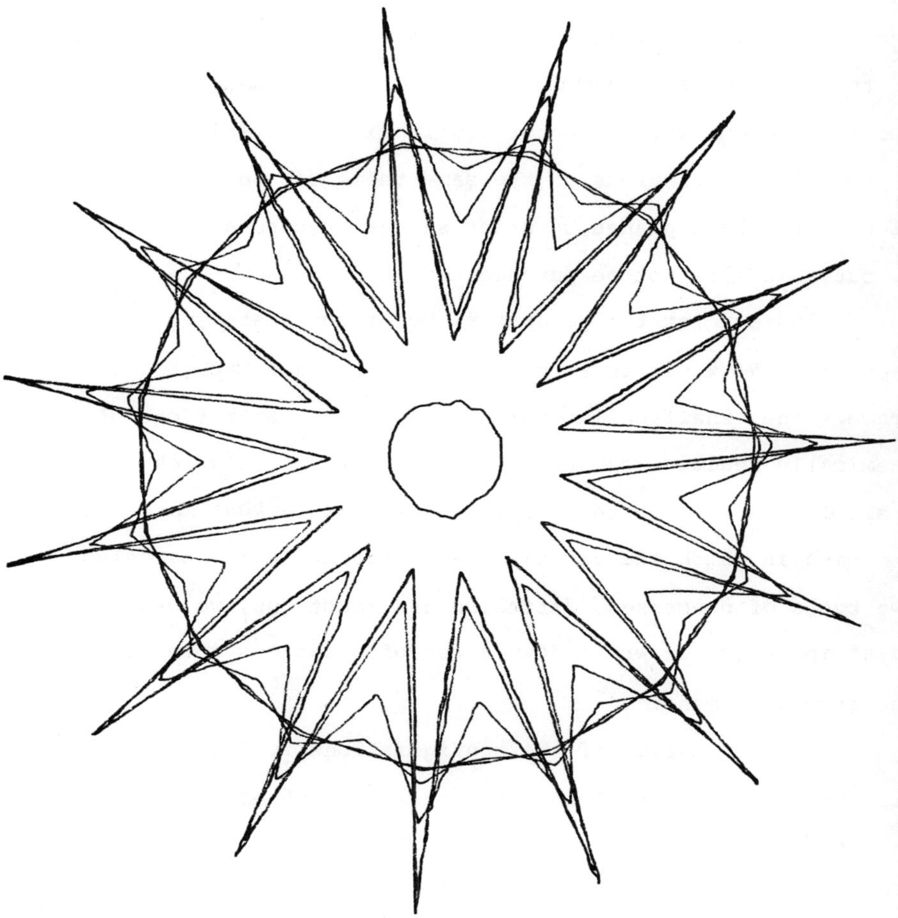

Figure 8.1b

asymptotically stable equilibrium.

This gave the appearance of two growth regimes: one towards circular shape, the other towards an image with tentacles. Note that no regimes were pre-programmed here in contrast to cases discussed later on in this section.

A weakness in these models is that they neglect biological noise except in the choice of the initial image. It is not known how the stability is affected if noise is introduced. Also, even in the deterministic case, a careful study of the whole phase portrait is needed, in particular one should find out whether other equilibria are possible.

Other variations on the same theme can be made by replacing the half-planes by other set generators as in Section 5, but this will not be pursued here.

Instead let the generators be of contrast type, so that each $g^{\alpha} = g^{\alpha}(x,y)$ is a function in the plane background space. They shall be interpreted as the concentrations of n chemically reactive substances, $\alpha = 1,2,\ldots n$, of which we shall let $\alpha = 1$ correspond to the substance that makes up the bath in which the organism is immersed. Assuming only two types of processes, diffusion and reactions, we can at least in principle write down a set of equations governing the growth process.

Identification of the configurations $c = (g^1, g^2, \ldots g^n)$ will be made by identifying the image by a threshold logic

$$I = \{(x,y) \mid \sum_{\alpha=2}^{n} g^{\alpha}(x,y) \geq \varepsilon\}. \tag{8.30}$$

Let the diffusion constant of substance α be D^{α}, assumed constant. It would seem natural to assume also a counteracting force representing cohesion but this will not be discussed here. The next step consists of specifying the reactions that take place between the substances, for example

$$\alpha+\beta \rightarrow 2\gamma \quad \text{at rate} \quad r \tag{8.31}$$

and so on. The first attempt to analyze biological morphogenesis in this way is due to Turing (1954), a pioneering paper that should be consulted by anyone interested in this topic. He arrived at equations of essentially the form

$$\frac{\partial g^{\alpha}}{\partial t} = D^{\alpha}\left(\frac{\partial^2 g^{\alpha}}{\partial x^2} + \frac{\partial^2 g^{\alpha}}{\partial y^2}\right) + \sum_{\beta=1}^{n} F_{\alpha\beta}(g^{\beta}) \tag{8.32}$$

where the first term on the right represents diffusion and the others the reaction kinetics.

Turing idealized the organism to be one-dimensional periodically bound together - the 1-torus - so that it can be looked at as a ring with fixed radius. The Laplacian in (8.32) whould then just be a second derivative. A more fundamental difference to (8.32) was that he dealt almost entirely with the linearized kinetics, corresponding to linear functions $F_{\alpha\beta}$. He studied the solutions of (8.32) (linearized) for large values of time t, especially the case of two morphogenes, and showed that certain patterns were formed, both stationary and oscillatory.

It has been argued that non-linear effects must be included; see Othmer-Soriven (1974). The mathematical analysis becomes harder, of course, but the last two authors have shown that at least in limited examples it is possible to obtain the symmetry properties of the patterns generated in this manner.

If the equations (8.32) are discretized in space we get a number of ordinary differential equations for the morphogenes. It becomes then a matter of studying the singular points in the phase portrait, especially the stable points, and to see what spatial patterns they represent. With parameters in the equations one is led to problems in structural stability and to Thom's theory of morphogenesis which will not be discussed here; the reader is referred to Thom (1973).

Instead we shall study growth patterns arrived at by rearrangements of the generators, so that only their coordinates vary, but otherwise they remain unchanged. As mentioned above we shall look for patterns such that the potential energy is minimized or some other criterion is made to take an extremium value.

Say that the generator classes are indexed by α and that two generators connected with each other correspond to the energy $t(g_1,g_2) = f(\alpha',\alpha")$ if $\alpha(g_1) = \alpha'$ and $\alpha(g_2) = \alpha"$. We consider for the moment only Σ = LINEAR with $\omega(g)$ = two double bonds with values $x - \frac{1}{2}$ and $x + \frac{1}{2}$ if x is the coordinate of the generator. Only integral x-values will be used. Here ρ should be EQUALITY.

The <u>total energy</u> of a configuration c will then be defined additively

$$t(c) = \sum_{i=1}^{\ell-1} f(a_i,a_{i+1}). \tag{8.33}$$

The <u>relative energy</u> is the average

$$r(c) = \frac{1}{\ell-1} t(c). \tag{8.34}$$

We also introduce the <u>minimal energies</u>, both total:

$$\begin{cases} t_\ell = \min\limits_{c\in S^\ell} t(c) \\ r_\ell = \dfrac{t_\ell}{\ell-1} \end{cases} \tag{8.35}$$

Of course, the minimum need not be uniquely attained.

Here we have assumed no agreement conditions ruling out some of the combinations of the elements of S as not regular. If, for example, the bonds of the ordered pair $a_i a_j$ do not agree, we make $f_{ij} = +\infty$ in the energy matrix.

More precisely, an ordered pair of generators ab is associated with the energy $t(ab) = f_{ij}$ if a has index i and b has index j. The <u>energy matrix</u> $\{f_{ij};\ i,j = 1,2,\ldots d\}$ is denoted by T and is assumed to have non-negative elements. T will <u>not</u> be assumed to be symmetric.

The problem before us is: how do minimal configurations look, especially if they are long? Do they exhibit any regularity?

We begin by some numerical experiments. To avoid solving the minimum problem by enumeration we apply a dynamic programming technique. Introduce

$$t_\ell(i,j) = \min\{t(c),\ c \in \mathscr{C}_\ell(\mathscr{R}) \text{ with } \alpha(g_1) = i,\ \alpha(g_\ell) = j \tag{8.36}$$

so that

$$t_{\ell+1}(i,j) = \min_k \{t_\ell(i,k) + t_{kj}\}. \tag{8.37}$$

We solve (8.37) successively keeping track of the resulting configurations (there may be others of equal total energy) for every ℓ up to some given value.

It is instructive to see what happens. With $d = 6$, $G = \{A,B,C,D,E,F\}$, and

$$T = \begin{Bmatrix} 3 & 1 & 14 & 18 & 13 & 15 \\ 15 & 20 & 18 & 5 & 7 & 8 \\ 11 & 12 & 17 & 9 & 17 & 6 \\ 9 & 11 & 10 & 6 & 4 & 4 \\ 12 & 17 & 1 & 11 & 10 & 20 \\ 15 & 12 & 18 & 13 & 17 & 4 \end{Bmatrix} \tag{8.38}$$

we get the minimal energy configurations in Table 8.1. Obviously the long chains will be made up almost exclusively of

A's. There is a boundary effect at the end of the configuration; a lonely B appears there.

For

$$T = \left\{ \begin{array}{rrrrrr} 4 & 7 & 13 & 3 & 14 & 13 \\ 17 & 5 & 10 & 8 & 5 & 1 \\ 19 & 9 & 3 & 19 & 9 & 3 \\ 18 & 2 & 4 & 5 & 8 & 6 \\ 3 & 16 & 10 & 7 & 10 & 17 \\ 19 & 14 & 5 & 14 & 19 & 6 \end{array} \right\} \tag{8.39}$$

the behavior is a bit more complicated, as indicated in Table 8.2. Note the transition in behavior at $\ell = 9$. Once the configuration contains a C it tends to repeat it, so that the large-scale behavior is again regular: mainly C's.

With T the same, except that t_{33} has been increased from 3 to the value 4, the generator pair CC no longer has such a favored role. Instead we get Table 8.3 with the tendency to repeat the cycle EADB with rare interruptions of AA and some boundary effects at the beginning and end.

This experiment points to the following conjecture. There is (at least) one cycle of length at most d such that a long configuration repeating this cycle is asymptotically minimal. That this is indeed so is borne out by the following analytical result.

<u>Theorem 8.1</u>. <u>Define the constant</u> τ <u>by</u>

$$\tau = \min\{\frac{1}{\ell-1}\, t(c),\ 2 \le \ell \le d+1,\ c \text{ a non-repeating cycle of length } \ell \text{ starting and ending in the same generator}\} \tag{8.40}$$

<u>Then</u>

$$\lim_{\ell\to\infty} r_\ell = \tau, \tag{8.41}$$

<u>so that minimum energy can be asymptotically attained by</u>

ℓ	r_ℓ	minimal configuration
2	1	AB
3	2	AAB
4	2.33	AAAB
5	2.5	AAAAB
6	2.6	AAAAAB
7	2.67	AAAAAAB
8	2.71	AAAAAAAB
9	2.75	AAAAAAAAB
10	2.78	AAAAAAAAAB

Table 8.1

ℓ	r_ℓ	minimal configuration
2	1	BF
3	1.5	DBF
4	2	ADBF
5	2.25	EADBF
6	2.60	EAADBF
7	2.67	DBEADBF
8	2.71	ADBEADBF
9	2.75	EADBEADBF
10	2.89	ADBFCCCCCC
11	2.90	ADBFCCCCCCC
12	2.91	ADBFCCCCCCCC
13	2.92	ADBFCCCCCCCCC
14	2.92	ADBFCCCCCCCCCC
15	2.93	ADBFCCCCCCCCCCC

Table 8.2

ℓ	r_ℓ	minimal configuration
2	1.00	BF
3	1.50	DBF
4	2.00	ADBF
5	2.25	EADBF
6	2.60	EAADBF
7	2.67	DBEADBF
8	2.71	ADBEADBF
9	2.75	EADBEADBF
10	2.89	EADBEAADBF
11	2.90	DBEADBEADBF
12	2.91	ADBEADBEADBF
13	2.92	EADBEADBEADBF
14	3.00	EADBEADBEAADBF
15	3.00	DBEADBEADBEADBF

Table 8.3

<u>repeating the cycle realizing</u> (8.40) <u>indefinitely</u>.

<u>Proof</u>: If c is a configuration of length ℓ we canwwrite

$$\frac{1}{\ell-1} t(c) = \sum_{i,j=1}^{d} t_{ij}\xi_{ij} \tag{8.42}$$

where

$$\xi_{ij} = \frac{1}{\ell-1} n_{ij} \tag{8.43}$$

with n_{ij} = number of transitions $g_i \to g_j$ in the configuration. We also have

$$\begin{cases} n_{ij} \text{ non-negative integer} \\ \sum_{j=1}^{d} n_{ij} = \sum_{j=1}^{d} n_{ji} + e_i \\ \sum_{i,j=1}^{d} n_j = \ell-1 . \end{cases} \tag{8.44}$$

In (8.44), $e_i = 1$ if the first generator in c is i but not the last one, it is -1 if the last generator in c is i but not the first one, and zero otherwise. On the other hand, if n_{ij} is a set of numbers satisfying (8.44) with all e_i equal to zero or one $+1$, another -1 and the rest zero, a configuration in $\mathscr{L}_\ell(\mathscr{R})$ exists with n_{ij} as the transition counts although there is no guarantee that this configuration is unique (see Notes). Asymptotically, we have to solve the (continuous-variable) linear programming problem

$$\begin{cases} \sum_{i,j=1}^{d} t_{ij}x_{ij} = \min \\ x_{ij} \geq 0 \\ \sum_{j=1}^{d} x_{ij} = \sum_{j=1}^{d} x_{ji}, \quad i = 1,2,\ldots,d \\ \sum_{i,j=1}^{d} x_{ij} = 1 \end{cases} \tag{8.45}$$

We now appeal to the following lemma which is perhaps known from the study of transportation polytopes.

Lemma 8.1. Consider the set $\mathscr{X}$ of $n \times n$ matrices $X = \{x_{ij};\ i,j = 1,2,\ldots,n\}$ with non-negative entries satisfying the relations

$$\begin{cases} \sum_{i,j=1}^{n} x_{ij} = 1 \\ \sum_{j=1}^{n} x_{ij} = \sum_{j=1}^{n} x_{ji}, \quad i = 1,2,\ldots,n \end{cases} \tag{8.46}$$

$\mathscr{X}$ is a bounded convex set whose extreme points can be written as

$$X = \begin{Bmatrix} \frac{1}{r} P_r & 0_{r\times n-r} \\ 0_{n-r\times n} & 0_{n-r\times n-r} \end{Bmatrix} \tag{8.47}$$

or as any matrix obtained from (8.47) by permuting indices. Here P_r is a permutation matrix over r objects with no invariant sets and $0_{\alpha\times\beta}$ is a zero matrix of dimensions α and β. The integer r satisfies $1 \leq r \leq n$.

Proof: It is obvious that $\mathscr{X}$ is a convex, bounded polyhedral set. Choose a matrix $X \in \mathscr{X}$ and assume it to be an extremal point of the convex set. Consider $n+1$ equations in n^2 real variables z_{ij}:

$$\Sigma : \begin{cases} \sum_{i,j=1}^{n} z_{ij} = 0 \\ \sum_{j=1}^{n} z_{ij} - \sum_{j=1}^{n} z_{ji} = 0, \quad i = 1,2,\ldots,n \end{cases} \tag{8.48}$$

Summing, the left-hand side of the n last equations of the system Σ vanishes because of the first equation so that Σ

has at most rank n. Consider a non-trivial linear combination of the left-hand sides of (8.48) with the weights c_o for the first relation and $c_1, c_2, \ldots, c_n$ for the rest. We then have to look for choices of the c's such that at least one $c_i \neq 0$, and we have the identity

$$c_o \sum_{i,j=1}^{n} a_{ij} + \sum_{i,j=1}^{n} c_i \sum_{j=1}^{n} [z_{ij} - z_{ji}] \equiv 0. \tag{8.49}$$

The coefficients of the z's must vanish and we get for z_{11}

$$c_o = 0, \tag{8.50}$$

for z_{12}

$$c_o + c_1 - c_2 = 0, \tag{8.51}$$

for z_{13}

$$c_o + c_1 - c_3 = 0, \tag{8.52}$$

and so on. This means that $c_o = 0$, c_i = constant for $i = 1,2,\ldots,n$. But this corresponds to the linear relation just mentioned. Since no other non-trivial linear relation of the left-hand sides of Σ vanishes identically, the rank is equal to n.

Now assume that X has at least $n+1$ positive elements. Solve Σ with the other entries made zero; there is a non-trivial solution and we use this to form

$$\begin{cases} X^1 = X + \varepsilon z \\ X^2 = X - \varepsilon z \end{cases} \tag{8.53}$$

where z is the solution matrix just mentioned. If ε is small enough $X^1, X^2 \in \mathscr{X}$ but $X = \frac{1}{2}(X^1+X^2)$, so that X cannot be an extreme point. This shows that any extreme X has at most n positive entries.

Call the number of positive entries r. If $r = 1$ the positive element must be 1 and appear in the same row and column.

If $r = 2$ we get one of two forms:

$$\begin{Bmatrix} 1/2 & 0 \\ 0 & 1/2 \end{Bmatrix} \quad \text{or} \quad \begin{Bmatrix} 0 & 1/2 \\ 1/2 & 0 \end{Bmatrix} \tag{8.54}$$

for the relevant submatrix and they are of the form $\frac{1}{2} P_2$. This indicates how to complete the proof. For a given value of r let s be the number of rows with positive sums (same columns, of course). If $s = r$ each row and each column will have a positive entry: this means that the submatrix is $\frac{1}{r} P_r$ or can be broken up into a direct sum of such matrices. If $s < r$ there will be one row (and the same column) in the submatrix without any positive entries, and the argument can now be completed by an inductive argument.

To see that a matrix X of the form (8.47) is actually an extremum point, assume the opposite so that there exist X^1, X^2 with $X = \frac{1}{2}(X^1+X^2)$. Both X^1 and X^2 must then have their positive elements confined to the first r columns and rows. Even more, they must be located in the cells where P_r has its ones. These positive elements must be all equal since P_r leaves no subset of r objects invariant. Hence, because of the first equation of (8.46), they must be equal to $1/r$.

Hence we can write an arbitrary X as a convex combination of extreme X's, $X^{(\nu)}$;

$$X = \Sigma\, c_\nu X^{(\nu)}, \quad c_\nu \geq 0, \quad \Sigma\, c_\nu = 1 \tag{8.55}$$

with

$$\sum_{i,j=1}^{d} t_{ij}x_{ij} = \Sigma\, c_\nu\, y_\nu \tag{8.56}$$

where

$$y_\nu = \sum_{i,j=1}^{d} t_{ij}x_{ij}^{(\nu)} \ . \tag{8.57}$$

The values of the minimum problem (8.45) is then obtained by making that $c_\nu = 1$, and the rest $= 0$, for which y_ν is the minimum. But this amounts to solving for τ in (8.44), as is seen by remembering the fact that the P_r are $1/r$ times permuted matrices over r signs. Hence an asymptotically minimal configuration can be obtained by repeating the cycle corresponding to the permutation of certain r generators, as stated in the theorem.

Note that we have not claimed that there are no non-periodic configurations that also are asymptotically minimal. Indeed, given an asymptotically minimal periodic configuration, we can always destroy its periodicity while preserving its asymptotic minimum property by changing a number of its generators if this number is $o(\ell)$.

We now turn to a two-dimensional variant of the problem. Consider configurations whose internal structure is a square lattice with generators having four (double) bonds. Each bond has a <u>direction value</u> E,N,W,S and <u>two location values</u> as in Figure 8.2. The bond relation says that the location values are equal and that the direction values appear in one of the combinations NS, SN, EW, WE.

The total energy of the configuration is defined additively over all the internal bonds

$$t(c) = [t(gg_1)+t(gg_2)+t(gg_3)+t(gg_4)] \tag{8.58}$$

with self-explanatory notation, summed over the finite lattice

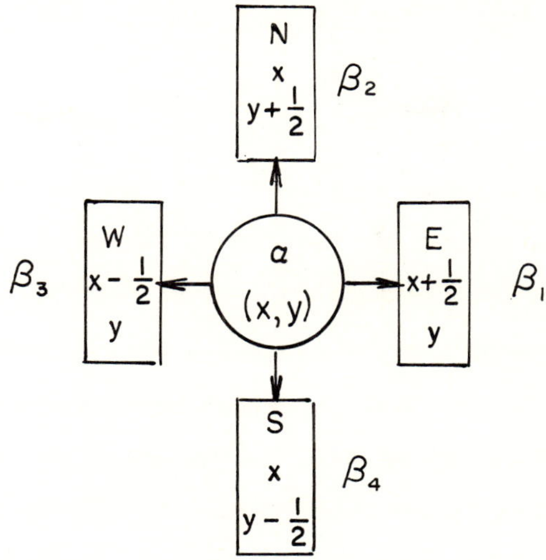

Figure 8.2

with obvious modifications at the boundary. Since T need not be symmetric, the directions of the arrows in the configuration are important. Asymptotically, this will not matter, however. In the total energy we will have one term $t(g_{ij}g_{i+1j})$ but there will also appear the term $t(g_{i+1j}g_{ij})$ except at the boundary of the lattice and of course the contribution of the boundary is asymptotically negligible. Hence we just have to get

$$\sigma = \min_{i,j} \frac{1}{2} (t_{ij}+t_{ji}). \tag{8.59}$$

Say that the minimum is attained for $i = i_o$, $j = j_o$, and call $g_{i_o} = a$, $g_{j_o} = b$. Then the very simple configuration in Figure 8.3 will have asymptotically minimal energy. The asymptotic minimum relative energy will then be 4σ.

```
a — b — a — b — ...
|   |   |   |
b — a — b — a — ...
|   |   |   |
a — b — a — b — ...
|   |   |   |
. . . . . . . . . .
```

Figure 8.3

The reason why this two-dimensional variant was so easy to deal with is that the neighborhood structure was simple. With an 8-neighbor definition, for example, it is not obvious what the result will be. Even numerical experiments seem difficult to carry out since there is no obvious dynamic programming approach available.

We assumed that *the generators of the configuration were coupled only locally, nearest-neighbor interaction*. It may be thought that this is what caused the asymptotically regular pattern.

To gain some insight into this we shall study configurations with *global interactions*. The model used was inspired by morphologial studies of the root system of plants and trees, but no claim is made that the following is even remotely concerned with biological reality. That is not the purpose here, although it is tempting to pursue this type of model in that direction. For the moment, however, we are concerned only with theoretical patterns.

If the arrangement of generators over a background space is governed by *an extremum principle guaranteeing minimum loss of energy flow, does this lead to a regular arrangement?* To study this question we shall assume that generators

$g_1, g_2, \ldots, g_n$, forming a free configuration, are distributed over a subset Q_L of the background space R^d. Here we write for the location of a generator $x = (x^1, x^2, \ldots, x^d)$ and let

$$Q_L\colon \{x \mid 0 \leq x^\nu \leq L;\quad \nu = 1,2,\ldots,d\}. \tag{8.60}$$

Energy is available but not necessarily used at each point $x \in Q_L$ and the maximum flow per time unit will be chosen as the unit. The n generators, which may represent botanical specimens or general ecological varieties, will not necessarily be able to absorb the maximum flow. Instead they screen each other according to a function $K = K(x-y)$. The loss of flow from the point x due to screening will be assumed to be, with x_ν as the νth generator,

$$e(x) = \prod_{\nu=1}^{n} K(x-x_\nu) \tag{8.61}$$

and the total loss due to screening is

$$E = \int_{Q_L} e(x)dx. \tag{8.62}$$

An example may be helpful in motivating this <u>multiplicative interaction</u>. Say that each generator g_ν is associated with a circle of radius r. If the whole system consisting of the configuration $c = (g_1, g_2, \ldots, g_n)$ can absorb energy from the area of the union of the circles we have a total absorbed flow equal to the area of the horizontally hatched region in Figure 8.4. The remainder of the total energy flow available, the loss, can be written as in equation (8.61) if

$$K(u) = \begin{cases} 1 & \text{if} \quad ||u|| > r \\ 0 & \text{if} \quad ||u|| < r \end{cases} \tag{8.63}$$

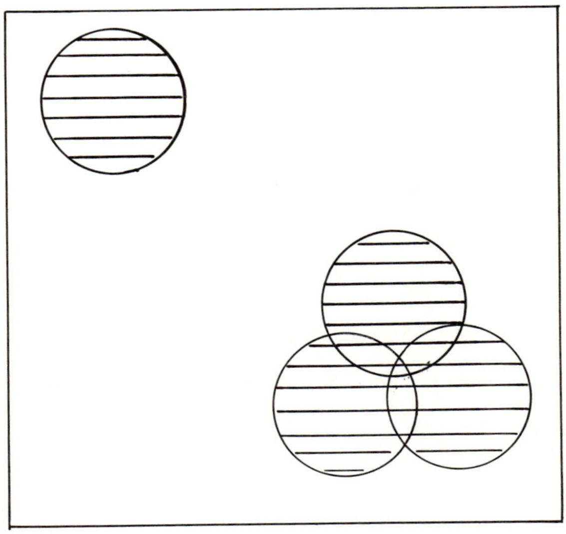

Figure 8.4

Of course, this is a degenerate case with K taking only the extreme values 0 and 1 and we shall be interested in smoother transitions from 0 (complete absorption) to 1 (complete loss). $K(u)$ will be assumed to be a non-negative function equal to zero only if $u = 0$ and tending to one if $||u|| \to \infty$; see below.

The multiplicative interaction has a simple probabilistic interpretation. Consider the available energy flow from a small neighborhood of the point x. The system represented by the configuration will absorb or lose it due to screening. Let the probability that it will be lost due to screening of x_i be $K(x-x_i)$. If these events are independent the total probability of its loss will be given by the expression (8.61). The integral E in (8.62) then represents an expected value.

The degenerate case with circles is connected with the problems of coverings in the plane; see Fejes Tóth (1953).

We do not assert that the minimum loss principle expressed through multiplicative interaction leads to a regular arrangement for a finite number of generators in the configuration. It is only asymptotically, for large configurations, that this is claimed to be true, in a way that will be made precise in the following theorem.

To express the asymptotics conveniently we shall study the average loss

$$A = \frac{1}{L^d} E \tag{8.64}$$

for a given <u>shape</u> ϕ of the configuration. The notion of shape of the configuration will be understood as follows. Introducing

$$k(u) = \log K(u), \tag{8.65}$$

we can write

$$A = \frac{1}{L^d} \int_{Q_L} \exp\{n \int_{Q_L} k(x-y)\ \Phi(dy)\}dx \tag{8.66}$$

where Φ is a discrete measure with the point mass $1/n$ in each of the generators g_ν. We shall now do the asymptotics, by assuming Φ to be absolutely continuous with a density that does not change its shape as n and L tend to infinity. The density will then be $\frac{1}{L^d}\phi(x/L)$. The more variable ϕ is, the less regular is the arrangement, the most regular one being associated with constant ϕ, $\phi \equiv 1$.

The result can be announced as follows.

<u>Theorem 8.2</u>. <u>Consider the average loss</u>

$$A = \frac{1}{L^d} \int_{Q_L} \exp\{n \frac{1}{L^d} \int_{Q_L} k(x-y)\phi(y/L)dy\}dx \tag{8.67}$$

<u>for</u> $L \to \infty$, $n/L^d = \alpha$ <u>and with the conditions</u>

$$\begin{cases} k \leq 0 \\ k \in L_1 \\ \phi \geq 0 \\ \int_{Q_L} \phi(y)dy = 1 \\ \phi \ \text{continuous} \end{cases} \tag{8.68}$$

<u>Then</u>

$$\min_{\phi} \lim_{L\to\infty} A = \exp(\alpha\kappa) \tag{8.69}$$

<u>where</u>

$$\kappa = \int_{R^d} k(u)du = \int_{R^d} \log K(u)du. \tag{8.70}$$

<u>The minimum can be attained for the most regular shape</u>

$$\phi(y) \equiv 1. \tag{8.71}$$

<u>Proof</u>: For a given ϕ satisfying the condition (8.68) we have

$$A = \frac{1}{L^d} \int_{Q_L} \exp\{\alpha H_L(x)\}dx \tag{8.72}$$

with

$$H_L(x) = \int_{Q_L} k(x-y)\phi(y/L)dy. \tag{8.73}$$

Changing variables

$$x = L\xi, \quad y = L\eta \tag{8.74}$$

so that ξ and η range over the unit cube Q_1, we get

$$A = \int_{Q_1} \exp\{\alpha H_L(L\xi)\}d\xi \tag{8.75}$$

and

$$H_L(x) = H_L(L\xi) = L^d \int_{Q_1} k[L(\xi-\eta)]\phi(\eta)d\eta. \tag{8.76}$$

If ξ is not on the boundary of Q_1 the non-positive kernel $L^d k[L(\xi-\eta)]$ has total integral over Q_1 equal to

$$\int_{L(\xi-Q_1)} k(u)du \to \kappa \quad \text{as} \quad L \to \infty \tag{8.77}$$

since the domain of integration $L(\xi-Q_1) \to R^d$. But if we only integrate the kernel over Q_1 excluding a neighborhood of ξ, the limiting value of that integral is zero. Hence it is a singular kernel and

$$H_L(L\xi) \to \kappa\, \phi(\xi) \tag{8.78}$$

for almost all ξ; the latter follows since ϕ is a continuous function. On the other hand,

$$H_L \leq 0, \tag{8.79}$$

the integrand in (8.75) is bounded by one and, by bounded convergence,

$$\lim_{L\to\infty} A = \int_{Q_1} \exp[\alpha\kappa\phi(\xi)]d\xi\ . \tag{8.80}$$

This integral represents a nonlinear functional in ϕ and we have to find its minimum. Remembering that α is positive and κ is negative, the function $\exp(\alpha\kappa\phi)$ is concave in ϕ so that

$$\int_{Q_1} \exp[\alpha\kappa\phi(\xi)]d\xi \geq \exp[\alpha\kappa \int_{Q_1} \phi(\xi)d\xi] = \exp(\alpha\kappa) \tag{8.81}$$

with equality for

$$\phi(\xi) \equiv 1. \tag{8.82}$$

This completes the proof of the theorem.

Remarks. (i) The configuration of generators was free: no restrictions were imposed on the generators except of course that they should fall inside Q_L. The extremum principle introduces a constraint, however, and it is this that forces the configuration to become systematic, at least asymptotically.

(ii) The assumptions in the theorem are stronger than needed. It is believed that ϕ need not be continuous and that integrability should be enough. Also, the form of the region Q_L is unnecessarily restricted. The proof goes through more generally, say when Q_1 is a sphere of given radius and $Q_L = LQ_1$. The only thing that is changed is the normalizing constant due to the volume of Q_1 being different from 1.

If the available energy flow is not uniform over Q_L but depends upon the location x, things can be expected to change. They do, but in a manner that may appear counterintuitive, so that it is worthwhile to examine such a case.

Say that the available energy flow is $a(x/L)$ where $a(\zeta)$ is a non-negative continuous function so that (8.62) will have to be replaced by

$$E = \int_{Q_L} a(x/L)e(x)dx. \tag{8.83}$$

To fix ideas, take $D = 2$ and let Q_L be the square of side L as in Figure 8.5, $x = (z_1, z_2)$ with coordinate axes as indicated in the figure.

With the same sort of asymptotic reasoning as above we let ϕ be a normalized density in Q_1 so that in Q_L we have the density of generators

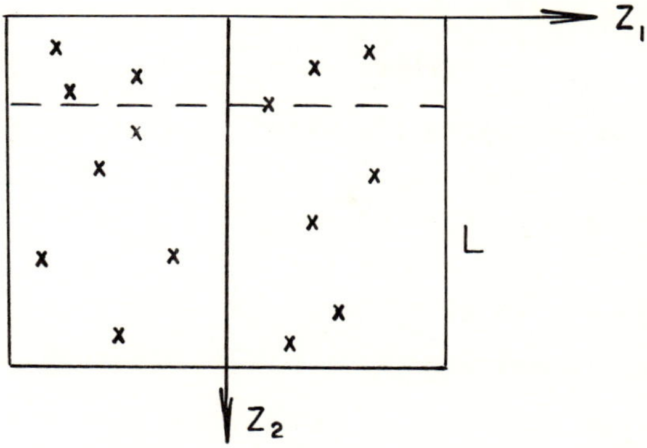

Figure 8.5

$$\frac{n}{L^2}\,\phi(x/L). \tag{8.84}$$

Repeating the asymptotics argument in the proof of the theorem, we are led to the functional

$$E(\phi) = \int_{\Omega_1} a(x)\exp[\alpha\kappa\phi(x)]dx. \tag{8.85}$$

We shall assume now that $a(z)$ depends only upon z_2, the depth below the z_1 axis, and decreases when z_2 increases. The reader may like to think of the system as a plane <u>root system</u> extending downwards with the available energy flow decreasing downwards, a soil gradient.

To minimize the functional $E(\phi)$ it is clear that the contribution from a horizontal line segment as the dotted one in Figure 8.5 has $a(z_2)$ = constant, so that we can let ϕ be constant along this horizontal line segment. This reduces our problem to a one-dimensional one.

After these reductions the problem has been condensed into the following, where the notation has been changed for simplicity. Find the minimum (if it is attained) of the functional

$$E(\phi) = \int_o^1 a(x)\exp[-\phi(x)]dx \tag{8.86}$$

with the non-negative function ϕ satisfying

$$\phi \in L_1(0,1) \quad \text{and} \quad \int_o^1 \phi(x)dx = -\alpha\kappa \tag{8.87}$$

while $a(x)$ is a given function which shall be assumed to satisfy the conditions

$$\begin{cases} \underline{a} \text{ decreasing on } [0,1] \text{ and } C_1[0,1] \\ a(0) = 1,\ a(0) = 0,\ 0 < a(x) < 1 \text{ for } 0<x<1 . \end{cases} \tag{8.88}$$

We need only consider non-increasing ϕ's. Indeed, if ϕ satisfies (8.87) but is decreasing somewhere we can always change it into another function ϕ' such that $E(\phi') \leq E(\phi)$ and ϕ' also satisfying (8.87). This can be done via a measure-preserving transformation of $[0,1]$ onto itself; see Hardy, Littlewood and Polya (1934), p. 276. Denote this class of functions by F.

Let $\phi_1, \phi_2, \phi_3, \ldots$ be a sequence in F such that

$$\lim_{i\to\infty} E(\phi_i) = \inf_{\phi\in F} E(\phi). \tag{8.89}$$

Using the bound

$$0 \leq \phi(x) \leq -\frac{\alpha\kappa}{x} \tag{8.90}$$

we can select a subsequence $\{\phi_i'\}$ from $\{\phi_i\}$ by the diagonal procedure such that $\phi_i'(r)$ converges to $\psi(r)$, with some $\psi \in F$, for any rational $r \in (0,1]$ and $\psi \in F$ and, using Fatou's lemma,

$$E(\phi_i') \to E(\psi) = \inf_{\phi \in F} E(\phi). \tag{8.91}$$

Thus the minimum is attained within F.

How does ψ look? Say that

$$x_o = \max\{u \mid 0 \leq u \leq 1,\ \psi(x) > 0 \text{ for } x < u,\ \psi(x) = 0 \text{ for } u \leq x \leq 1\}. \tag{8.92}$$

Note that the value of ψ at the single point x_o does not influence $E(\psi)$. A straightforward variational argument on the support $[0,x_o)$ of the function ψ shows that $a(x)\exp[-\psi(x)]$ is constant there, so that we can write

$$\psi(x) = \begin{cases} \lambda + \log a(x), & 0 \leq x < x_o \\ 0, & x_o \leq x \leq 1 \end{cases} \tag{8.93}$$

where λ is a constant. This constant is certainly not arbitrary but must satisfy

$$\lambda + \log a(x_o) \geq 0, \tag{8.94}$$

since $\psi(x)$ should be non-negative, and

$$\lambda x_o + \int_0^{x_o} \log a(x)dx = -\alpha\kappa \tag{8.95}$$

since ψ should have total integral equal to $-\alpha\kappa$. These two conditions are enough to guarantee that $\psi \in F$, the only possible exception arising if $x_o = 1$. We shall see that the latter case does not occur.

For ψ as in (8.93) our functional takes the value

$$E(\psi) = x_o e^{-\lambda} + \int_{x_o}^{1} a(x)dx. \tag{8.96}$$

Solving (8.95) for λ the inequality in (8.94) gives us

$$R(x_0) = -\alpha\kappa + \int_0^{x_0} \log \frac{a(x_0)}{a(x)}\, dx \geq 0 \tag{8.97}$$

The function R is continuously decreasing with $R(0) = -\alpha\kappa > 0$ and $R(1) = -\infty$. The last relation follows from

$$\begin{aligned} R(x_0) &\leq -\alpha\kappa + \int_0^{\varepsilon} \log \frac{a(x_0)}{a(x)}\, dx \\ &= -\alpha\kappa+\varepsilon \log a(x_0) - \int_0^{\varepsilon} \log a(x)dx \downarrow -\infty \end{aligned} \tag{8.98}$$

for any fixed $\varepsilon \in (0,1)$. Hence there is a unique root ρ of the equation

$$R(\rho) = 0. \tag{8.99}$$

Therefore we must have x_0 between 0 and ρ and $\rho < 1$. (8.95) gives us, after differentiation with respect to x_0,

$$\lambda + x_0 \frac{d\lambda}{dx_0} + \log a(x_0) = 0. \tag{8.100}$$

Hence

$$\frac{dE}{dx_0} = e^{-\lambda}[1+\lambda+\log a(x_0)] - a(x_0). \tag{8.101}$$

Writing $a(x_0) = e^{-u}$, $u > 0$, we get

$$e^{\lambda} \frac{dE}{dx_0} = 1+\lambda-u-e^{\lambda-u}. \tag{8.102}$$

Because of (8.94), $\lambda-u \geq 0$, so that

$$dE/dx_0 \leq 0. \tag{8.103}$$

Hence to make E small we should push x_0 as far as possible to the right, that is to the value ρ. Combining these results, we have

<u>Theorem 8.3</u>. <u>The minimum of</u> $E(\phi)$ <u>under the given conditions</u> (8.88) <u>and</u> (8.89) <u>is attained for the generator density</u>

$$\psi(x) = \begin{cases} \lambda + \log a(x), & 0 \le x < \rho \\ 0, & \rho \le x \le 1 \end{cases} \tag{8.104}$$

<u>where</u> ρ <u>is the solution of</u>

$$R(x_o) = -\alpha\kappa + \int_o^{x_o} \log \frac{a(x_o)}{a(x)}\, dx = 0, \quad x = \rho, \tag{8.105}$$

<u>and</u>

$$\lambda = \frac{-\alpha\kappa - \int_o^{\rho} \log a(x)dx}{\rho} = -\log a(\rho). \tag{8.106}$$

<u>The resulting minimum value is</u>

$$\min_{\phi\in F} E(\phi) = \rho e^{-\lambda} + \int_\rho^1 a(x)dx = \rho a(\rho) + \int_\rho^1 a(x)dx. \tag{8.107}$$

Let us take a particular case to get some idea of how the extremal configuration will look. Say that we have a linear soil gradient upwards

$$a(x) = 1 - x, \quad 0 \le x \le 1. \tag{8.108}$$

Then R will be given by

$$R(x_o) = -\alpha\kappa + x_o + \log(1-x_o). \tag{8.109}$$

To get simple numerical values put $\alpha = 1$, $\kappa = -1/e$. Then

$$\begin{cases} \rho = 1 - 1/e \\ \lambda = 1 \\ \phi(x) = 1 + \log(1-x), & 0 \le x \le \rho \\ \phi(x) = 0, & \rho \le x \le 1 \end{cases} \tag{8.110}$$

The actual pattern will then appear as a finite collection of generators generated by the measure over the square with ϕ as the vertical marginal density and uniform conditional densities along the horizontal line segments. A configuration of size $n = 50$ has been drawn from this probability measure and is represented in Figure 8.6.

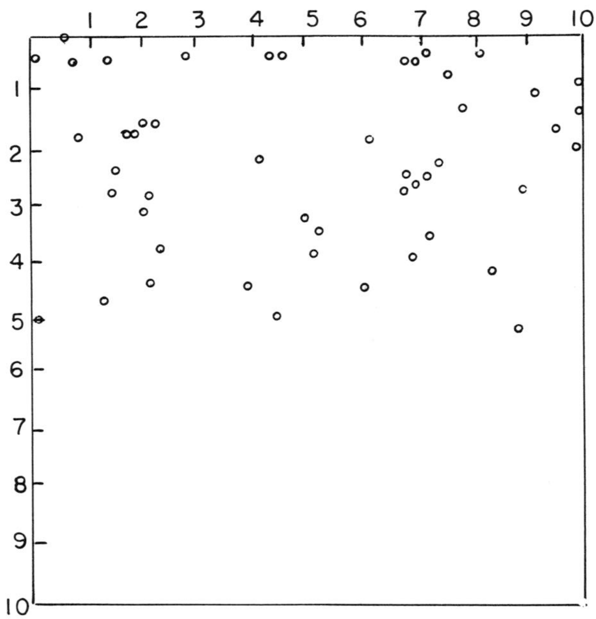

Figure 8.6

As described by equation (8.110), the resulting configuration shows a fairly regular pattern and will have its largest density at the top and the density decreasing downwards. This could be expected from the original formulation of the model. What is surprising, however, is the fact that the density is not only small further down but is <u>equal to zero</u> from $x = \rho$ onwards. This counter-intuitive <u>depth effect</u> was not expected initially and seems remarkable even now when a formal proof of it has been given.

We now turn to another aspect of the same extremum problem. In the previous formulation we considered a configuration of generators distributed over a given area. The number of generators was fixed and large. Asymptotically this was described via a plane density.

Suppose now instead that we use as generators, not points in Q_L, but directed line segments with $\omega(g) = 1$, and the bond value equal to the first end point. Let ρ = EQUALITY. On these line segments we select points (absorbing elements) with a constant linear density μ (see Figure 8.7). The idea behind this is that roots are idealized as straight, all with the same apex, and with points of energy absorption distributed uniformly along the straight roots. Let now n, the given fixed but large natural number, mean the number of roots and use a continuous description for the asymptotics. If ψ stands for the density of root ends, what will the resulting density ϕ be for the points of energy absorption? If we can answer this question we can go ahead and look at the same sort of extremum problem as before but with restrictions on the number of root ends instead.

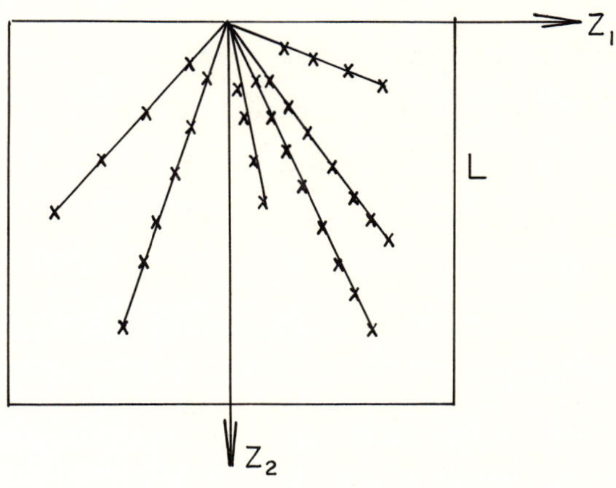

Figure 8.7

To find the density ϕ expressed in ψ we consider a small annular region $A = (\rho,\rho+d\rho)\times(\alpha,\alpha+d\alpha)$ in polar coordinates. In order that a point of absorption fall in A a root must pass through it and a point of absorption on this root must fall in A. This gives us the product

$$\phi(\rho,\alpha)\rho d\rho d\alpha = \int_{r=\rho}^{b(\alpha)} \psi r \, dr d\alpha \cdot \mu d\rho \tag{8.111}$$

where $b(\alpha)$ is the distance from the origin to the boundary of the region along the direction α. Hence

$$\phi(\rho,\alpha) = \frac{\mu}{\rho}\int_{\rho}^{b(\alpha)} \psi r dr. \tag{8.112}$$

Given a desired density ϕ, when can it be realized by a root system? It is clear from (8.112) that $\phi\rho$ must be absolutely continuous with the non-positive derivative

$$\partial(\phi\rho)/\partial\rho = -\mu\psi\rho \leq 0 \tag{8.113}$$

where of course all the statements shall be qualified by "almost everywhere". This equation (8.113) enables us to find the root and density for the system that leads to a given point density, assuming of course that such a system exists.

This need not be the case, not even in simple cases. We showed for the model with no soil gradient that a uniform distribution of points was due to the extremal principle. That means that $\phi \equiv$ constant ϕ_o and (8.113) gives us only the contradictory relation

$$\psi = -\phi_o/\mu\rho < 0: \tag{8.114}$$

<u>There is no root system distribution leading to a uniform point distribution</u>.

What is the way out of this dilemma? We have used absolutely continuous measures with integrable densities ψ.

This is closed in L_1-norm but what we really need is closure under the weak topology of measures. We must therefore approach the problem this time with more general ψ-measures.

Going back to the proof of Theorem 8.2, we should find a measure Φ minimizing the functional

$$E(\psi) = \int_{Q_1} \exp[-\frac{c}{\rho} p(\alpha) \int_{\rho}^{b(\alpha)} r\Phi_{\alpha}(dr)]\rho d\rho d\alpha \qquad (8.115)$$

where c is a positive constant and the integrand in the exponent is along a ray in the α-direction out to the boundary, and p is a marginal density over values of the angle α. As we shall see, this class of measures is sufficiently wide to make the minimum attained. Along α the measure Φ_{α} is normed to one; this is done by $p(\alpha)$.

To minimize (8.115) we first keep p fixed and vary the distribution Φ_{α} along a ray. The inner integral should be made as large as possible, which means that we should put all the Φ_{α} measure at the boundary point with $r = b(\alpha)$.

When this is done we get after change of variable

$$E(\psi) = \int_{0<\alpha<\pi/2} H[cp(\alpha)]b^2(\alpha)d\alpha \qquad (8.116)$$

where H is non-negative, decreasing and differentiable. We shall carry out the analysis for the case $H(x) = e^{-x}$, which should be qualitatively similar, but it should be remarked that general H can be dealt with by the same method. Then of course the exponential and logarithmic functions will have to be replaced by H and its inverse.

Now $p \in L_1(0,\pi/2)$, non-negative, and integral 1. But (8.116) is similar to (8.84) and we can apply the same technique with some modifications. Here we have used the symmetry of Q_1 around the vertical $\alpha = 0$; for a non-symmetric region a small modification is needed in the following derivation. First, apply a measure-preserving transformation from $(0,\pi/2)$ onto itself which makes $b(\alpha)$ non-increasing without affecting the conditions imposed on p. This done, we follow the previous argument and are led to

$$p(\alpha) = \begin{cases} \dfrac{\lambda+\log b^2(\alpha)}{c}, & \alpha \le \alpha_o \\ 0 & \alpha_o < \alpha \le \pi/2 \end{cases} \tag{8.117}$$

with

$$\begin{cases} \lambda + \log b^2(\alpha_o) \ge 0 \\ \lambda\alpha_o + \displaystyle\int_o^{\alpha_o} \log b^2(a)da = c \end{cases} \tag{8.118}$$

and

$$E(\psi) = \alpha_o e^{-\lambda} + \int_{\alpha_o}^{\pi/2} b^2(a)d\alpha. \tag{8.119}$$

Again, we get by differentiating (8.118) and (8.119) with respect to α_o

$$\begin{aligned} e^{\lambda} \frac{dE(\psi)}{d\alpha_o} &= 1 + \lambda + \log b^2(\alpha_o) - b^2(\alpha_o)e^{\lambda} \\ &= 1 + \lambda - u - e^{\lambda-u} \end{aligned} \tag{8.120}$$

putting $b^2(\alpha) = e^{-u}$. But (8.118) implies that $\lambda-u \ge 0$ so that the above derivative in (8.119) is non-positive and we should push α_o as much as possible to the right without violating the constraints.

But now things change from the previous case since $b(\alpha)$ is bounded away from zero in contrast to $a(x)$ in (8.84). The function

$$S(\alpha_o) = c + \int_o^{\alpha_o} \log \frac{b^2(a_o)da}{b^2(\alpha)} \qquad (8.121)$$

takes the value $c > 0$ for $\alpha_o = 0$ and decreases. We must then put α_o equal to $\pi/2$ if $S(\pi/2) \geq 0$ and otherwise $= \rho$ where ρ is the root of the equation $S(\rho) = 0$.

For the square region the boundary distance $b(\alpha)$ is

$$b(\alpha) = \begin{cases} \dfrac{1}{\cos\alpha}, & 0 \leq \alpha < \beta \\ \dfrac{1}{2\sin\beta}, & \beta < \alpha \leq \dfrac{\pi}{2} \end{cases} \qquad (8.122)$$

with tan $\beta = 1/2$, and it depends upon c, i.e. on μ, the linear density, where the breakpoint b corresponding to α_o after transformation by the measure-preserving transformation occurs. We get three possibilities:

$$\begin{cases} \text{I} & b = \sqrt{5}/2 \\ \text{II} & 1/2 < b < 1 \\ \text{III} & 1 < b < \sqrt{5}/2 \end{cases} \qquad (8.123)$$

These three cases lead to root systems as indicated in Figure 8.8, only schematically to show the qualitative behavior.

We meet another phenomenon here: a distance effect. The root system tends to emphasize long roots. In Case I this is not so obvious although equation (8.117) indicates that this is so even in this, the least extreme case. In Case II it becomes more apparent and the whole system has been pushed downwards, leaving part of the region unoccupied. In the most interesting case, the third one, this tendency has even led to the breaking up of the root system into two components, connected with each other only at the apex.

Consider again a configuration consisting of point

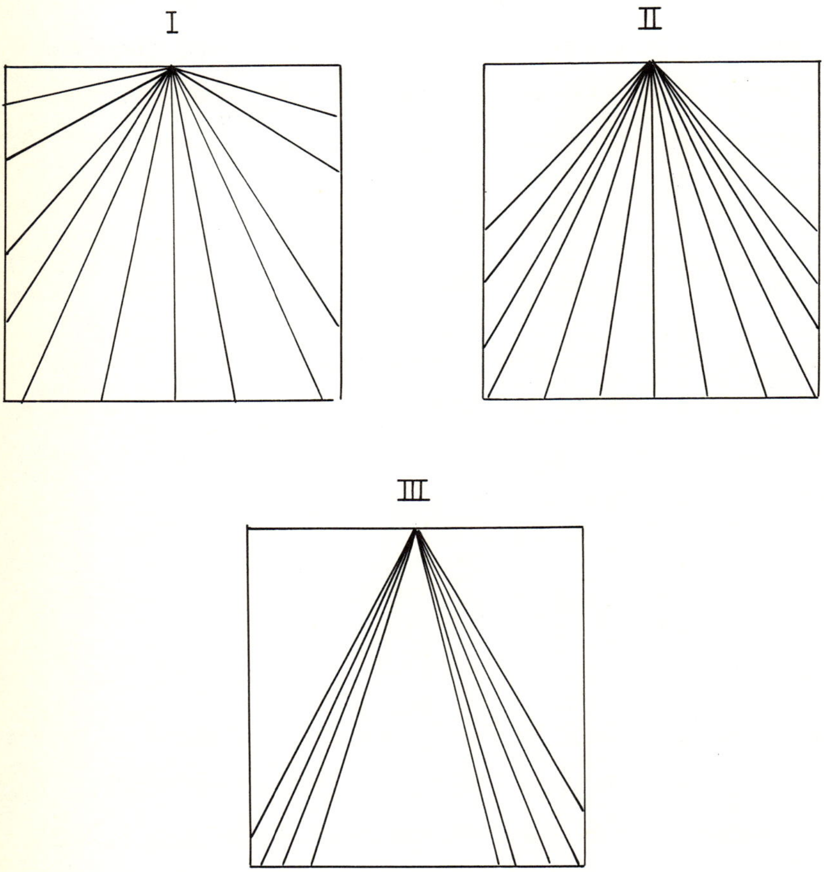

Figure 8.8

generators spread out over Q_L. In the previous case the generators screened each other but now we shall assume that they balance each other so as to lead to a static equilibrium. What configuration has minimum potential energy?

Assume that the total potential energy is made up of all terms of the type $P(x_\nu - x_\mu)$ where P is a given function of the vector difference between the locations x_ν and x_μ of two generators. As before, we shall study the problem asymptotically and use a continuous density shape ϕ just as before. Approximate the sum

$$E = \sum_{\nu,\mu=1}^{n} P(x_\nu - x_\mu) \tag{8.124}$$

by an integral. We get two types of regular patterns, a uniform one and one with all the generators lumped into a single point.

We shall assume that Q_1 is a bounded convex subset of R^d which is actually more than what is needed. We have the following result.

<u>Theorem 8.4. Assuming that the potential energy function</u> $P(z)$ <u>is</u> $L_1(R^d)$ <u>with</u>

$$P = \int_{R^d} P(z)dz \neq 0. \tag{8.125}$$

<u>For a configuration of particles, the point generators, distributed over</u> Q_L <u>with a non-negative and continuous density of shape</u> ϕ, <u>the total potential energy</u> E_L <u>is then given asymptotically by the expression</u>

$$\lim_{L\to\infty} L^d E_L(\phi) = P \cdot \int_{Q_1} \phi^2(z)dz \tag{8.126}$$

<u>which is minimized for a uniform shape</u>

$$\phi(z) \equiv 1/m(Q_1) \tag{8.127}$$

<u>if</u> $P > 0$, <u>and can be made arbitrarily small by lumping into a single point if</u> $P < 0$.

Proof: We have

$$E_L(\phi) = \frac{1}{L^{2d}} \int_{u\in Q_L} \int_{v\in Q_L} P(u-v)\phi(u/L)\phi(v/L)dudv \qquad (8.128)$$

so that

$$L^d E_L(\phi) = \int_{\xi\in Q_1} K_L(\xi)\phi(\xi)d\xi \qquad (8.129)$$

with the integral kernel

$$K_L(\xi) = \int_{Q_L-\xi L} P(z)\phi(\xi + \frac{z}{L})dz. \qquad (8.130)$$

But ϕ is bounded and P integrable so that $K_L(\xi)$ is uniformly bounded. On the other hand, unless $\xi \in \partial Q_1$, the domain of integration in (8.130) tends to the entire space R^d as $L \to \infty$, so that again appealing to the same properties

$$K_L(\xi) = \int_{R^d} P(z)\phi(\xi + \frac{z}{L})\, dz + o(1) \qquad (8.131)$$

for almost all ξ. Now $\phi(\xi + \frac{z}{L}) \to \phi(\xi)$ for any fixed z, and by dominated convergence of the integrand in (8.131)

$$\lim_{L\to\infty} K_L(\xi) = \phi(\xi)\int_{R^d} P(z)dz = \phi(\xi)P, \quad \text{a.e. } \xi\ . \qquad (8.132)$$

This, together with bounded convergence of the integrand in (8.129), proves (8.126). The last statement of the theorem then follows immediately. In particular, if P is negative it is clear that inf of the limit is $-\infty$. We can get arbitrarily close by having the function ϕ close to a δ-function so that the particles tend to lump in a single point.

The important fact here is that the <u>asymptotically regular configuration of generators follows from the extremum principle whatever the form of the region</u> Q_1 <u>or the interaction potential</u> $P(z)$, with mild restrictions only.

One could ask what the real reason is for this tendency to one of these two simple patterns? The total potential energy consists of interaction energy only, with no external force field applied. It would be interesting to see what would happen if a potential well is placed along the boundary of Q_L and if its influence would gradually tend not to affect the shape of the extremum configuration as L tends to infinity.

If $P = 0$ the right-hand side of (8.126) is zero irrespective of the shape ϕ of the configuration. To say then that the choice of ϕ does not matter is really to avoid the issue. Instead one should look more carefully into the asymptotics, getting the leading term of the asymptotic expansion of E_L and then trying to maximize E_L with respect to ϕ.

A similar problem but with a different motivation has been studied in a discrete formulation (see Grenander and Vitale (1969)) for weights that satisfy

$$\begin{cases} \sum_{\mu=1}^{n} \phi_\nu = 1 \\ \sum_{\nu,\mu=1}^{n} P(\nu-\mu)\, \phi_\nu \phi_\mu = \min \end{cases} \tag{8.133}$$

where P is a non-negative definite function. If the sum

$$\sum_{-\infty}^{\infty} P(\nu) = 0, \tag{8.134}$$

which corresponds to the integral above, then it was shown that the ϕ_ν sequence took on a parabolic shape as n increased. This tendency is shown graphically in Figure 8.9 for a typical case. In (a), $n = 30$ and in (b) $n = 50$. The tendency to parabolic shape was also proven analytically by the same authors.

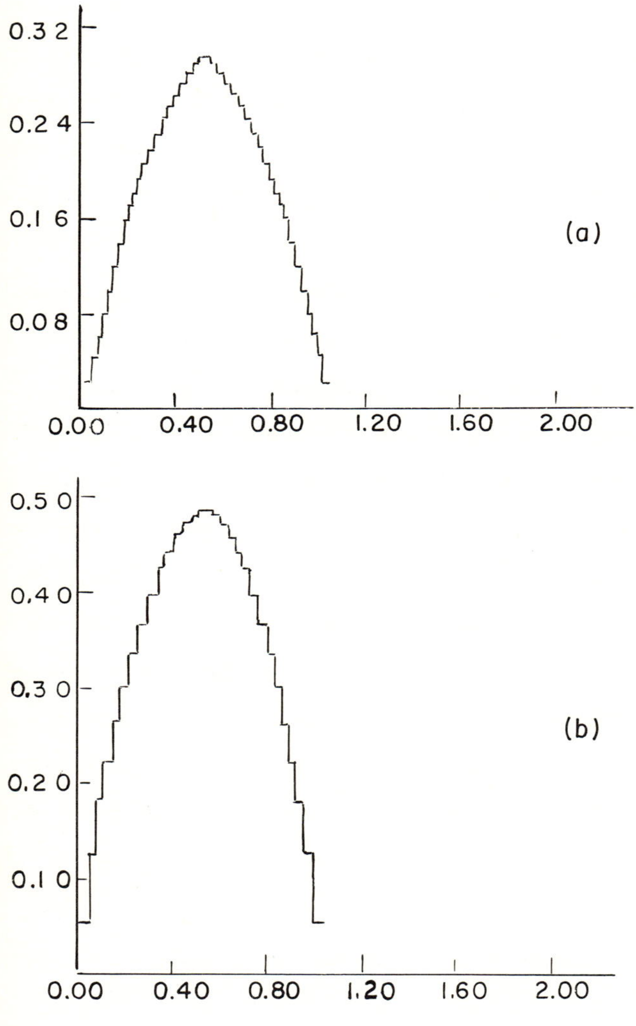

Figure 8.9

If this result carries over to our continuous problem, it would mean that even with only one generator class other asymptotic patterns than the uniform and lumped ones can occur.

We have demonstrated in some cases how extremum principles force the configuration to assume a regular pattern. It should be noted, however, both that the extrema need not be uniquely attained, and that the statements are all asymptotic in character. <u>It is only for large configurations that a regular appearance is caused by the extremum principles</u>.

Our finding agrees well with the statement by Fejes Tóth (1953): "Die reguläre Gestalt der Extremalfigur ist hier oft eine Folgerung der Extremalforderung" - the regular appearance of the extremal figure is often a consequence of the extremum demand.

So far we have assumed that the number of generators remains constant, generator growth. The picture changes drastically if this assumption is dropped.

In configuration growth, that will be discussed now, the configuration changes in a drastic manner, for example by the addition of more generators. If we think of the generators as cells, each of which occupies a certain volume and excluding each other, growth will depend upon the mechanism that produces more cells. Let us make time discrete, $t = 0,1,2,...$, and space two-dimensional and discrete consisting of the lattice points (x,y), both x and y integers.

Most mathematicians - professional ones as well as amateurs - seem to enjoy games. One very enjoyable mathematical game was introduced some years ago, called "game of life", see Gardner (1970, 1971). The inventor of the game, J. H. Conway, suggested the following rules for birth and death. To each lattice point (x,y) one associates the eight neighbors $(x+1,y)$, $(x+1,y+1)$, $(x,y+1)$, $(x-1,y+1)$, (x-1,y), (x-1,y-1), (x,y-1), (x+1,y-1).

If the image at time t consists of the set $I(t)$ of lattice points then it changes into a new set $I(t+1)$ according to rules depending upon the number $n(x,y)$ of neighbors of (x,y) belonging to $I(t)$

$$\begin{cases} \text{death:} & \text{if } (x,y) \in I(t), \text{ then } (x,y) \notin I(t+1) \\ & \text{iff } n(x,y) = 0,1,4,5,6,7 \text{ or } 8 \\ \text{birth:} & \text{if } (x,y) \notin I(t), \text{ then } (x,y) \in I(t+1) \\ & \text{iff } n(x,y) = 3 \\ \text{survival:} & \text{if } (x,y) \in I(t), \text{ then } (x,y) \in I(t+1) \\ & \text{iff } n(x,y) = 2 \text{ or } 3. \end{cases} \tag{8.135}$$

The rules were chosen in order to lead to complex growth patterns, not to model biological growth. Actually some remarkable patterns appear, some of which settle down to stable ones, some oscillate, and some actually grow without limit.

As an illustration we show the growth patterns resulting from original configurations of four cells, see Figure 8.10. For a,b,c, and d stable configurations are reached rapidly. In d no limit exists, the "traffic light" oscillates periodically.

A reader inclined towards mathematical games will find it rewarding to trace the life histories of more involved initial configurations. It is remarkably difficult to guess how they will evolve in time and some surprising cases turn up often.

Perhaps this is more than a sophisticated parlor game. Of course the rules in (8.135) would appear artificial in a

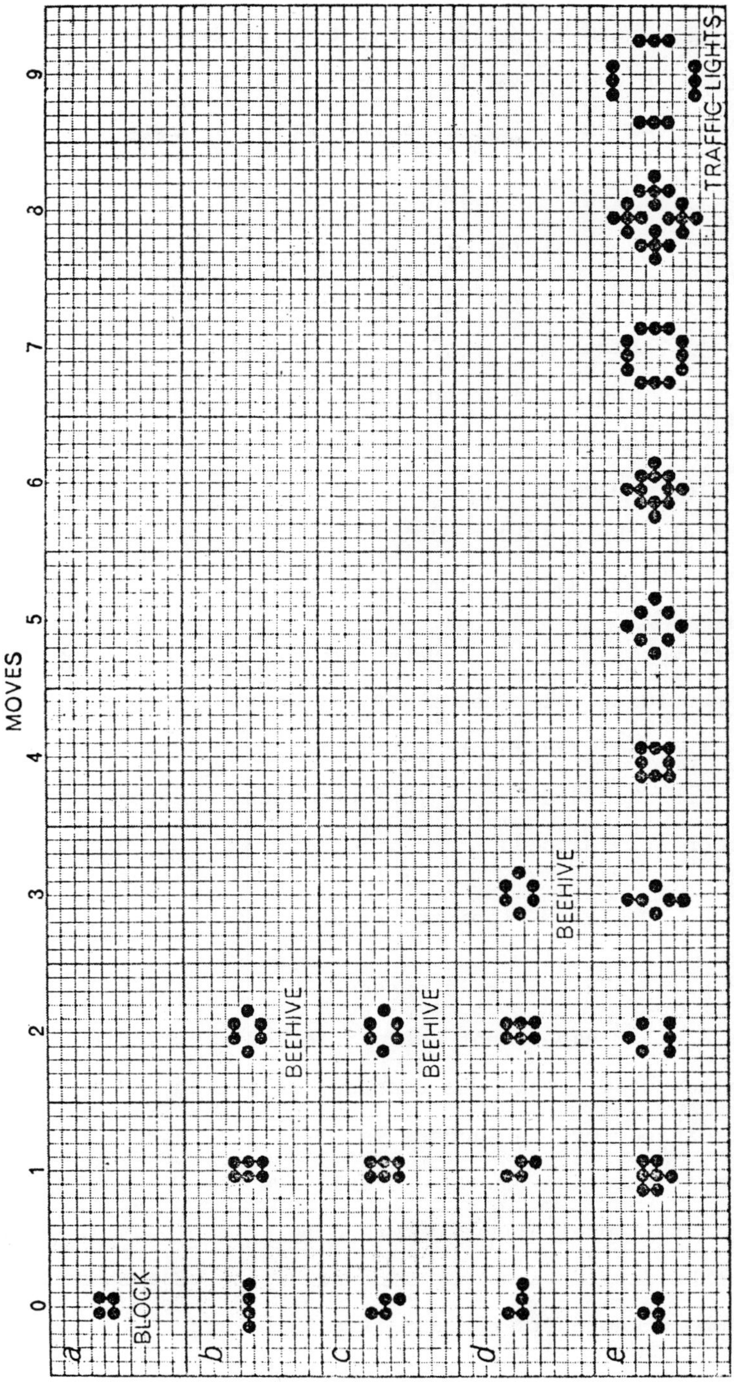
MOVES
0
1
2
3
4
5
6
7
8
9
a
BLOCK
b
BEEHIVE
c
BEEHIVE
d
BEEHIVE
e
TRAFFIC LIGHTS

biological context, but possibly alternative rules could give rise to models of promising capability. We shall look at one such model which differs from the above in that

(i) only four neighbors are counted: $(x+1,y)$ $(x-1,y), (x,y+1)$, and $(x,y-1)$

(ii) it is probabilistic with the probability of birth at $(x,y) \in I(t)$ given by $p_{n(x,y)}$, where the vector $p = (p_0,p_1,p_2,p_3,p_4)$ is constant

(iii) no decay

(iv) the previous history of $I(s)$, $s < t$, does not affect $I(t+1)$. (8.136)

In other words: I_t is a <u>Markovian stochastic process taking sets as values</u>. At time $t = 0$ we start with some simple shape I_0 and want to study how it grows.

How fast does I_t grow? We consider the behavior of the stochastic process

$$N(t) = \#\{\text{points in } I_t\} \tag{8.137}$$

and in particular how this depends upon the initial size $N(0)$. A pilot experiment was carried out by simulating $I(t)$ on the computer. This was done several times and we printed out the shapes of the successive I_t's and values of $N(t)$. We also printed out boundary statistics for the obvious reason that the model allows growth only at the boundary and in a way that depends upon the fine structure of the boundary. The numbers chosen to characterize this were

$$B(t) = \{\# \text{ points in boundary } \partial I_t\} \tag{8.138}$$

and

$$B_i(t) = \frac{1}{B(t)} \#\{\text{points in } I_t \text{ with } i \text{ connections to } I_t\}; \quad i = 1,2,3,4. \tag{8.139}$$

It was found empirically that

$$B(t) \tilde{=} G\sqrt{N(t)} \quad \text{for large } t \tag{8.140}$$

where G was close to the value 4 (this should be related to the definition of neighborhood used) and when $p = (0, 1/2, 3/4, 7/8, 15/16)$

$$\left\{\begin{array}{l} B_1(t) \cong .60 = B_1 \\ B_2(t) \cong .35 = B_2 \\ B_3(t) \cong .05 = B_3 \\ B_4(t) \cong .01 = B_4 \end{array}\right\} \text{ for large } t \tag{8.141}$$

These numbers are given only to show the typical behavior of $B_i(t)$. There is a good deal of random variation around these values, but they seemed to stabilize slowly for large values of t.

The particular choice of the <u>boundary sensitive</u> birth parameters was motivated by assuming that each neighbor can give rise to a new cell, where there is none, and each of these have probability one-half. If the events are independent and we add cells by logical OR this results in the given p-vector.

Other choices of p are of interest and have also been tried; see below.

A completely analytic treatment of the model was not attempted at this stage. Instead, it was suggested that the behavior of the boundary should be exploited by a heuristic approach as follows.

Given $N(t)$, what increase

$$\Delta N(t) = N(t+1) - N(t) \tag{8.142}$$

could we expect? If we can trust (8.140) and (8.141) we would expect approximately

$$\Delta N(t) \cong FB(t) \cong GF\sqrt{N} \tag{8.143}$$

where G characterized the global behavior of the boundary and we have used

$$F = \frac{1}{2} B_1 + \frac{3}{4} B_2 + \frac{7}{8} B_3 + \frac{15}{16} B_4 \tag{8.144}$$

as a characterization of the fine structure of ∂I_t. This would give us

$$\sum_{N(0)}^{N(t)} \Delta N/\sqrt{N} \cong GFt \tag{8.145}$$

or

$$2(\sqrt{N(t)} - \sqrt{N(0)}) \cong GFt \tag{8.146}$$

so that

$$N(t) \cong (\sqrt{N(0)} + GFt/2)^2. \tag{8.147}$$

To get a bit closer to the real behavior of $N(t)$ we approximate it by a birth process with the birth intensities

$$\lambda_k = GF\sqrt{k}. \tag{8.148}$$

This means that $N(t) = N$ if $S_N \le t$ and $S_{N+1} > t$, where

$$S_N = x_1+x_2+\cdots+x_N \tag{8.149}$$

with x_k exponentially distributed with mean values $1/\lambda_k$.

The sequence $\{x_k\}$ satisfies Liapunov's condition for the central limit theorem (see e.g. Cramér (1945), p. 215).

Indeed, the variance sum behaves like

$$\sigma_N^2 = \sum_1^N \text{const}/k \sim \text{const. } \log N \tag{8.150}$$

while the corresponding sum of absolute third moments

$$\rho_N^3 = \sum_1^N \text{const}/k^{3/2} \tag{8.151}$$

converges so that ρ/σ tends to zero as wanted. Note that the convergence is quite slow, however.

This implies that if $P_N(t)$ stands for the probability of the event $\{S_N \leq t < S_{N+1}\}$

$$P_N(t) = P\{S_N \leq t\} - P\{S_{N+1} \leq t\} \tag{8.152}$$

and

$$P\{N' < N(t) \leq N^N\} = P\{S_{N''} \leq t\} - P\{S_{N'} \leq t\}. \tag{8.153}$$

Let us choose N' and N'' such that

$$\begin{cases} (t-m_{N''})/\sigma_{N''} = A \\ (t-m_{N'})/\sigma_{N'} = -A \end{cases} \tag{8.154}$$

with

$$m_N = \sum_1^N 1/\lambda_K \cong \frac{2}{GF}\sqrt{N} \tag{8.155}$$

and A is such that $\Phi(A) = 1 - \frac{\varepsilon}{2}$. Then we can state with probability $1-\varepsilon$ that $N(t)$ falls between N' and N''. But since $\sigma_N/m_N \to 0$ it follows that

$$t/m_{N(t)} \to 1 \text{ in probability} \tag{8.156}$$

so that $N(t)$ grows approximately as the function

$$(\tfrac{1}{2}\, GFt)^2. \tag{8.157}$$

We can look at this as <u>a law of large numbers for the</u>

growth model. After this heuristic excursion we returned to the computational experiment. An example is shown in Figure 8.11 where we have plotted $N(t)$ and (8.147) using two plotting symbols. The agreement is seen to be close.

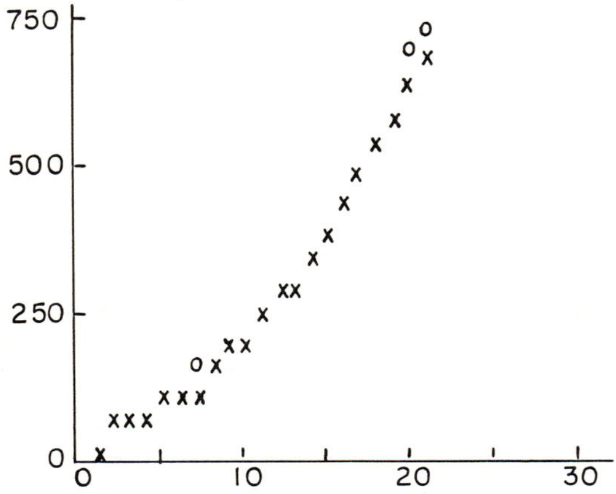

Figure 8.11

We have also looked at the shape of $I(t)$, in particular for large values of t. Initially we had conjectured that the normalized shape would tend to a circle, but this was not supported by the computational experiment. It seems to depend upon the p_i, how they vary with i.

If the p_i increase with i it can be expected that irregularities on ∂I will be filled in rapidly and that the boundary will be smooth with almost no holes in the interior.

On the other hand, if p_i decreases with i the opposite will happen: ∂I becomes irregular and holes in the interior fill in only slowly.

Two printouts are shown in Figure 8.12. Interior holes have not been shown. In (a) there were few, but in (b) a moderate number of holes. Of these (a) corresponds to the

Figure 8.12a

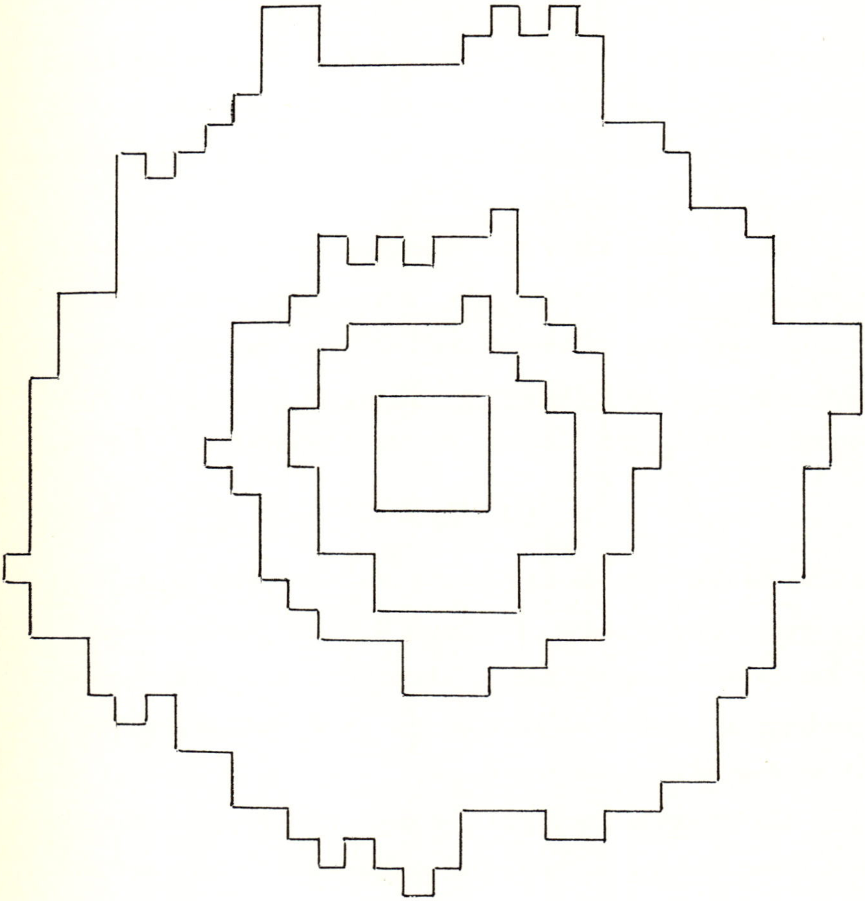

Figure 8.12b

first case and (b) to the second. Notice, in addition to what has already been said, that in the first case there is a strong tendency toward diamond shape. We do not know the reason for this but it may be due to the choice of neighborhood with only four elements in the east, north, west, and south directions. Possibly one could attempt to carry the heuristic analysis one step further by assuming that the stochastic vector

$$V(t) = \{N_1(t), N_2(t), N_3(t), N_4(t), N(t)\} \qquad (8.158)$$

forms a Markov chain, with $N_i(t) = B(t)$. This is not strictly true, but it is believed that the analysis would lead to approximations of high accuracy, since already a crude attempt leading to (8.147) was good.

Now we shall allow the configuration growth to affect not only the number but also the index of its generators. Let the generators g^α be of contrast type and represent <u>fields with rotational symmetry around its center</u> $\zeta = (\xi,\eta)$. Then we can write the field as a scalar quantity in $z = (x,y)$

$$g^\alpha(z) = \phi^\alpha(||z - \zeta||). \qquad (8.159)$$

The idea behind this is to provide the cells with <u>positional information</u> in the sense of Wolpert (1970). Assuming that the fields generated by the g's add, then this will define a gradient assumed to be felt by the individual cells, who will move in this direction.

The growth pattern will be generated by two mechanisms, one for the <u>motions</u> of cells and centers of fields, and another for <u>transitions</u> from one generator to one or several others. The second one involves change of generator index.

Consider a configuration at time t consisting of <u>neutral</u> generators g_z^1, $z \in A(t)$, and a finite number of <u>active</u> generators with generator index α_i and position $\zeta_i = \zeta_i(t)$, $i = 1,2,\ldots n$. The total field $F(z,t)$ at time t is

$$F(z,t) = \sum_{i=1}^{n} \phi^{\alpha_i}(||z-\zeta_i(t)||) \qquad (8.160)$$

and this makes a neutral generator at $z = z(t)$ move according to the law

$$\dot{z}(t) = \text{grad } F(\cdot,t). \tag{8.161}$$

The active generators on the other hand move according to

$$\dot{\zeta}_i(t) = \text{grad } F_i(\cdot,t) \tag{8.162}$$

where

$$F_i(\zeta,t) = \sum_{j\neq i} \phi^{\alpha_i\alpha_j}(||\zeta-\zeta_j||) \tag{8.163}$$

summed over all $j = 1,2,\ldots n$ different from i. A field $\phi^{\alpha\beta}$ need not be symmetric in the two generator indices α and β.

The transitions depend upon the age of the active generator. Say that the age of g^{α} is t. Then we will have a transition

$$g^{\alpha} \to g^{\beta_1}, g^{\beta_2}, \ldots g^{\beta_r} \tag{8.164}$$

at time t_{α}, where $r,\beta_1,\ldots\beta_r$ depend upon α and where the positions of the new active generators on the right hand side of (8.164) are specified relative to the position of the center of g^{α} at the time of transition. The new generators are of course of age zero at the moment of their birth.

Because of the transitions these growth patterns will exhibit regimes - intervals of time-homogeneous behavior. In contrast to the case of starshaped growth the regimes are now explicitly built into the organism.

The expression (8.163) for the cell velocities implies that

$$\text{curl } z = 0 \quad \text{and} \quad \text{div } z = \Delta F \tag{8.167}$$

so that the motion is irrotational with the local rate of

growth given by the Laplacian of the F-field. It should be noted that it may be necessary, to arrive at reasonable growth patterns, to allow the ϕ's to have singularities at small values of the argument. This is to prevent the generators from penetrating and overtaking each other.

The $\phi^{\alpha\beta}$-functions may be increasing or decreasing. The first case means that generators repel each other, the second one that they attract each other. For ϕ^{α} we have only considered increasing functions. We have also assumed that the fields have limited range so that $\phi = 0$ for large values of the argument.

Simulations of this growth model have resulted in a variety of growth patterns, a few of which are shown in Figure 8.13.

In (a) there are now transitions and the initial configuration has three generators. In (b) and (c) the initial number of generators is four and two respectively but transitions occur. The rather strange boundary behavior is caused by the fact that the ϕ-functions used were zero for arguments larger than some threshold value.

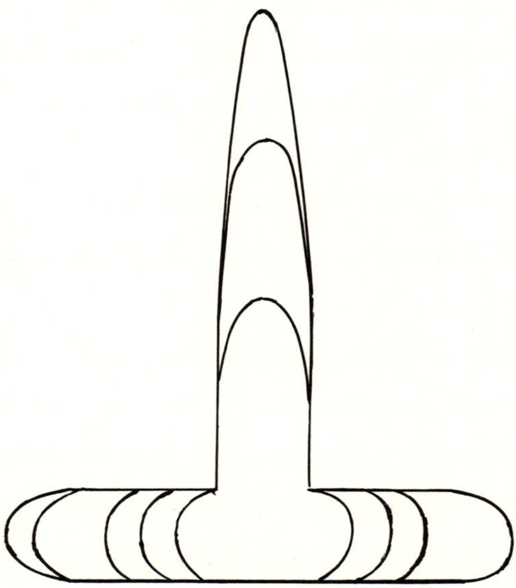

Figure 8.13a

Figure 8.13b

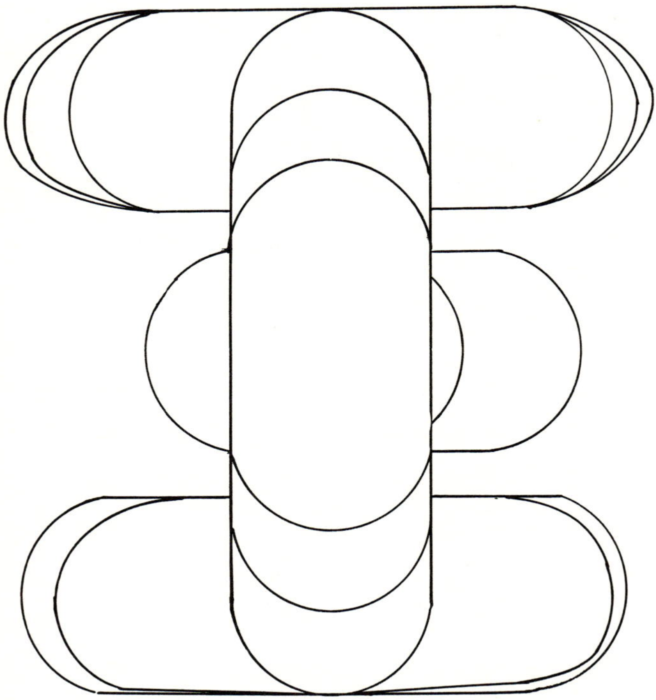

Figure 8.13c

9. Boolean Patterns

Instead of classifying pattern types according to the space by which they are supported, as in sections 4 - 8, we can consider the way the images are obtained from the generators. There is one such type that deserves to be singled out: the Boolean patterns.

Starting from a set G of (simple) features g these are combined through propositional calculus. The features themselves are binary variables taking two possible truth values, say TRUE and FALSE. These features are the generators in this section and we shall form images as certain well-formed Boolean expressions.

Given any Boolean function ϕ in terms of the f's, it can be written in the conjunctive normal form

$$\phi = \bigwedge_{b \in B} \left(\bigvee_{a \in A} f^{e_{ab}} \right). \tag{9.1}$$

The exponents take the values 1, 0, -1 with the convention $f^1 = f$, $f^{-1} = \sim f$, f^o = not occurring. Correspondingly there is a dual form: the disjunctive normal form where conjunctions and disjunctions have switched roles. This dual representation will not be explicitly mentioned below.

This fits into the pattern formalism by letting G consist of features $g = g(x)$ defined on some background space X. On X a group S of similarity transformations induces mappings $s: G \to G$ for any $s \in S$, see Section 3.3.

Generator classes G^α shall consist of pairwise comparable features: if g_1 and $g_2 \in G^\alpha$ then one implies the other, so that either $g_1 \Rightarrow g_2$ or $g_2 \Rightarrow g_1$.

The connection type Σ shall have connectivity as in Figure 9.1. Any $\sigma \in \Sigma$ shall consist of one or several isolated subgraphs, each with at most p generators and the subgraphs with full graph structure.

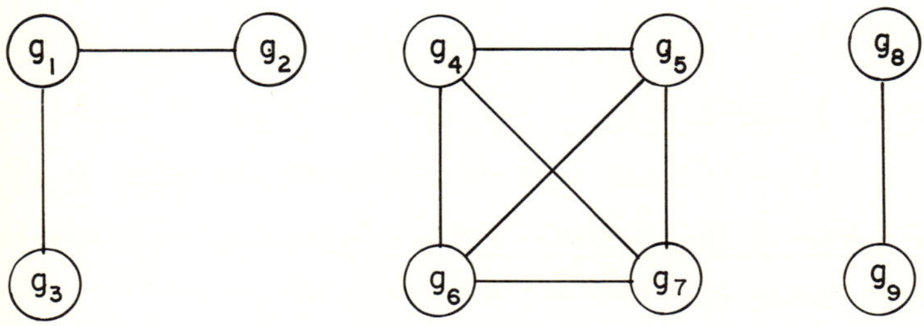

$\omega(c)$ = 7 double arrows

$n = 9$

$p = 4$

Figure 9.1

The arity of any g is $p-1$ double arrows, all its bonds are equal to the generator index α that corresponds to the generator.

The bond relation ρ will be chosen as "NOT IMPLY" to guarantee that no superfluous specification occurs in the configuration.

Let the regular configuration $c = \mathscr{C}(\mathscr{R})$ be written as $\{g_{ij};\ i = 1,2,\ldots m;\ j = 1,2,\ldots a_i\}$ with $a_i \leq p$ so that c is the combination of $c_i = \{g_{ij};\ j = 1,2,\ldots a_i\}$, each c_i with connection type FULL. We do not actually insist that the indices are denumerable later on.

Given two configurations c (as above) and $c' = \{g'_{ij};\ i = 1,2,\ldots m';\ j = 1,2,\ldots a'_i\}$, $a_i \leq p$, we shall identify them, cRc', if

$$\bigwedge_{i=1}^{m} \left(\bigvee_{j=1}^{a_i} g_{ij}(x)\right) = \bigwedge_{i=1}^{m'} \left(\bigvee_{j=1}^{a_i'} g_{ij}(x)\right), \quad \forall x \in X. \tag{9.2}$$

In other words we identify the configurations according to their logical function.

Case 9.1. (Logical feature images). The set $\mathcal{T}$ consisting of all resulting logical feature images is called a logical feature image algebra of structural complexity p over the generator space G.

We now specify the properties of G a bit further. Let us consider a family G of (simple) features g: each g is a function on X taking as values TRUE and FALSE. As usual, such a binary-valued function can be identified with its truth set in X and we shall not distinguish between these two interpretations.

If for two elements x and y from X we have $g(x) = g(y)$, $\forall g \in G$, this shall imply $x = y$. Otherwise we would have to divide X into equivalence classes in the usual way; this is assumed to have been done already.

For patterns allowing a group S of similarity transformations $s: X \to X$ we shall ask that G be closed under the transformation induced by the s's: if $f \in G$ then $sg \in \mathcal{F}$, $\forall s \in S$.

The specification space will often be finite, in which case we do not need any topology on it. Otherwise a topology will be introduced via all neighborhoods of the form

$$N = \{x \mid \bigwedge_{g \in M} g(x) = \text{FALSE}\} \tag{9.3}$$

where M is an arbitrary finite subset of G. It will be sufficient for the following that X is a regular topological space.

In the present image algebra we have many operators $I \to sI$ and a binary operator $(I_1, I_2) \to I_1 \wedge I_2$.

A measure μ will be given on G. Little will be assumed about it except that it is invariant under the mappings $f \to sf$; $s \in G$. The total μ-measure of G may be infinite or finite; in the latter case it will be normalized to 1.

Many cases exhibit some symmetry with respect to the negation operator "~". If G is closed with respect to "~" and if μ is ~-invariant some of the calculations would be simplified in the following. In discrete cases this may very well be so, but if X is not discrete, topological peculiarities would arise.

Instead the following, weaker, property will often be used. The measure μ will be said to be negation invariant if the μ-measure is the same for the sets of features satisfying $g(E) \equiv \text{TRUE}$ and $g(E) \equiv \text{FALSE}$, for any closed set $E \subset X$.

Occasionally we shall have to condition the measure μ by some constraint C on the features used. We then use the notation μ_C.

Sets in specification space X will be described in terms of the logic given through $\mathscr{F}$. The simplest descriptions we shall use have the form

$$E = \{x \mid \bigwedge_{g \in G_o} g(x) = \text{TRUE}\}, \tag{9.4}$$

where E is the given subset and G_o is some collection of features. If $g_o(x) = \text{TRUE}$ for all x in E we shall write

$$g_o(E) \equiv \text{TRUE}. \tag{9.5}$$

Now we are ready to define clustering quantitatively. Given a subset E of the background space, how can we express the clustering of it? It may be that we have some natural volume or distance in X. Occasionally invariance (S) will single out one particular measure m or one particular family of measures (or distances). Then we could use *ad hoc* criteria to express the clustering (or, rather, lack of clustering), such as

$$m(E) \text{ or diameter } (E). \tag{9.6}$$

Our point of departure will be quite different.

Principle of logical homogeneity: *A set will be said to cluster if it is possible to make many statements that are true for all points in* E *or false for all points in* E. Put another way: the set will be said to cluster if it is homogeneous under the logic introduced through the feature family G. The word "many" will get its meaning by using the fact that on G we have a measure μ. If μ is chosen so that it really expresses the importance that we attribute to the different features, we have in the above principle the embryo of an operative definition of clustering.

For a given subset E of X consider the set of all features that are identically constant in E:

$$E^G = \{g \mid g(E) \equiv \text{TRUE or } g(E) \equiv \text{FALSE}\}. \tag{9.7}$$

Definition 9.1. *The clustering of* E *is defined as*

$$\text{clust}(E) = \mu(E^G). \tag{9.8}$$

We shall also use the partial clustering measures

$$\begin{aligned} \text{clust}^T(E) &= \mu\{g \mid g(E) \equiv \text{TRUE}\} \\ \text{clust}^F(E) &= \mu\{g \mid g(E) \equiv \text{TRUE}\} \end{aligned} \tag{9.8a}$$

It will be convenient to use the complementary quantity when $\mu(G) = +\infty$ as follows:

<u>Definition 9.2</u>. <u>The size of a set</u> E <u>is defined as</u>

$$\text{size}(E) = \mu\{g \mid g(E) \not\equiv \text{TRUE nor FALSE}\}. \tag{9.9}$$

Frequently μ is negation invariant and then it is enough to compute clust^T or only clust^F and multiply by two.

Actually, one could also define size <u>in general</u> by an expression like (9.9) but only using those g's for which $g(E) \not\equiv$ TRUE. Both possibilities seem of interest and we shall have occasion to use both; which one is being used should be clear from context.

If $E_1 \subset E_2$ then $E_1^G \supset E_2^G$, so that $\text{clust}(E_1) \geq \text{clust}(E_2)$ and $\text{size}(E_1) \leq \text{size}(E_2)$. In particular, if E consists of a single point we always get $\text{clust}(S) = \mu(G)$, the largest possible value, while for $E = X$ we have $\text{clust}(X) = 0$ (this assumes that G does not contain the constant features or, at least, that they have μ-measure zero). It is obvious that size is not additive in general for disjoint sets; more about this below.

Given E and the set of features that are identically true on E,

$$E_+^G = \{g \mid g(E) \equiv \text{TRUE}\}, \tag{9.10}$$

we can ask what set in background space is described by E_+^G.

In other words, we introduce the _feature closure_

$$[G] = \{x \mid g(x) = \text{TRUE}, \ \forall g \in E_+^G\}$$
$$= \{x \mid g(S) = \text{TRUE} \Rightarrow g(x) = \text{TRUE}\}. \quad (9.11)$$

It follows directly from (9.11) that $E \subset [E]$. In case of equality we shall say that E is _feature closed_.

Theorem 9.1. If E is feature closed it is also topologically closed.

Proof: Let x be a point not belonging to E and such that all its neighborhoods N_x intersect E; we want to show that it must then also belong to E. Since E is feature closed, this is the same as saying that $x \in [E]$. If $x \notin [E]$ there must be some feature g_o such that

$$g_o(E) \equiv \text{TRUE}, \quad g_o(x) = \text{FALSE}. \quad (9.12)$$

But x has the neighborhood

$$N_x = \{x \mid g_o(x) = \text{FALSE}\} \quad (9.13)$$

and this is disjoint from E by (9.12) contrary to assumption, and a contradiction is arrived at; S must be topologically closed; as stated.

Any set described by a single feature g_o is automatically feature closed. Indeed, in such a case the set E_+^G consists of all the _implicants_ of g_o: all the g such that $g(x) = \text{TRUE}$ implies that $g_o(x) = \text{TRUE}$. Let $x \in [E]$; since $g_o(E) \equiv \text{TRUE}$ we must have $g_o(x) = \text{TRUE}$ so that $x \in E$: the set is feature closed.

Consider now an arbitrary intersection of feature closed sets E_a

$$E = \bigcap_{a \in A} E_a. \tag{9.14}$$

Then E is also feature closed $[E] = E$. To see this assume the opposite so that there is a point $x \in [E]$, $x \notin E$. Then there must be an <u>a</u> such that $x \notin E_a = [E_a]$ so that there exists a g_o such that $g_o(E_a) \equiv \text{TRUE}$, $g_o(x) = \text{FALSE}$. But since $x \in [E]$ it follows that for any g with $g(E) \equiv \text{TRUE}$, we must have $g(x) = \text{TRUE}$. But $E \subseteq E_a$ so that $g_o(E) \equiv \text{TRUE}$, implying $g_o(x) = \text{TRUE}$, which contradiction proves the assertion.

Hence conjunctions of simple features form feature closed sets. This gives us

<u>Theorem 9.2</u>. <u>A set in</u> X <u>is feature closed if and only if it can be written as</u>

$$\{x \mid g(x) = \text{TRUE}, \ \forall g \in G_o\}. \tag{9.15}$$

<u>where</u> $G_o \subset G$.

This means that feature closed is an idempotent operator. Indeed since $[E]$ can be written as a conjunction of simple features it is clear that it is closed so that $[[E]] = [E]$, the idempotent property.

<u>Remark 9.1</u>. If size is defined in terms of identically true features, then the size of a set E is equal to the size of the feature closed set $[E]$. This follows since the set of identically true features is the same for E and $[E]$. Also, if μ is negation invariant we get

$$\text{size}(E) = 1 - \mu\{g|g(E) \equiv \text{TRUE}\} - \mu\{g|g(E) \equiv \text{FALSE}\}$$
$$= 1 - 2\mu\{g|g(E) \equiv \text{TRUE}\} = \text{size}([E]) \tag{9.15a}$$

Remark 9.2. If a volume m is given in the background space, one is naturally led to ask what sets of given m-volume exhibit maximum clustering; more about this later; this leads to a sort of isoperimetric question.

Remark 9.3. Sometimes we will introduce the feature logic in a product space, say $X = Y \times Y$. It is then of interest to study the relations between size (and cluster) in the two marginals Y and in the product X. The following describes two possibilities.

Given a set $E \subset X$ we shall denote its projection upon the two marginal spaces by E_1 and E_2 respectively. Given the values of $\text{size}(E_1)$, $\text{size}(E_2)$, what can we say about the value of $\text{size}(E)$? The answer depends upon what μ-measure we introduce in G in terms of the measures μ_1 and μ_2 we have in G_Y.

Theorem 9.3. a) Let $G = G_1 \cup G_2$ where G_1 consists of all features $g(y_1)$ and G_2 of all features $g(y_2)$, $g \in G$. If μ is simply given by μ on G_1 and on G_2 then

$$\text{size}(E) = \text{size}(E_1) + \text{size}(E_2). \tag{9.16}$$

b) Assume $\mu(G) = 1$ and let G consist of all features $g(x)$ = truth value of $\{g_1(y_1) = g_2(y_2)\}$; $g_1, g_2 \in G$. If $\mu(dg) = \mu(dg_1)\mu(dg_2)$ then

$$\text{size}(E) \le \text{size}(E_1) + \text{size}(E_2) - \text{size}(E_1)\cdot\text{size}(E_2). \tag{9.17}$$

Proof of (a): The features from G that cut E form a set

$$\{g \in G_1 | g \text{ cuts } E\} \cup \{g \in G_2 | g \text{ cuts } E\}. \tag{9.18}$$

But in order that an $g(x)$ from G_1 cut E it is necessary and sufficient that $g(y_1)$ cut E_1, since the feature does not depend upon the other coordinate. Similarly for the second term, and since the terms in (9.18) are disjoint the relation (9.16) follows.

Proof of (b): Assume for the moment that E is a product set, $E = E_1 \times E_2$. In order that a feature g cut E there must be points $x = (y_1, y_2)$ and $x' = (y_1', y_2')$ such that $g(x) \neq g(x')$. But this means that

$$\begin{aligned} &\text{truth value } \{g_1(y_1) = g_2(y_2)\} \\ \neq\ &\text{truth value } \{g_1(y_1') = g_2(y_2')\}. \end{aligned} \tag{9.19}$$

But this is the same as saying that either

$$\begin{cases} g_1(y_1) = g_2(y_2) \\ g_1(y_1') \neq g_2(y_2') \end{cases} \tag{9.20}$$

or

$$\begin{cases} g_1(y_1) \neq g_2(y_2) \\ g_1(y_2') = g_2(y_2') \end{cases} . \tag{9.21}$$

Both (9.20) and (9.21) say that either g_1 separates y_1 from y_1' or g_2 separates y_2 from y_2'. Hence either g_1 cuts E_1 or g_2 cuts E_2. On the other hand, say that g_1 cuts E_1. Then obviously g = truth value $\{g_1(y_1) = g_2(y_2)\}$ cuts E (remember that E is a product set). To calculate size(E) we just have to get the μ-measure of the event that

g_1 or g_2 cut E_1 or E_2 respectively, so that

$$\text{size}(E) = \text{size}(E_1) + \text{size}(E_2) - \text{size}(E_1)\cdot\text{size}(E_2) \quad (9.22)$$

where we have used the normalization of μ-measure, $\mu(G) = 1$. To get rid of the assumption that E be a product set, we note that in general $E \subset E_1 \times E_2$ which, together with the observation above that size is monotonic, gives us the inequality (9.17).

Remark 9.4. We always have $[A \cup B] \supset [A] \cup [B]$, as the reader may verify.

To familiarize ourselves with the concepts introduced above, we shall examine a few examples.

Example 1. The simplest one is where background space is completely unstructured. Say it is finite, $X = \{1,2,3,\ldots,n\}$ and that our logic is given through all features of the form: x belongs to a certain subset. Such a feature is determined by the vector $\phi = (\phi_1,\phi_2,\ldots,\phi_n)$ where ϕ_i is its indicator function. How many g's (i.e., we take μ as counting measure) are identically true or false for a given non-empty set E with ν elements? Well, we must have $\phi_i = 1$, $i \in E$; the remaining ϕ's are arbitrary so that we get, if we leave out the two trivial features,

$$\text{clust}(E) = 2^{n-\nu+1} - 2, \quad (9.23)$$

or normalizing $\mu(G) = 1$

$$\text{clust}(E) = \frac{2^{n-\nu+1}-2}{2^n-2} \quad (9.24)$$

Here every set is trivially feature closed.

Example 2. The above logic is too discriminating to be of interest. We therefore modify the setup slightly to exhibit a more special case with more structure. Let the logic be determined by the set G consisting of the features $g_1, g_2, \ldots, g_{n-\ell+1}$, where

$$g_i(x) = \text{TRUE if } x \neq i, i+1, i+2, \ldots, i+\ell-1 \tag{9.25}$$

and ℓ is a fixed natural number less than n. G separates the points of the background space. Given a subset $E \subset X$, how is the closure formed? Look at Figure 9.2 where $n = 18$, $\ell = 3$, $E = \{3,4,6,10,11,14\}$. The set E_+^G consists of the features g_7, g_{15}, g_{16}, and the corresponding sets of the form (2.4) are indicated in the figure. This leads to the closure $[E] = \{1,2,3,4,5,6,10,11,12,13,14\}$. It is obvious that the closure operation in this case consists simply of filling in gaps of length less than ℓ. To determine clust(E) we just have to find the sum of the lengths of the remaining gaps minus twice the number of such gaps. The interested reader is advised to look at the corresponding case of a two-dimensional lattice where the intervals $(i, i+1, i+2, \ldots, i+\ell-1)$ are replaced by all the translates of a given fixed set. Again, closure leads to a regularization or cleaning of the original picture.

Example 3. Let $X = [0,1]$ and with the features

$$\begin{cases} g_\xi^+(x) = \text{TRUE iff } \xi \leq x \leq 1 \\ g_\xi^-(x) = \text{TRUE iff } 0 \leq x \leq \xi \end{cases} \tag{9.27}$$

G is not closed under negation. The measure μ is defined by

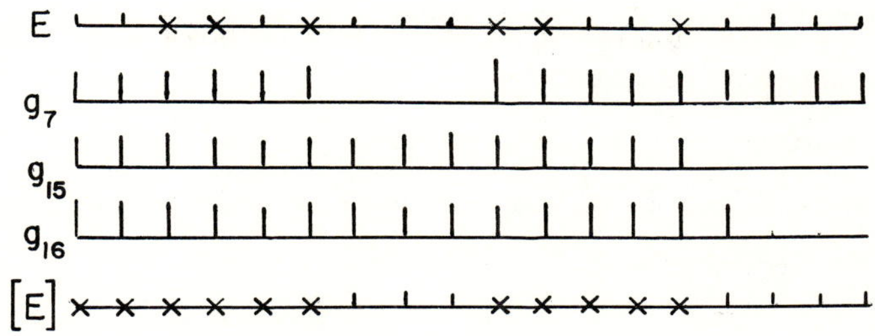

$$\text{Clust}(E) = \frac{3}{16}$$

Figure 9.2

$$P\{g_\xi^+ \mid \xi \le u\} = P\{g_\xi^- \mid \xi \le u\} = \frac{u}{2} \tag{9.28}$$

so that ξ is uniformly distributed as well as the value of the two-valued superscript. It is negation invariant.

Any conjunction of g's determines a closed interval in X so that feature closed sets in X are intervals. The feature closure of a set is the circumscribed interval.

To find what size means here it is enough to assume that E is feature closed = an interval $[a,b]$. Then in order that g_ξ^+ or g_ξ^- cut E it is necessary and sufficient that $a \le \xi \le b$, so that $\text{size}(E) = b-a$ = length, or for a general closed set

$$\text{size}(E) = \text{length of smallest circumscribed interval.} \tag{9.29}$$

Example 4. In our fourth example we take the Euclidean plane R^2 as our specification space and consider all features of the form

$$g(x) = \text{TRUE} \Leftrightarrow \ell(x) \ge 0 \tag{9.30}$$

where $\ell(x)$ is an arbitrary linear expression

$$\ell(x) = ax + by + c. \tag{9.31}$$

In other words, we use a <u>linear threshold logic</u>. For such a feature g the set of the form $\{x|g(x) = \text{TRUE}\}$ is a closed half plane which can be identified with a directed line. It is obvious that a set is feature-closed in this logic if and only if it is (topologically) closed and convex.

What is the size of a set E in this feature logic? If we want the model to be invariant with respect to S = the group of Euclidean transformations of the plane, the measure μ is essentially unique, as is well known in integral geometry (see e.g., Blaschke (1955)). To calculate the size of E we just have to find the μ-measure of the set of directed lines that intersect E. But this is also a classical result in integral geometry: it is proportional to the length of the boundary of the convex closure of E.

The reader may be interested in examining a modification of the last case when the features have the form

$$g(x) = \text{TRUE} \equiv x \in C(a,b,r) \tag{9.32}$$

where $C(a,b,r)$ is an open circle whose center has the coordinates a,b (arbitrary) and whose radius is r (fixed); or where the circle is replaced by translates of other geometric shapes.

<u>Example 5</u>. Let X be the unit square $0 \leq x,y \leq 1$ in the plane with the features

$$\begin{cases} g_\xi^{1+}(x,y) = \text{TRUE iff } \xi \le x \le 1 \\ g_\xi^{1-}(x,y) = \text{TRUE iff } 0 \le x \le \xi \\ g_\eta^{2+}(x,y) = \text{TRUE iff } \eta \le y \le 1 \\ g_\eta^{2-}(x,y) = \text{TRUE iff } 0 \le y \le \eta \end{cases} \tag{9.33}$$

and with the negation invariant μ measure determined by having the superscripts and subscript uniformly distributed over their respective ranges. The superscripts are the generator indices.

Any conjunction of g's determines a rectangle (here understood to have sides parallel to the coordinate axes) so that feature closed sets are just rectangles. The feature closure of a set is formed by taking the smallest circumscribed rectangle.

A feature g_ξ^{1+} or g_ξ^{1-} cuts E iff E has a point (ξ,y) and similarly for the other features. Hence

$$\text{size}(E) = \frac{1}{4}\text{ perimeter of the circumscribed rectangle.} \tag{9.34}$$

Let E_1 and E_2 be two subsets of background space that we want to separate using that particular logic G that happens to be available to us. In the same spirit as above we are led to define the separation as the number of features that have one truth value on E_1 and the opposite one on E_2, or, more formally,

Definition 9.4. *By the separation of two sets* E_1 *and* E_2 *we mean the number*

$$\text{sep}(E_1,E_2) = \mu\{g|g \text{ identically constant on } E_1 \text{ and on } E_2 \text{ and } g(E_1) \ne g(E_2)\}. \tag{9.35}$$

The following statements hold:

1. $\text{sep}(E_1,E_2) \geq 0$; $\text{sep}(E_1,E_2) = 0$ if the two sets intersect but not necessarily the converse; $\text{sep}(E_1,E_2) = \text{sep}(E_2,E_1)$.

2. If $E_1' \subseteq E_1$ and $E_2' \subseteq E_2$ then $\text{sep}(E_1',E_2') \geq \text{sep}(E_1,E_2)$.

3. The triangle inequality does not necessarily hold for separation.

Statement 1 is obvious. To see that 2 holds we notice that if g is identically constant on E_1 and E_2 with different truth values this is also true on E_1' and E_2' respectively. Hence the g-set appearing in (9.35) is not smaller for the E_1' and E_2' and their separation is at least as large as the separation between E_1 and E_2. It will be seen later that statement 3 is true.

In Example 3 $\text{sep}(E_1,E_2)$ means simply the length of the greatest interval such that E_1 is on one side of it and E_2 on the other.

It will be instructive to see what separation means in the fourth example above. We can let E_1 and E_2 be closed convex sets (see Figure 9.3). It is then a direct consequence of a classical result in integral geometry (see Blaschke (1955, p. 16)) that

$$\begin{aligned}\text{sep}(E_1,E_2) &= \text{length of } AB + \text{length of } CD \\ &\quad - \text{length of } CA - \text{length of } BD\end{aligned} \tag{9.36}$$

where AB and CD are the two common tangents (support lines) separating E_1 and E_2 and CA and BD are the arcs indicated along the boundaries of E_1 and E_2.

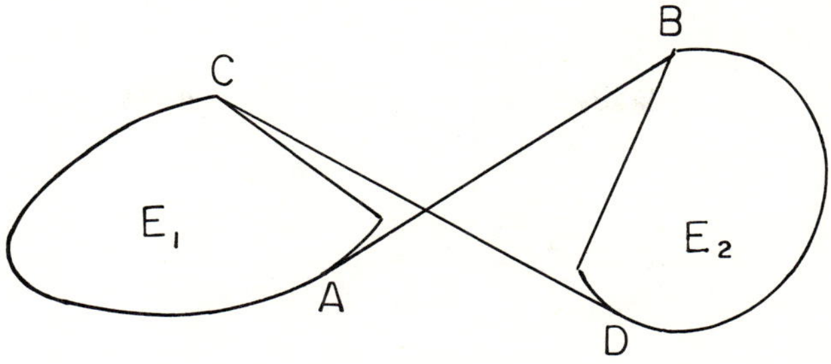

Figure 9.3

Returning to the general case, let g be one of the features and consider its truth values on E_1 and E_2. We have to distinguish 9 disjoint possible cases (see Table 9.1). A feature in case 1 is uniformly true on both E_1 and E_2 while one in case 3 is uniformly true on E_1 but neither uniformly true nor uniformly false on E_2; it <u>cuts</u> E_2. Note the last case, number 9, of features that cut both E_1 and E_2; the μ-measure of class 9 will be denoted $\text{cut}(E_1 \& E_2)$.

We now combine these cases as follows.

(a) features that cut $E_1 \cup E_2 = (2)+(3)+(4)+(6)+(7)+(8)+(9)$

(b) features that cut $E_1 = (7)+(8)+(9)$

(c) features that cut $E_2 = (3)+(6)+(9)$

(d) features that cut both E_1 and $E_2 = (9)$

(e) features that separate E_1 and $E_2 = (2)+(4)$.

Using the additive property of μ-measure and Definition 9.2, we get

<u>Theorem 9.4</u>. <u>For two sets E_1 and E_2 of finite size and finite separation we have</u>

$$\text{size}(E_1 \cup E_2) = \text{size}(E_1) + \text{size}(E_2) + \text{sep}(E_1,E_2) - \text{cut}(E_1 \& E_2). \tag{9.37}$$

case	E_1	E_2
(1)	TRUE	TRUE
(2)	TRUE	FALSE
(3)	TRUE	TRUE and FALSE
(4)	FALSE	TRUE
(5)	FALSE	FALSE
(6)	FALSE	TRUE and FALSE
(7)	TRUE and FALSE	TRUE
(8)	TRUE and FALSE	FALSE
(9)	TRUE and FALSE	TRUE and FALSE

Table 9.1

Remark 1. Let E_1 and E_2 be two congruent and parallel rectangles in the plane and use the linear threshold logic of the fourth example. If the rectangles touch each other the term $\text{sep}(E_1,E_2)$ in (9.30) vanishes but not $\text{cut}(E_1 \& E_2)$. On the other hand, if the rectangles are far apart the difference $\text{sep}(E_1,E_2)-\text{cut}(E_1 \& E_2)$, which is easily calculated, becomes large. Therefore the value of $\text{size}(E_1 \cup E_2)$ can be smaller, equal to or larger than the sum $\text{size}(E_1) + \text{size}(E_2)$.

Remark 9.5. If C stands for the constraint that a feature g cut E_1, then we can write

$$\text{cut}(E_1 \& E_2) = \text{size}(E_1)\cdot\text{size}(E_2|E_1) \tag{9.38}$$

where the last factor has been calculated using the conditional measure μ_C. The separation of two clusters can therefore be expressed in terms of four calculations of size (see (9.31)).

So far we have completely neglected an important aspect of clusters, almost always present in practical cluster analysis. This is <u>the aspect of frequency</u>: all the members of the cluster do not appear with the same frequency. While some of the peripheral members of E may be very rare, others forming the core of E may occur with high frequency. It is often this circumstance that makes it possible to achieve meaningful separation of overlapping clusters and we must take this possibility into account.

We shall now see how what has been said above can be modified to take frequency into account and deal with diffuse Boolean patterns. When defining cluster (and size) we can no longer ask for the set of all features that are uniformly true; that set may be empty even in non-trivial situations. Instead, we will have to be satisfied if the feature is true with high probability.

<u>Definition 9.5. Given a probability distribution P over background space X and a number α between 0 and 1, by the feature closure of P on the level α we shall understand the subset of X</u>

$$[P]^{\alpha} = \{x \mid P\{g = \text{TRUE}\} \geq \alpha \Rightarrow g(x) = \text{TRUE}\}. \qquad (9.39)$$

<u>The cluster of P on level α is</u> $\mu\{f \mid P(g = \text{TRUE}) \geq \alpha\}$.

Obviously $[P]^{\alpha}$ is automatically a feature closed subset of X. The values of α that we are likely to be concerned with are close to 1; we may ask, e.g., for the 95%

closure of the cluster, and so on. Note that $[P]^{\alpha}$ may well be empty.

It will be instructive to see what this concept reduces to in the example above.

Consider example 1 and let P be defined <u>via</u> the probabilities $p_1, p_2, \ldots, p_n$ associated with the n members of X. A feature g, given as a subset $\phi \subset X$, satisfies the relation in (9.39) if and only if

$$\sum_{i=1}^{n} \phi_i p_i \geq \alpha. \tag{9.40}$$

But this implies that an element i will be excluded from $[P]^{\alpha}$ if and only if $p_i < 1-\alpha$, so that in this unstructured case we have simply

$$[P]^{\alpha} = \{i \mid p_i \geq 1 - \alpha\}, \tag{9.41}$$

and

$$-\log_2 \{\text{cluster}_{\alpha}(P)\} = \#\{i \mid p_i \geq 1-\alpha\}. \tag{9.42}$$

Turning to example 2, we see that a feature $g_i(x)$ satisfies the relation (9.39) if

$$\sum_{\nu=i}^{i+\ell-1} p_{\nu} < 1-\alpha. \tag{9.43}$$

Therefore the α-level closure of P is obtained by removing all gaps of length less than ℓ that contain less than $1-\alpha$ of the probability mass.

To examine example 4, let P be a bivariate normal distribution in $X = R^2$. Since features are now given in terms of half spaces we can subject X to any non-singular linear transformation. In this way we reduce P to a bivariate normal probability distribution P' centered at the origin, with standard deviation 1 and correlation coefficient 0. But in this rotation-symmetric case $[P']$ is simply a closed circle. Transforming P' back to P we get, if $\alpha > 1/2$,

$$[P]^{\alpha} = \text{concentration ellipse containing the mass } 1-e^{-r^2} \tag{9.44}$$

where $\Phi(r) = \alpha$, and, if $\alpha < 1/2$,

$$[P]^{\alpha} = \text{empty set.} \tag{9.45}$$

The features should have the property that they separate points. For a probability measure P the following question was posed by R. Vitale. If we know $P(g(x) = \text{TRUE}) = P(g)$ for all g, does this determine P?

Let us look at Example 4. Then we would know the mass in any closed half plane. But then it is known (see Cramér-Wold (1936)) that P is uniquely determined.

On the other hand, if we define features by

$$g^{xy}(x) = \text{TRUE} \quad \text{iff} \quad x \in E_{xy} \tag{9.46}$$

where E_{xy} is the translation (x,y) of a given fixed Borel set E, does $P(g^{xy})$ determine P if it is given for all x,y?

A partial answer can be given as follows. Assume E to be closed and bounded. Then

$$P(g^{xy}) = m(E) \iint \phi(x-u,y-v)P(du,dv) \tag{9.47}$$

where

$$\phi(\xi,\eta) = \frac{1}{m(E)} \times \text{indicator function of } -E \tag{9.48}$$

so that ϕ is a frequency function with compact support. We can then calculate the Fourier transform

$$\iint e^{i(tx+sy)} P(g^{xy})dxdy = \hat{\phi} \cdot \hat{P}. \tag{9.49}$$

However, $\hat{\phi}$ is an entire function so that its zeroes cannot cluster. That implies that we can divide in the above equation by $\hat{\phi}$ except at isolated points and $\hat{P}$ is uniquely determined by continuity, and also P.

We have not studied the case when E is not bounded. Note that the half plane case corresponds to the last one only if rotations are included in S.

The descriptions of clusters used up till now were all expressed *via* a number of features from G combined by the logical "and" operation. If the clusters cannot be described in this way with sufficient accuracy, we must use more powerful descriptions. This will be done successively in a number of steps, the first of which is the following.

Definition 9.6. *By a feature logic of the second order we mean all descriptions of the form*

$$\bigwedge_{a,b} (g_a \vee g_b) \tag{9.50}$$

where

$$g_a, g_b \in G.$$

To avoid superfluous specification we would delete the appearance of one of the generators when g_a and g_b belong to the same generator class.

This means that we replace our features G by the set G_2 consisting of all features of the form $g_a \vee g_b$. We can then ask for the feature closure under G_2 of a set $E \subset X$. This 2-closure will be denoted by $[E]_2$.

Similarly, we introduce the second-order quantities $size_2(E)$ and $cluster_2(E)$. For simplicity we shall assume μ to be normalized, $\mu(G) = 1$. Here some caution is needed to get convincing and logically consistent definitions. Now we consider a pair h of features $h = (g_a, g_b)$ from the product space $G \times G$ on which we have the product measure,

$$\mu_2(dh) = \mu(dg_a) \cdot \mu(dg_b). \tag{9.51}$$

Note that if $h = g_a \vee g_b$ cuts E then either g_a or g_b (or both) must also cut E since otherwise h would be identically constant on E. Hence

$$\begin{aligned} size_2(E) = {} & \mu_2\{h \text{ and } g_a \text{ cut } E \text{ but not } g_b\} \\ & + \mu_2\{h \text{ and } g_b \text{ cut } E \text{ but not } g_a\} \\ & + \mu_2\{h \text{ and } g_a \text{ and } g_b \text{ cut } E\}. \end{aligned} \tag{9.52}$$

The three sets appearing on the right hand side of (9.52) can be rewritten as follows. The first set has g_b identically constant, but it could not be identically true since the h would also be identically true and not cut E. Similarly in the second set $g_a(E) \equiv \text{FALSE}$, so that using the fact that μ_2 is a product measure over G_2

$$\begin{aligned} size_2(E) = {} & 2\mu\{g_b(E) \equiv \text{FALSE}\}\mu\{g_a \text{ cuts } E\} \\ & + \mu_2\{h \text{ and } g_a \text{ and } g_b \text{ cut } E\} \end{aligned} \tag{9.53}$$

and, if μ is negation invariant,

$$\begin{aligned} size_2(E) = {} & [1 - \mu\{g_b \text{ cuts } E\}]\mu\{g_a \text{ cuts } E\} \\ & + \mu_2\{h \text{ and } g_a \text{ and } g_b \text{ cut } E\}. \end{aligned} \tag{9.54}$$

Rearranging (9.54) we get

$$\text{size}_2(E) = \text{size}(E) - \mu_2\{g_a \text{ and } g_b \text{ cut } E \text{ but not } h\}. \tag{9.55}$$

This shows the inequality $\text{size}_2(E) \leq \text{size}(E)$ so that

<u>Theorem 9.5</u>. <u>We always have</u>

$$E \subset [E]_2 \subset [E] \tag{9.56}$$

<u>and if</u> μ <u>is negation invariant</u>,

$$\text{size}_2(E) \leq \text{size}(E). \tag{9.57}$$

On the other hand if we explore E using only probes g_a and g_b that cut E we would get a different measure

$$\text{size}_2'(E) = \mu_2\{(g_a, g_b) \mid g_a, g_b \text{ and } g_a \vee g_b \text{ cut } E\} \tag{9.58}$$

which is not larger than $\text{size}_2(E)$.

To make the discussion concrete, let us take a look at example 2; what is G_2 and 2-closure then? If we combine the features g_i and g_j from G by "or", we will get the feature $g = g_i \vee g_j$

$$\begin{aligned} g(x) = \text{TRUE if } & x \notin i, i+1, \ldots, i+\ell-1 \\ \text{or if } & x \notin j, j+1, \ldots, j+\ell-1. \end{aligned} \tag{9.59}$$

But this implies that

$$g(x) = \text{TRUE if } x \notin \{i, i+1, \ldots, i+\ell-1\} \cap \{j, j+1, \ldots, j+\ell-1\} \tag{9.60}$$

so that g corresponds to the intersection of the (discrete) intervals we started from. In particular, we then have the feature, for any h,

$$g(x) = \text{TRUE if } x \neq h. \tag{9.61}$$

But it follows from (9.61) that we can then separate any point $k \in E$ from E by a feature in G_2: all sets are 2-closed in this special case.

What happens in example 3? It is enough to notice that a disjunction of two features determines a set in X consisting of one or two closed intervals or is the empty set. The 2-closed sets then turn out to consist of topologically closed but otherwise arbitrary sets (if only finite configurations are allowed we get finite unions of closed intervals) and 2-feature closure coincides with topological closure.

Say now that E is the union of the intervals $i_k = [a_k, b_k]$, $k = 1,2,\ldots n$, assumed to be disjoint and ordered in increasing order on $[0,1]$. We can then compute $\text{size}_2(E)$ which turns out to be a function of two quantities

$$\begin{cases} \text{circumscribed length} = b_n - a, \\ \text{sum of gaps} = \sum_{k=1}^{n-1} (a_{k+1} - b_k). \end{cases} \tag{9.62}$$

On the other hand to compute $\text{size}_2'(E)$ say that $g_1 \& g_2$ cut E and $(g_1 \vee g_2)(E) \equiv \text{TRUE}$. If g_1 and g_2 come from the same generator class $g_1 \vee g_2 = g_1$ or $= g_2$ which is impossible. If they come from different generator classes, say $g_1 = g_{x_1}^{+}$ and $g_2 = g_{x_2}^{-}$ we must have either $a_1 \leq x_1 \leq x_2 \leq b_n$ or $a_1 \leq x_2 < x_1 \leq b_n$ with the condition that the interval (x_2, x_1) falls in one of the gaps. Hence

$$\begin{aligned} &\mu_2\{g_1 \& g_2 \text{ cut } E, (g_1 \vee g_2)(E) \equiv \text{TRUE}\} = \\ &= \frac{1}{2}\left[\frac{(b_n - a_1)^2}{2} + \frac{1}{2}\sum_{k=1}^{n-1} (a_{k+1} - b_k)^2\right]. \end{aligned} \tag{9.63}$$

If $g_1 \& g_2$ cut E and $(g_1 \vee g_2)(E) \equiv$ FALSE we find with a similar reasoning that this corresponds to the μ_2-measure zero. Hence

$$\begin{aligned} \text{size}_2'(E) &= \mu_2\{g_1 \& g_2 \text{ cut } E\} \\ &\quad - \mu_2\{g_1 \& g_2 \text{ cut, } (g_1 \vee g_2)(E) \equiv \text{constant}\} \qquad (9.64) \\ &= \frac{3}{4}(b_n - a_1)^2 - \frac{1}{4}\sum_{k=1}^{n-1}(a_{k+1} - b_k)^2 \end{aligned}$$

with the sudden appearance of Hausdorff measure.

Turning to Example 5 it is clear that a disjunction of two features will give rise to a set as the shaded ones in Figure 9.4. In (a) and (b) the two generators combine to a single one modulo R.

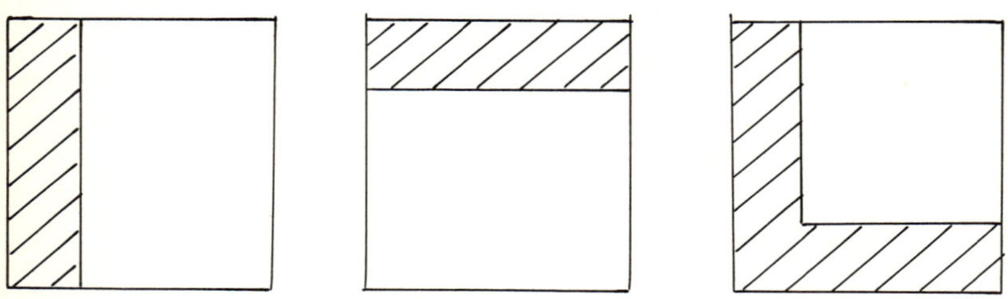

(a) (b) (c)

Figure 9.4

In (c) they do not reduce to a single one modulo R.

Combining such 2-features we get sets as in Figure 9.5 that have been discussed in Section 5.

To find $\text{size}_2'(E)$ in this example consider the set shown in Figure 9.6 where $[a_1,b_1]$, $[a_2,b_2]$ are the projections of E on the X_1 and X_2 axes respectively. The

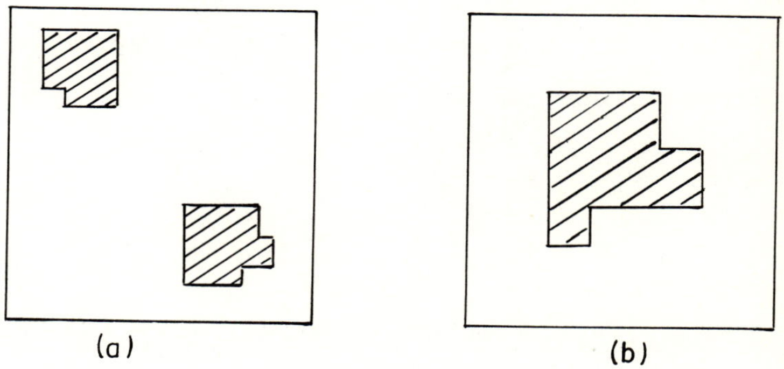

Figure 9.5

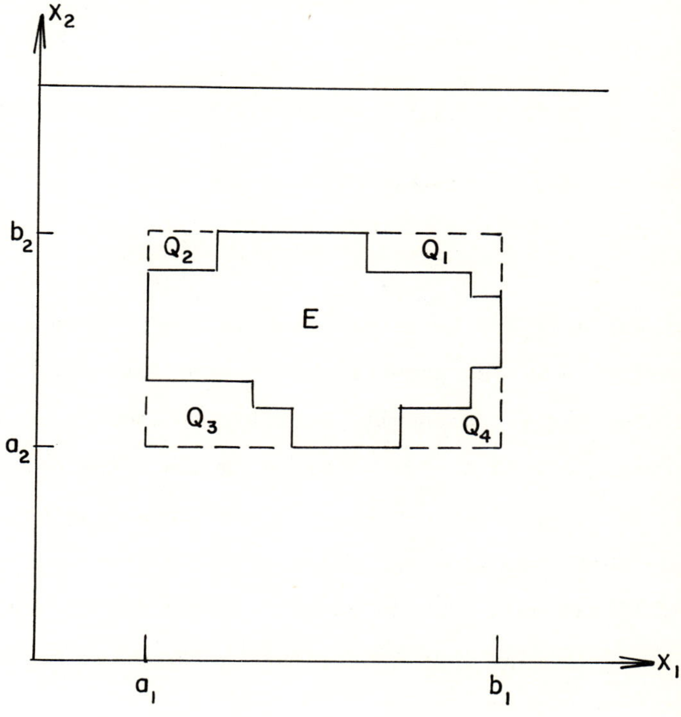

Figure 9.6

corresponding lengths are denoted by $\ell_1 = b_1 - a_1$ and $\ell_2 = b_2 - a_2$.

Now if we want to find two features g_1 and g_2 such that they both cut E and also $h = g_1 \vee g_2$ we separate cases. First consider the case with

$$\begin{cases} g_1 = \xi_{\xi_1}^{1+} \\ g_2 = \xi_{\xi_2}^{2+} \end{cases} \tag{9.65}$$

The truth set of the disjunction consists of three quadrant pieces joined together and in order that $h = g_1 \vee g_2$ also cut E we must have the point (ξ_1, ξ_2) in the set $E \cup Q_2 \cup Q_3 \cup Q_4$ with the μ_2-measure

$$\frac{1}{8}\,[m(E) + m(Q_2) + m(Q_3) + m(Q_4)] \tag{9.66}$$

Combining all possibilities we get

$$\text{size}_2'(E) = \frac{3}{8}[m(Q_1)+m(Q_2)+m(Q_3)+m(Q_4)] + \frac{1}{2}\,m(E). \tag{9.67}$$

Note that while in the 1-feature case we were led to length, the perimeter, now area appears in the size-measure.

Higher-order logics will be able to describe in a more detailed way the sets of specification space: the closures $[E]_1, [E]_2, [E]_3, \ldots$ will form a non-increasing sequence. It may happen that this sequence converges in a finite number of steps. This has been seen in example 2; the reader may be interested in looking into this question for other feature logics for plane sets.

An image I that is feature-closed of order p can be written as a conjunction of (at most) p disjunctions.

$$I = \bigwedge_{\alpha=(\alpha_1,\alpha_2,\ldots,\alpha_p)\in A} (g_{\alpha_1} \vee g_{\alpha_2} \vee \cdots \vee g_{\alpha_p}) = \bigwedge_{\alpha} g_\alpha. \tag{9.68}$$

It is natural to use as a measure of its <u>numerical complexity</u> (see section 4.1) the power of the index set A over which the conjunction ranges, and to use p as a measure of its structural complexity. The disjunctions $g_\alpha = g_{\alpha_1} \vee g_{\alpha_2} \vee \ldots \vee g_{\alpha_p}$ are called (composite) <u>features of order</u> p.

With the linear threshold logic the set E_1 in Figure 9.7 is 2-closed but not E_2, while $[E_2] = E_3$.

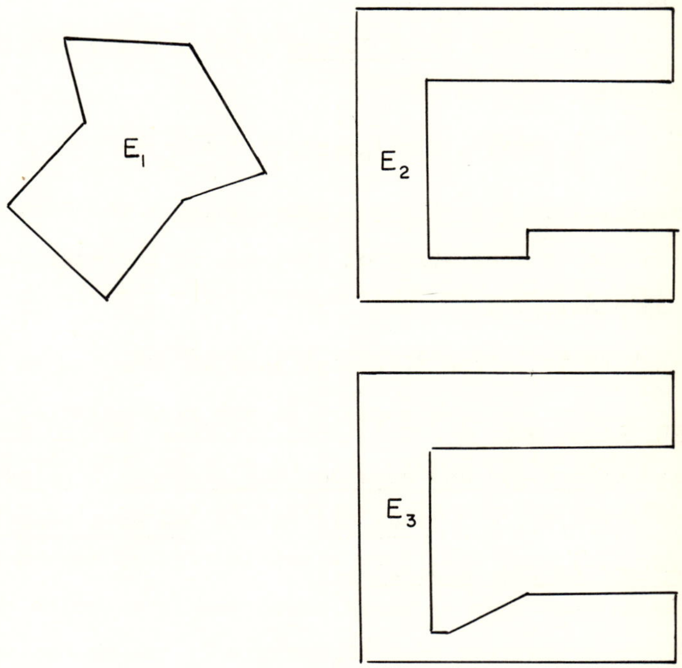

Figure 9.7

<u>Theorem 9.6. The p-closure operations satisfy the cancellation law</u>

$$[[E]_r]_s = [E]_t, \quad t = \min(r,s). \tag{9.69}$$

<u>Proof</u>: If $r = s$ the relation is obvious since p-closure is an idempotent operation. If $r < s$ the set $[E_r]$ can be written as a conjunction of disjunctions with at most r simple features. It is then s-feature closed so that $[[E]_r]_s = [E]_r$. Finally, if $r > s$ we can apply the monotonic operation s-closure to the inclusion

$$E \subseteq [E]_r \subseteq [E]_s \tag{9.70}$$

and get

$$[E]_s \subseteq [[E]_r]_s \subseteq [E]_s \tag{9.71}$$

which proves (9.69).

Introduce the <u>outer layers</u> L_p of a set E in background space

$$L_p = [E]_p - [E]_{p+1}, \quad p = 1,2,\ldots \tag{9.72}$$

The outer layers describe how accessible is the "boundary" of E using probes consisting of features of increasing structural complexity.

In general we can state the following

<u>Theorem 9.7.</u> <u>Let</u> E <u>be a (topologically) closed subset of specification space and consider the sequence of closures</u> $[E]_p$, $p = 1,2,3,\ldots$. <u>Then, as</u> $p \to \infty$, <u>we have monotonic convergence to the original set</u>

$$[E]_p \downarrow E. \tag{9.73}$$

<u>Proof</u>: Since $[E]_p$ is non-increasing and $E \subset [E]_p$, $\forall p$, it is clear that the sequence converges to some set $E' \supset E$. If $E' \neq E$ take a point $x_o \in E'$ and $x_o \notin E$. Since E is (topologically) closed we can find a neighborhood N_{x_o} of it

such that $N_{x_o} \cap E = \phi$ (remember that X is regular). Such a neighborhood is defined through a finite number of features from G. If we take p greater than this number the p-closure of E will not contain x_o so that $x_o \notin E'$ is contrary to the assumption, and the assertion is proved.

It should be possible to extend the discussion above about higher order features to the case when frequencies are taken into account. This has not yet been made.

CHAPTER 4

THE DEFORMATIONS

1. Deformations on Generators, Configurations, Images.

It can happen that the images we actually observe are exactly the images in the algebra $\mathcal{I}$. This will simplify the analysis. A few situations like that will be discussed in Part III (see Appendix).

It is fair to say, however, that in most instances we can only observe corrupted versions of the ideal images in $\mathcal{I}$, and we then encounter the fundamental question of how such deformations occur. For a complete pattern synthesis it will be necessary to specify the deformation mechanism. This specification is also indispensable for the analysis stage.

Let us denote by d a mapping from the image algebra $\mathcal{I}$ into a set $\mathcal{I}^{\mathcal{D}}$ of objects that can be observed. The elements

$$I^{\mathcal{D}} \in \mathcal{I}^{\mathcal{D}} \tag{1.1}$$

will be spoken of as the deformed images.

Usually there will be many transformations d and we will not know in advance which one will be applied. We use the symbol $\mathscr{D}$ to denote the set of all of them.

So far we have not said anything about the nature of the deformed images. The simplest case is when $\mathscr{I}^{\mathscr{D}} \subset \mathscr{I}$, in other words when the objects we observe are of the same type as the pure images in the image algebra. We shall then speak of <u>automorphic</u> deformations and d then maps the image algebra into itself.

In the opposite case, <u>heteromorphic</u> deformations, the set $\mathscr{I}^{\mathscr{D}}$ can be of many different types, as we shall see in this chapter. It may be that $\mathscr{I}^{\mathscr{D}}$ also has image algebra structure, although a different one from I. It should be pointed out that even then the structures can be radically different so that I and $I^{\mathscr{D}}$ are conceptually far apart. Quite often we will have $\mathscr{I} \subset \mathscr{I}^{\mathscr{D}}$ so that the pure images are special instances of deformed ones. Typically d loses structure so that $\mathscr{I}^{\mathscr{D}}$ will be less structured than $\mathscr{I}$.

In the case when $\mathscr{I} \subset \mathscr{I}^{\mathscr{D}}$ we shall often extend the domain of d from $\mathscr{I}$ to $\mathscr{I}^{\mathscr{D}}$, still with the range equal to $\mathscr{I}^{\mathscr{D}}$. We can then repeatedly apply a sequence of d's and it will then be natural to extend $\mathscr{D}$ to be a semigroup of transformations.

It will also often be possible to extend the domain of the similarity transformations s from $\mathscr{I}$ to $\mathscr{I}^{\mathscr{D}}$, still in $\mathscr{I}^{\mathscr{D}}$. We combine the above statements into a condition that will be satisfied in most cases in what follows. In this section S will be assumed to form a group.

<u>Definition 1.1</u>. <u>The deformation mechanism is said to be</u>

<u>regular if</u>

$$\left.\begin{array}{ll} \text{(i)} & \mathcal{T} \subseteq \mathcal{T} \\ \text{(ii)} & \underline{\text{each}}\ \ d\ \ \underline{\text{is defined on}}\ \ \mathcal{T}^{\mathcal{D}}\ \underline{\text{and}}\ \mathcal{D} \\ & \underline{\text{still forms a semigroup}} \\ \text{(iii)} & \underline{\text{each}}\ \ s\ \ \underline{\text{is defined on}}\ \ \mathcal{T}^{\mathcal{D}},\ \underline{\text{with}} \\ & \underline{\text{values in}}\ \mathcal{T}^{\mathcal{D}}\ \underline{\text{with}}\ \ S\ \ \underline{\text{forming a group}}. \end{array}\right\} \quad (1.2)$$

Automorphic deformations is a very special case of a regular $\mathcal{D}$. The two types of transformations, d and s, would then be defined on the same set. They play very different roles, however. The similarity transformations usually change the image in a systematic and easily understood manner. When S is a group they do not destroy any information since the inverse s^{-1} will recover the original image. The deformations, on the other hand, can corrupt the image to such an extent that it is impossible to recover it exactly: they lose information.

The interaction between similarity transformations and deformations is playing an important part, and we shall introduce two properties that will simplify the pattern analysis considerably when they hold.

<u>Definition 1.2</u>. <u>Consider a regular deformation mechanism on an image algebra</u>. <u>We shall say that it is</u>

(i) <u>covariant if</u> $sd = ds$ <u>holds for all</u> $s \in S$, $d \in \mathcal{D}$

(ii) <u>homomorphic if for</u> $I = I_1 \sigma I_2$, <u>all in</u> $\mathcal{T}$, <u>we have</u> $d(I) = d(I_1)\sigma d(I_2)$.

It should be observed that these are strong conditions that will not be satisfied very often. Of course, if $\mathcal{D}$ is a commutative semigroup with $S \subset \mathscr{D}$ then deformations are certainly covariant. In another simple case, $\mathscr{T}^{\mathscr{D}}$ = vector space and d consisting of linear operators on it, then deformations are homomorphic.

Let $I^{\mathscr{D}}$ be a metric space with a distance δ satisfying the conditions

(i) $\delta(I_1^{\mathscr{D}}, I_2^{\mathscr{D}}) \geq 0$ with equality if $I_1 = I_2$

(ii) δ is symmetric and satisfies the triangle inequality. (1.3)

If $\delta(I_1^{\mathscr{D}}, I_2^{\mathscr{D}}) = 0$ implies $I_1^{\mathscr{D}} = I_2^{\mathscr{D}}$ the distance is definite but this will not be assumed throughout.

It is natural to ask that the metric be tailored to the similarity relations in $\mathscr{T}^{\mathscr{D}}$ and this will be done in two ways.

<u>Definition 1.3</u>. <u>For a distance</u> δ <u>defined on a regular</u> $\mathscr{T}^{\mathscr{D}}$ <u>we shall say that it is</u>

(i) <u>invariant if</u> $\delta(sI_1^{\mathscr{D}}, sI_2^{\mathscr{D}}) = \delta(I_1^{\mathscr{D}}, I_2^{\mathscr{D}})$ <u>for all</u> $s \in S$, $I_1^{\mathscr{D}}$ <u>and</u> $I_2^{\mathscr{D}} \in \mathscr{T}^{\mathscr{D}}$.

(ii) <u>completely invariant if</u> $\delta(sI_1^{\mathscr{D}}, tI_2^{\mathscr{D}}) = \delta(I_1^{\mathscr{D}}, I_2^{\mathscr{D}})$ <u>for all</u> s <u>and</u> $t \in S$, $I_1^{\mathscr{D}}$ <u>and</u> $I_2^{\mathscr{D}} \in \mathscr{T}^{\mathscr{D}}$.

Starting from a given distance δ_0 define

$$\delta_1(I_1^{\mathscr{D}}, I_2^{\mathscr{D}}) = \inf_{s \in S} \delta_0(sI_1^{\mathscr{D}}, sI_2^{\mathscr{D}})$$

$$\delta_2(I_1^{\mathscr{D}}, I_2^{\mathscr{D}}) = \inf_{s,t \in S} \delta_0(sI_1^{\mathscr{D}}, tI_2^{\mathscr{D}}) \tag{1.4}$$

It is then easy to verify that δ_1 is invariant and δ_2 completely invariant.

The deformation will sometimes be based on some physical mechanism requiring power, energy or some similar quantity to transform a pure image into its observed form. To use a more neutral term we shall speak of the <u>effort</u> required.

<u>Definition 1.4</u>. <u>On a regular deformation space</u> $\mathscr{D}$ <u>consider a non-negative function</u> e <u>with the properties</u>

(i) $e(d_1 d_2) \leq e(d_1) + e(d_2)$

(ii) $e(sd) = e(ds)$, <u>all</u> $d \in \mathscr{D}$, $s \in S$.

<u>The function</u> e <u>is called an invariant effort function</u>. <u>If we have</u> (i) <u>and</u>

$$\text{(iii)} \quad e(sd) = e(dt), \quad \text{all } d \in \mathscr{D}, \ s \text{ and } t \in S \tag{1.5}$$

e <u>is called a completely invariant effort function</u>.

If $\mathscr{D}$ is covariant the condition in (ii) holds automatically. We have

<u>Theorem 1.1</u>. <u>Let</u> e <u>be a completely invariant effort and put</u>

$$\delta(I_1^{\mathscr{D}}, I_2^{\mathscr{D}}) = \min_{d_1, d_2} [e(d_1)+e(d_2); \ I_1^{\mathscr{D}} = d_1 I_2^{\mathscr{D}} \text{ and } I_2^{\mathscr{D}} = d_2 I_1^{\mathscr{D}}] \tag{1.6}$$

<u>assumed to be attained</u>. <u>Then</u> δ <u>is a completely invariant distance</u>.

<u>Remark</u>: We have tacitly assumed that the relation $I_1^{\mathscr{D}} = dI_2^{\mathscr{D}}$ viewed as an equation in d always has at least one solution. If this is not so, the corresponding value of e(d) should be replaced by $+\infty$, and it may be necessary to allow the value $+\infty$ for the resulting distance. This would only

slightly affect the proof.

<u>Proof</u>. The function δ is symmetric in its two arguments and, to prove the triangle inequality, we consider fixed $I_1^{\mathcal{D}}$, $I_2^{\mathcal{D}}$, and $I_3^{\mathcal{D}}$. If there exist d_1, d_2, d_3, d_4 such that

$$\begin{aligned} I_1^{\mathcal{D}} &= d_2 I_2^{\mathcal{D}} \\ I_2^{\mathcal{D}} &= d_1 I_1^{\mathcal{D}} \\ I_2^{\mathcal{D}} &= d_3 I_3^{\mathcal{D}} \\ I_3^{\mathcal{D}} &= d_4 I_2^{\mathcal{D}} \end{aligned} \tag{1.7}$$

we write $d^{(1)} = d_2 d_3$, $d^{(2)} = d_4 d_1 \in \mathcal{D}$, so that

$$\begin{aligned} I_1^{\mathcal{D}} &= d^{(1)} I_3^{\mathcal{D}} \\ I_3^{\mathcal{D}} &= d^{(2)} I_3^{\mathcal{D}} . \end{aligned} \tag{1.8}$$

Hence, using property (ii),

$$e(d^{(1)}) + e(d^{(2)}) \leq e(d_1) + e(d_2) + e(d_3) + e(d_4) \tag{1.9}$$

from which follows

$$\begin{aligned} \delta(I_1^{\mathcal{D}}, I_3^{\mathcal{D}}) &\leq \min[e(d^{(1)}) + e(d^{(2)});\ d^{(1)} = d_2 d_3,\ d^{(2)} = d_4 d_1] \\ &\leq \min[e(d_1) + e(d_2);\ I_1^{\mathcal{D}} = d_2 I_2^{\mathcal{D}} \text{ and } I_2^{\mathcal{D}} = d_1 I_1^{\mathcal{D}}] \\ &\quad + \min[e(d_3) + e(d_4);\ I_2^{\mathcal{D}} = d_3 I_3^{\mathcal{D}} \text{ and } I_3^{\mathcal{D}} = d_4 I_4^{\mathcal{D}}] \\ &= \delta(I_1^{\mathcal{D}}, I_2^{\mathcal{D}}) + (I_2^{\mathcal{D}}, I_3^{\mathcal{D}}). \end{aligned}$$

Finally, complete invariance follows from condition (iii) since $I_1^{\mathcal{D}} = dI_2^{\mathcal{D}}$ implies $s_1 I_1^{\mathcal{D}} = (s_1 d s_2^{-1}) s_2 I_2^{\mathcal{D}}$ so that $e(s_1 d s_2^{-1}) = e(d)$. This proves that δ is completely invariant.

If we had started with an effort function e that was

only invariant we could only claim that the resulting distance was invariant.

Now let us also introduce a probability measure P on some σ-algebra $\mathscr{A}(\mathscr{D})$ of subsets of $\mathscr{D}$. This means that we will be able to talk about some deformations as more likely than others. In $\mathscr{T}$ and $\mathscr{T}^{\mathscr{D}}$ we also need σ-algebras $\mathscr{A}(\mathscr{T})$, $\mathscr{A}(\mathscr{T}^{\mathscr{D}})$ such that for any subsets E of $\mathscr{T}$ and F of $\mathscr{T}^{\mathscr{D}}$ with $E \in \mathscr{A}(\mathscr{T})$, $F \in \mathscr{A}(\mathscr{T}^{\mathscr{D}})$ we have

$$\{d \mid dE \in F\} \in \mathscr{A}(\mathscr{D}). \tag{1.11}$$

For fixed $I \in \mathscr{T}$ the deformed image $\mathscr{T}^{\mathscr{D}} = dI$ will have a probability measure P_I over $\mathscr{T}^{\mathscr{D}}$.

We now introduce a wider and more useful version of covariant deformations.

<u>Definition 1.3</u>. <u>The regular deformations</u> $\mathscr{D}$ <u>with a probability measure</u> P <u>will be said to be covariant in probability if for any similarity transformation</u> $s \in S$ <u>the transformations</u> ds <u>and</u> sd <u>have the same probability distribution over</u> $\mathscr{D}$.

In case the deformation restricts a contrast pattern to a random subset E, but not its values, we shall interpret covariance in probability to mean that the random set sE has the same probability distribution as E.

If this definition applies, we have for any fixed s and $E \in \mathscr{A}(\mathscr{D})$,

$$P(d \in E) = P(sd \in sE) = P(ds \in sE) = P(d \in sEs^{-1}). \tag{1.12}$$

Conversely, if (1.12) holds for any s and E then the deformations are covariant in probability.

An important consequence of covariance in probability is given by

Theorem 1.2. Assume that the deformations are covariant in probability and let $\mathscr{P}$ be a pattern consisting of an equivalence class modulo S

$$\mathscr{P} \in \mathscr{T}/S. \tag{1.13}$$

Then, if E is an S-invariant set in $\mathscr{A}(\mathscr{T}^{\mathscr{D}})$, the conditional probabilities $P(E|\mathscr{P})$ are well defined: $P(E|I)$ does not depend upon I as long as $I \in \mathscr{P}$.

Proof: Consider the conditional probability

$$f(s) = p(I^{\mathscr{D}} \in E | I = sI_0) \tag{1.14}$$

where I_0 is a fixed prototype (see (5.1.14). Then

$$f(s) = P(dsI_0 \in E) = P(dsI_0 \in E) \tag{1.15}$$

because of covariance in probability. But

$$P(sdI_0 \in E) = P(dI_0 \in s^{-1}E) = P(dI_0 \in E) \tag{1.16}$$

since E was S-invariant. Hence f is a constant so that the conditional probability $P(E|\mathscr{P})$ is indeed well defined since it does not depend upon what image gI_0 we start from in the pattern $\mathscr{P}$.

If this had not been so we could not speak of $P(E|\mathscr{P})$ unless, of course, we impose a probability measure on the pure image algebra $\mathscr{T}$ too.

For the preceding discussion of this paragraph we should add that it is desirable to choose the algebraic, topological, and probabilistic structures so that they fit together in a natural manner. A reader interested in how this can be done in the standard algebraic-topological settings should consult Grenander (1963).

The actual choice of P is more difficult than the measure-theoretic precautions that are needed. It has to be made from case to case, using the available subject-matter knowledge in order to arrive at the usual compromise: the model should approximate the phenomenon studied reasonably well and at the same time be analytically or numerically tractable. Nevertheless, we can state some general principles that may help us in constructing the model of deformations.

First, we should try to separate $\mathcal{D}$, which may be quite a complicated space, into simple factors $\mathcal{D} = \mathcal{D}_1 \times \mathcal{D}_2 \times \ldots$. The product may be finite, denumerably or non-denumerably infinite, as we shall see below. Sometimes we are given such a separation directly, for example when the deformations consist of a topological transformation of the background space followed by a mask deformation. We can also expect some help from the way in which the image algebras are formed from elementary parts. If we consider images whose configurations have n generators, all identifiable, we may try writing

$$\begin{aligned} I^{\mathcal{D}} &= dI = (d_1 g_1, d_2 g_2, \ldots, d_n g_n) \\ I &= (g_1, g_2, \ldots, g_n) \end{aligned} \tag{1.17}$$

in the hope that the properties of the factors d_v should be tractable. But this would work only when the generators can be uniquely determined from the image. Instead we could try the corresponding separation, but for canonical configurations whose generators are determined in the image algebra under consideration.

Once we have been able to separate $\mathscr{D}$ into reasonably simple factors we must decide upon what probability measure to introduce on $\mathscr{D}$. To be able to do this it is important that the way in which we have separated the deformations into factors is such that the factors of d appear to behave independently of each other. The complete specification of P cannot be made without empirical information and in order to arrive at sufficiently accurate estimates we must have enough structure in the axiomatic model. This is a critical step in the determination of P and requires such understanding of the deformation mechanism that the ensuing analysis does not force our data into an inadequate mould. If we can actually carry out the separation so that the factors are stochastically independent we are left with the problem of determining the marginal distributions of the factors. To exemplify this, let us think of pure generators being generated by a mechanism of the form $L_0x = 0$, where the reader may think of L_0 as a difference operator, and the deformed generators are given by $L_0x = \varepsilon$. The first thing we would try would then be to assume that the values of ε (for different arguments) are independent. If this could not be accepted as an adequate approximation we could try to remove the dependence by working not with x itself but with some transformation (possibly linear) of it. In other words, we could choose the model so that the deformations have a simple probabilistic form. Or, as another example, if we deal with contrast patterns (see Section 3.3) with a discrete background space X we may try to model P by assuming that different points of X are mapped independently of each other, but with different distributions, into the background space of $\mathcal{I}^{\mathcal{D}}$.

To narrow down the choice of the marginal distributions that remain to be specified, we appeal to the role played by the similarity transformations. Again, if our $\mathcal{D}$ has been well chosen we could hope that P should have a corresponding invariance property. Thus, if I and I' are similar pure images, I' = sI, we would first examine the possibility that dI and dI' = dsI have the same probability distribution. Or we may try the model postulating that sdI has the same probability distribution as dsI: this leads us to covariance in probability.

Applying these principles, we may be able to specify the analytic form of P and the remaining free parameters must be estimated empirically.

Deformation mechanisms will be classified according to two criteria: <u>level</u> and <u>type</u>.

By the level of a deformation mechanism we shall mean that stage in the pattern synthesis of images at which $\mathcal{D}$ is defined. The highest level, images, is when $\mathcal{D}$ is defined directly for each I, irrespective of the way the (pure) image has been synthesized from configurations, rules, restrictions, generators, attributes. The lowest level is when $\mathcal{D}$ is defined in terms of the generators that make up a configuration in I. The intermediate level is when $\mathcal{D}$ is defined on $\mathcal{L}(\mathcal{R})$.

The type of $\mathcal{D}$ describes how different $\mathcal{I}^{\mathcal{D}}$ is from $\mathcal{I}$. We already have one crude classification according to type:

automorphic vs. heteromorphic. We shall also subclassify heteromorphic deformations into five classes that are neither disjoint nor exhaustive: changing the contrast, changing the background space, incomplete observations, indirect observations, and a few others.

These criteria are not intended to lead to a complete classification of deformations. Nevertheless, they will be helpful in describing the nature of the $\mathcal{D}$'s to be dealt with.

All through this section we have assumed that the images go through a single deformation. We may also study what happens if we have a set $\{d(t)\}$ of primitive deformation mechanisms that are applied successively. The simplest case is when t takes only a finite number of values, say $t = 1,2,\ldots,k$, and we apply the total deformation

$$\mathcal{D} = d(k)d(k-1)\ldots d(1). \tag{1.18}$$

The set of t's can also be denumerably infinite or have the power of the continuum. In the latter case we shall write $\mathcal{D}(t)$ for the total deformation from $t = 0$ to t and assume the semigroup property

$$\mathcal{D}(t_1+t_2) = \mathcal{D}(t_1)\,\mathcal{D}(t_2). \tag{1.19}$$

Furthermore, if the $\mathcal{D}$'s are probabilistic we shall assume that partial deformations over disjoint t-intervals are stochastically independent.

2. Automorphic Deformations.

To study automorphic deformations, $\mathcal{T}^{\mathcal{D}} \subseteq \mathcal{T}$, it is convenient to start at the lowest deformation level. That is when $\mathscr{D}$ is constructed in terms of the generators.

Definition 2.1. A mapping $D: G \longrightarrow G$ such that generator classes are preserved, $DG^{\alpha} \equiv G^{\alpha}$, is called a generator deformation.

Given D, which is defined on G only, we try to extend it to $\mathscr{C}(\mathscr{R})$ and $\mathscr{T}$. In order to do this we must take care that the deformed configurations and images do not violate the image algebra structure.

A case of some importance is when $D = S$ or a power of it, and we separate two cases.

Definition 2.2. For an image algebra with some (arbitrary) Σ but with $\rho \equiv$ TRUE and R given through equality of configurations (see Definition 3.4.1), we define shift deformations on $\mathscr{C}$ and $\mathscr{T}$ by

$$(g_1,g_2,\ldots,g_n) \to (sg_1,sg_2,\ldots,sg_n); \quad s \in S \tag{2.1}$$

and jittering deformations by

$$(g_1,g_2,\ldots,g_n) \to (s_1g_1,s_2g_2,\ldots,s_ng_n); \; s_i \in S, \quad i \tag{2.2}$$

Occasionally we shall use these terms "shift" and "jittering" even when s-transformations in (2.1) and (2.2) are not necessarily similarity transformations but more general deformations of the generators.

For probabilistic deformations we also have to specify the probability measure over S and S^n respectively. We do this by a probability measure μ over S in the first case, and by μ^n in the second, where μ^n stands for the product measure with identical marginal measures equal to μ.

To illustrate how this definition works we shall consider a few cases.

<u>Case 2.1</u> (ciphers). Let G be a finite set and S the permutation group for $\#(G)$ objects and form the shifted configurations as in (2.1).

If we interpret G as an alphabet then the above deformation is just an ordinary cipher.

<u>Case 2.2</u> (shifted Euclidean patterns). Let $\mathcal{T}$ be given as in Case 2.3.3, R = identifying unions of sets, and deformations by (2.1).

This case does not fall strictly under Definition 3.2.2 since identification is not equality of configurations. This is not serious, however, since we can verify directly that if c and c' belong in the same image I then their deformed versions are equivalent modulo R so that the definition in (2.1) extends immediately from $\mathcal{C}$ to $\mathcal{T}$.

We can also use the deformations in (2.2) when the bond relation ρ is no longer $\equiv$ TRUE but then of course we cannot expect automorphic deformations. The following two cases illustrate this.

<u>Case 2.3</u> (jittered crystal). Consider images as in Case 3.5.1, and let S be the translation group acting as in (2.2).

This means that instead of observing the regular lattice (3.5.2) we will only be able to see a point set of the form

$$I^{\mathcal{D}} = \{z_{\nu\mu} + n_{\nu\mu}\} \tag{2.3}$$

where all the $n_{\nu\mu}$ are two-dimensional stochastic vectors with the same distribution and independent of each other. A difficulty arises here that could easily be missed at first glance. Note that we observe the <u>set</u> $I^{\mathscr{D}}$, not points with a certain labelling with subscripts ν,μ. This means that different points in $I^{\mathscr{D}}$ will appear on equal footing (no labelling) which will make the analysis more difficult (see Chapter 10).

<u>Case 2.4</u> (corruption of terminal symbols). On the set of terminal symbols V_T a set D of mappings is given. For an image with the terminal string $(X_1,X_2,\ldots,X_n)$ we define the deformed image as

$$I^{\mathscr{D}} = (X_1',X_2',\ldots,X_n') \tag{2.4}$$

where

$$x_i' = d_i x_i,\ d_i \in D$$

It is clear that this is still a generator deformation since only the local nature of the original string matters, not the global, syntactic properties.

This characteristic feature of Case 2.4 that it is <u>syntax-independent</u> is clearly quite limiting. More severe deformations that also affect the derivation tree will be considered later on. In rare cases, say when $\mathscr{R}$ consists of very mild restrictions, it can happen that a non-generator D maps admissible configurations into admissible ones. As an example of this we give the following example.

<u>Case 2.5</u> (non-generator wave form changes). Consider Case 2.4.1 with the generator classes given through

$$g \in G^{\alpha} \quad \text{iff} \quad L_{\alpha} g = 0 \tag{2.6}$$

where L_{α} is a second-order time-independent differential operator and $p = 1$. Deformations are given for an arbitrary generator g defined on $[a,b]$ by

$$\alpha \to \beta \,, \quad g \to g^{D} \tag{2.7}$$

where g^D is defined on the same interval, satisfies the equation $L_{\beta} g^D = 0$ and has the boundary values

$$\begin{aligned} g_1^D(a+0) &= g_2^D(a-0) \\ g_1'^D(a+0) &= g_2'^D(a-0) \end{aligned} \tag{2.8}$$

where g_1 and g_2 are two generators joined together by Σ at the point a.

We now go ahead to the next level when $\mathcal{D}$ is expressed in terms of the configuration. Let us look at two illustrative cases.

<u>Case 2.6</u> (randomized response). Let $\{g_1, g_2, \ldots, g_k\}$ be a set of truth variables and put S = the operation of selecting ℓ out of k objects with equal probabilities without replacement.

This corresponds to a device used in sample surveys to prevent invasion of privacy. One is interested in the occurrence of certain properties with the truth variables $g_1, g_2, \ldots, g_{\ell}$, $\ell < k$. If these are sensitive questions, for example relating to sexual practice or criminal activities, the respondents may refuse to answer or may lie. To make them feel more secure one adds $k-\ell$ safe questions, not of a sensitive nature, and permutes the order of all k questions at random.

The respondent is asked to answer these or usually a fixed subset (a mask deformation), say the first one, in the permuted order. For obvious reasons this is called randomized response.

Case 2.7 (reordering). Let $\mathscr{C}$ have structure LINEAR but with $\rho \equiv$ TRUE. Let $\mathscr{D}$ consist of the permutation group so that $\mathscr{D}$ maps $c = (g_1, g_2, \ldots, g_n)$ into $c' = (g_1', g_2', \ldots, g_n')$ where c' is just a reordering of c.

Note that this case is different from Case 2.1 where the permutation applied to individual generators but here instead to the entire configuration.

This preserves the number of generators and we can talk of the changes $g_i \rightarrow g_i'$ but these are not true generator deformations since they depend upon the entire configuration that we start from.

Case 2.8 (noisy sequential machine). Let the configurations be generated by a sequential machine. We now change the input string $i_1, i_2, \ldots, i_n$ into $i_1', i_2', \ldots, i_n'$ in accordance with one probability distribution over the input set and another over the initial symbol. This defines a probabilistic deformation mechanism on $\mathscr{C}$.

These new configurations are still regular so that $\mathscr{D}$ maps $\mathscr{C}$ into $\mathscr{C}$ so that $\mathscr{D}$ is automorphic. It also preserves the number of generators in c. A different situation occurs in the following case.

Case 2.9 (missing generators). With Σ = LINEAR consider generators with $\omega_{in} = \omega_{out} = 1$ and with the same bond value in and out. If ρ is a complete order relation consider a configuration $c = (g_1, g_2, \ldots, g_n)$. $\mathscr{D}$ maps c into $c' = (g_{i_1}, g_{i_2}, \ldots, g_{i_m})$ where $m \leq n$ and $(i_1, i_2, \ldots, i_m)$ is an ordered subset of $(1, 2, \ldots, n)$.

This is also automorphic but here we can drop some of the generators so that their number is not preserved.

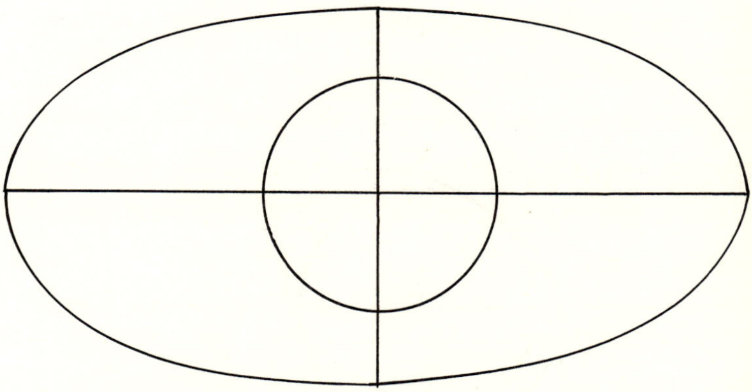

(a)

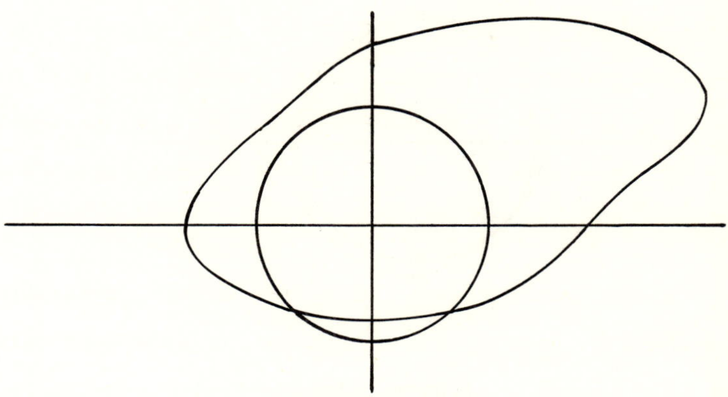

(b)

Figure 2.1

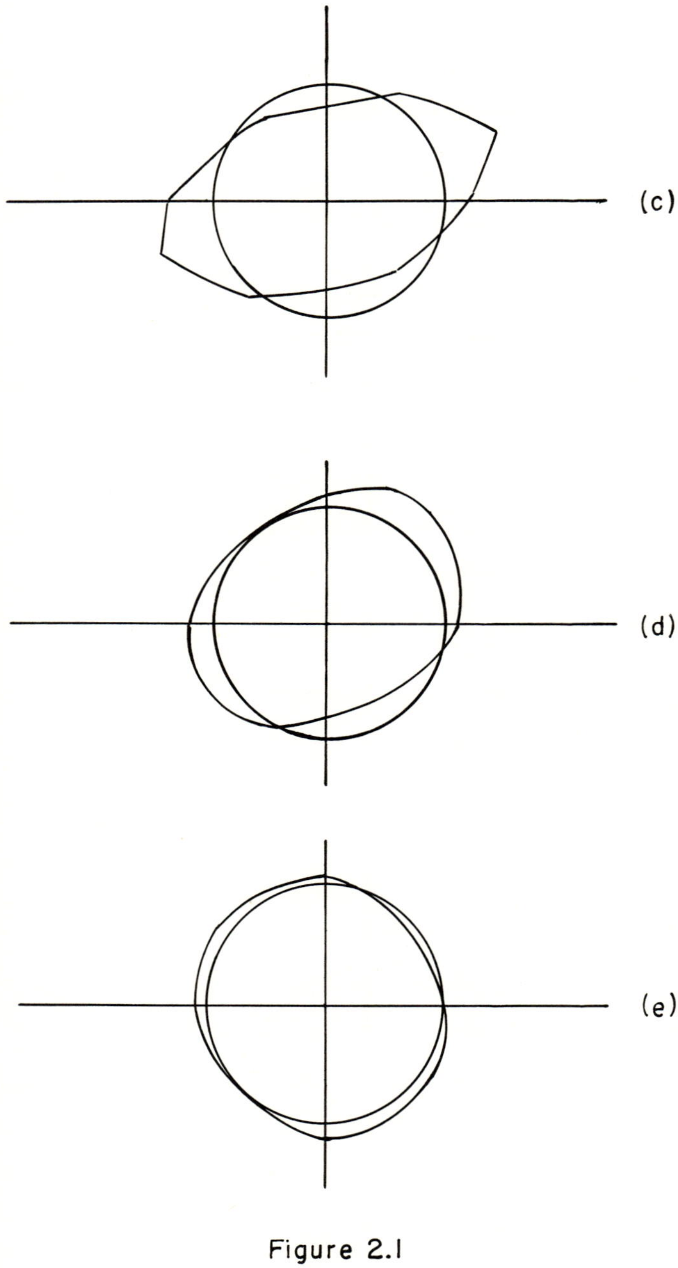

Figure 2.1

If R is defined simply through equality of configurations any $\mathscr{D}$ that maps $\mathscr{C}^{\mathscr{R}}$ into itself gives rise to automorphic deformations of $\mathscr{T}$. In general, when we have more general identification rules, we must make sure that $\mathscr{D}$ satisfies the consistency condition in order to be automorphic:

$$\text{for any R-equivalent } c_1 \text{ and } c_2 \text{ we must have } (dc_1)R(dc_2) \text{ for all } d \in \mathscr{D}. \tag{2.9}$$

Then we can directly extend $\mathscr{D}$ from $\mathscr{C}^{\mathscr{R}}$ to the image algebra.

We now come to deformations defined directly on $\mathscr{T}$. It could be said that all deformation mechanisms must ultimately be extended to $\mathscr{T}$, but as pointed out in Section 1 we always try to decompose them into simple parts, say on G or $\mathscr{C}^{\mathscr{R}}$. Often it will be more convenient to introduce $\mathscr{D}$ directly on the image algebra, however. The simplest case of this part is additive noise.

Case 2.10 (generalized additive noise). If $\mathscr{T}$ is a semigroup under its binary operation '∘' we define

$$\mathscr{D}: \quad I \to dI = I^{\mathscr{D}} = I \circ n \tag{2.10}$$

where n is a stochastic image taking values in $\mathscr{T}$.

This is more general than one may think at first glance. It certainly contains the usual type of additive noise when $\mathscr{T}$ is a topological vector space with vector addition as the binary operator.

To fix ideas let us consider a separable, real Hilbert space $\mathscr{H}$ with the dual space $\mathscr{H}^*$ with elements denoted h^*. The generators g are arranged into classes G^{α} where each G^{α} is a one-dimensional subspace of $\mathscr{H}$ and $G^{\alpha} \perp G^{\beta}$ if

$\alpha \neq \beta$. S consists of multiplication by real scalars.

Let Σ = FULL and ρ denote that two connected generators should have different generator index. We allow infinite configurations and define the identification rule $(g_1,g_2,\ldots)R(g_1',g_2',\ldots)$, assuming that $g_1+g_2+\ldots = g_1'+g_2'+\ldots$ and we must of course assume that these sums converge for regular configurations.

Hence $\mathcal{T}$ will be a closed subspace of $\mathcal{A}$ and if we let P be a probability measure on $\mathcal{A}$ with d standing for the corresponding stochastic element, we define

$$\mathcal{D}: \quad I \to I^{\mathcal{D}} = I+d. \tag{2.11}$$

In order that $I^{\mathcal{D}} \subseteq \mathcal{T}$, automorphic deformations, we must demand that $d \in \mathcal{T}$ a.c., in other words that the support of P satisfies $s(P) \subseteq \mathcal{T}$.

The technique of handling stochastic elements in a Hilbert space is conveniently handled via the properties of $h^*(I)$, $h^* \in \mathcal{A}^*$. Indeed, for any bounded functional h^* the quantity $h^*(I)$ is just a real-valued stochastic variable. If we prescribe the distribution function for $h^*(I)$ for all h^* this defines P uniquely (see, e.g., Grenander (1963), p. 128).

A particularly important sort of additive noise is the <u>Gaussian</u> one where each $h^*(I)$ has a normal distribution with the parameters

$$\begin{aligned} &\text{mean} = E[h^*(I)] = m(h^*) = h^*(d_m) \\ &\text{variance} = \text{Var}[h^*(I)] = V(h^*) = (Qh^*,h^*). \end{aligned} \tag{2.12}$$

The element $d_m \in \mathcal{A}$ plays the role of the mean of the Gaussian distribution and the linear operator Q should be Hermitian, non-negative, bounded and of Hilbert-Schmidt type with finite trace.

In order that additive Gaussian noise be automorphic the mean d_m and the covariance operator Q must satisfy certain conditions.

Theorem 2.1. Consider the deformation mechanism in (2.11) in a separable Hilbert space and d given by a Gaussian distribution with parameters as in (2.12). Then $\mathscr{D}$ is automorphic if and only if

$$d_m \in \mathscr{T} \tag{2.13}$$

and

$$Qz = 0 \quad \text{for any } z \perp \mathscr{T}. \tag{2.14}$$

Proof: We can write with strong convergence a.c.

$$d = d_m + \sum_\nu \sigma_\nu \xi_\nu e_\nu$$

where $\sigma_\nu^2 \geq 0$ are the eigenvalues of Q, $\{e_\nu\}$ are normed eigenvectors associated with the σ_ν^2, and the ξ_ν are independent normally distributed stochastic variables with mean zero and standard deviation 1 (see, e.g., Grenander (1963), p. 140). But that means that, with probability one, the stochastic element d spans the linear subspace

$$\mathscr{A}_{sub} = d_m \oplus e_1 \oplus e_2 \oplus \dots . \tag{2.16}$$

To be sure that $\mathscr{A}_{sub} \subseteq \mathscr{T}$ we must have all the vectors d_m, $e_1, e_2, \dots$ in $\mathscr{T}$ and this would also be sufficient. But to claim that all $e_\nu \in \mathscr{T}$ is the same as to demand (2.14), which completes the proof.

For additive noise we choose the effort formation $e(d) = \frac{1}{2}||d||$ which satisfies Definition 1.4 since vector addition in $\mathscr{A}$ is commutative, so that e is invariant. We get the usual distance

$$d(I_1^{\mathscr{D}}, I_2^{\mathscr{D}}) = ||I_1^{\mathscr{D}} - I_2^{\mathscr{D}}|| \tag{2.17}$$

which is of course invariant (S).

Additive noise deformations are trivially covariant since they have commutative addition.

We get a family $\mathscr{D}(t)$ of additive Gaussian noise deformations (see (1.19)) if we define $\mathscr{D}(t)$ through the parameters

$$\mathscr{D}(t): m_t = t \cdot m, \quad Q_t = t \cdot Q \tag{2.18}$$

where m and Q are fixed and with properties as before.

Moving away from topological vector spaces, let $\mathscr{T}$ now be a subset of the space B^n consisting of all n-tuples of 0 or 1 with addition defined component-wise modulo 2. Let the vector $d \in B^n$ take values according to some probability measure P and define $I^{\mathscr{D}}$ as in (2.9).

Consider the νth component of the image

$$I_\nu \to I_\nu^{\mathscr{D}} = I_\nu + d_\nu . \tag{2.19}$$

This means that I_ν will be unchanged with probability $P(d_\nu = 0)$ and changed with the complementary probability. The changes themselves have probabilities that do not depend on the pure image, or in terms of communication engineering we have a noisy, symmetric <u>binary channel</u>. Thus $\mathscr{D}$ will be automorphic if the support $s(P)$ satisfies

$$s(P) + \mathcal{T} \subseteq \mathcal{T}. \tag{2.20}$$

Let I be a Boolean image, $I = I(x)$ where x is a vector of truth variables, and define $I^{\mathcal{D}}$ as in Case 2.10 with n as a randomly chosen Boolean image from $\mathcal{T}$. If the semigroup operation is logical "and", we speak of a <u>conjunctive contrast deformation</u>; if it is logical "or", it is a <u>disjunctive contrast deformation</u>.

An unconventional form of additive noise appears in the study of plane convex sets (see Case 5.5.9). Assume given a probability measure P on the closed, bounded convex sets in R^2. This will induce a distribution on the set of support functions s. Define <u>Minkowsky deformation</u> through the equation

$$\mathcal{D}: I \to I^{\mathcal{D}} = I+d \quad \text{with Minkowsky addition} \tag{2.21}$$

and d is the random set associated with the support function s. Rewriting this in terms of the corresponding support functions, we get

$$s^{\mathcal{D}}(\phi) = s^{I}(\phi) + s^{d}(\phi), \quad 0 \leq \phi < 2\pi. \tag{2.22}$$

This implies that the set of deformed images always contains the pure ones. With δ as the Hausdorff distance we get the expected size of the deformation as

$$E[\delta(I^{\mathcal{D}},I)] = E[\sup_{\phi}|s^{d}(\phi)|] \tag{2.23}$$

and the expected increase in perimeter (see (3.5.69)) equal to

$$\int_0^2 E[s^d(\phi)]d\phi = E[(\text{perimeter of } I^{\mathscr{D}})-(\text{perimeter of } I)] \qquad (2.24)$$

In Figure 2.1(a) we show the pure image as an ellipse (disregard the circle for the moment). The deformations s^d consist of adding a randomly oriented unit interval. In Figures 2.1(b), (c), (d), (e), we display the deformed images for $n = 2,10,30,40$ normalized as $\frac{1}{n} I^{\mathscr{D}}$. All the figures also contain the limiting shape, a circle.

To explain the tendency of the rescaled deformed image to the shape of a circle we prove a limiting theorem.

<u>Theorem 2.2</u> (<u>Law of large numbers for convex sets</u>; R. Vitale). <u>Given a pure image</u> $I \in \mathscr{I}$ (as above), <u>we expose it to a sequence of deformations following</u> (2.21)

$$\begin{aligned} I_n^{\mathscr{D}} &= \mathscr{D}_n I_{n-1}^{\mathscr{D}}, \quad n = 1,2,\ldots \\ I_0^{\mathscr{D}} &= I \end{aligned} \qquad (2.25)$$

<u>and normalize the deformed images via the scale change</u>

$$C_n = \frac{1}{n} I_n^{\mathscr{D}}. \qquad (2.26)$$

<u>Then</u>

$$C_n \to C \quad \underline{\text{in Hausdorff norm}} \qquad (2.27)$$

<u>with probability one, where</u> C <u>is a convex set with support function</u>

$$s(\phi) = E[s^d(\phi)]. \qquad (2.28)$$

<u>Proof</u>: The deformed image I_n^d has the support function

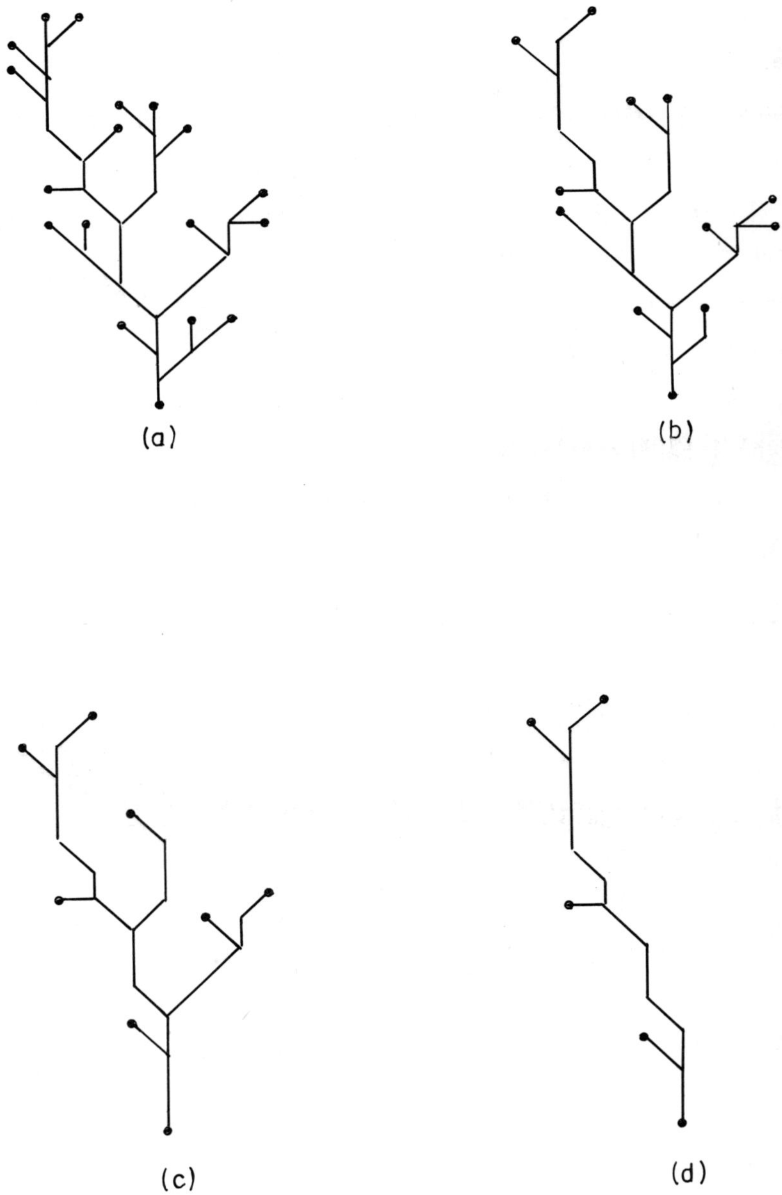

Figure 2.2

$$s_n(\phi) = s^I(\phi) + s_1^d(\phi) + s_2^d(\phi) + \dots + s_n^d(\phi) \qquad (2.29)$$

where the n last terms are independently and identically distributed as $(\frac{1}{n})s_n$. Now a support function can be viewed as belonging to the Banach space $C[0,2\pi]$ with sup norm. But then apply the strong law of large numbers (see Grenander (1963), p. 144) to the average $(\frac{1}{n})s_n$ and we find that, with probability one, it converges to the mean given by (2.7) in the sup norm. But sup-norm topology for the support function is the same as Hausdorff distance topology for the convex sets, and this completes the proof.

Note that the limiting shape does not depend upon the shape of the original pure image.

Another mechanism involving repeated deformations occurs in the study of channel networks; see Section 5.5.

Case 2.11 (pruning deformation). Let the probabilistic family $\mathscr{D}(t)$ of deformations satisfy

i) $\mathscr{D}(t) = d(t_n,t)\dots d(t_1,t_2)d(0,t_1)$ with $0 < t_1 < t_2 < \dots < t_n < t$.

ii) the d's corresponding to disjoint intervals are stochastically independent.

iii) $d(t,t+\varepsilon)$ destroys one at random of the m links of magnitude one with probability $\lambda(t)\varepsilon+o(\varepsilon)$ where λ is a non-negative continuous function.

iv) $d(t,t+\varepsilon)$ leaves the network unchanged with probability $1-\lambda(t)\varepsilon+o(\varepsilon)$.

The meaning of this definition is that the small links, in terms of magnitude, are deleted first and that this is done successively in terms of the deformation parameter t and with independence. As t increases the value of $\lambda(t)$ may change, say decrease, so that the remaining links are less likely to be affected.

The size of the deformation can be expressed by $N(I,I^{\mathcal{D}})$ = the number of links that have been deleted.

To show how this will work we have simulated $\mathcal{D}(1)$, $\mathcal{D}(2)$, $\mathcal{D}(3)$ for $\lambda(t) \equiv .4$ on an image I of magnitude $m = 19$. The results are shown in Figure 2.2.

As the channel network is pruned the number $N(I,I^{\mathcal{D}})$ decreases as a function of the deformation parameter. The exact manner in which this happens is easy to derive.

<u>Theorem 2.3</u>. <u>The probability distribution of</u> $N(I,I^{\mathcal{D}})$, <u>where</u> I <u>has magnitude</u> m, <u>is binomial</u> $N(m,p)$ <u>with</u>

$$p = 1 - \exp(-\int_0^t \lambda(s)ds).$$

<u>Proof</u>: The proof is a standard application of death processes. Put $I^{\mathcal{D}}(t) = \mathcal{D}(t)I$ and introduce the probabilities

$$P_k(t) = P(\text{magnitude of } I^{\mathcal{D}}(t)=k);\ k = 0,1,\ldots,m \tag{2.31}$$

Because of the conditions in Definition 2.11 we get for $k = 0,1,2,\ldots,m-1$

$$\begin{aligned} P_k(t+\varepsilon) &= P_k(t)[1-\lambda(t)\varepsilon k] \\ &\quad + P_{k+1}(t)\lambda(t)\varepsilon(k+1)+o(\varepsilon) \end{aligned} \tag{2.32}$$

and

$$P_m(t+\varepsilon) = P_m(t)[1-\lambda(t)\varepsilon m] + o(\varepsilon) \tag{2.33}$$

leading to the system of differential equations

$$\dot{P}_k = -\lambda(t)kP_k + \lambda(t)(k+1)P_{k+1}, \; k = 0,1,\ldots,m-1$$
$$\dot{P}_m = -\lambda(t)mP_m. \tag{2.34}$$

These equations can be recognized as belonging to a pure death process. We can solve this directly starting with $k = m$, solve for $k = m-1$, and so on and get finally

$$P_k(t) = \binom{m}{k}[1 - \exp(-\int_0^t \lambda(s)ds)]^{m-k} \exp(-k\int_0^t \lambda(s)ds). \tag{2.35}$$

But this is a binomial distribution $B(m,p')$ with

$$p' = \exp(-\int_0^t (s)ds). \tag{2.36}$$

Since $N(I,I^{\mathscr{D}}(t)) = m-h$ the result follows with $p = 1-p'$.

Note that this deformation mechanism affects only the topology of the channel, not its metric properties.

When the bond relation is identically true and with some symmetric graph structure and identification by permutation (see Section 5.3) we meet another deformation mechanism.

<u>Case 2.12</u> (nexus sampling). Given an image with n nodes and with certain binary relations, $I^{\mathscr{D}}$ is obtained by sampling a fraction r of the n nodes. $I^{\mathscr{D}}$ will show those and only those connections whose two nodes both belong to $I^{\mathscr{D}}$.

The sampling operation can be performed by simple random sampling as in Figure 2.3, where $n = 35$ and $r = 4/7$. When we go from I to $I^{\mathscr{D}}$, not only is the number of nodes reduced by the factor r, but one also notices a drastic change in the coarse topology (see Section 5.5). If the average connectivity is denoted by α, so that α denotes the number of

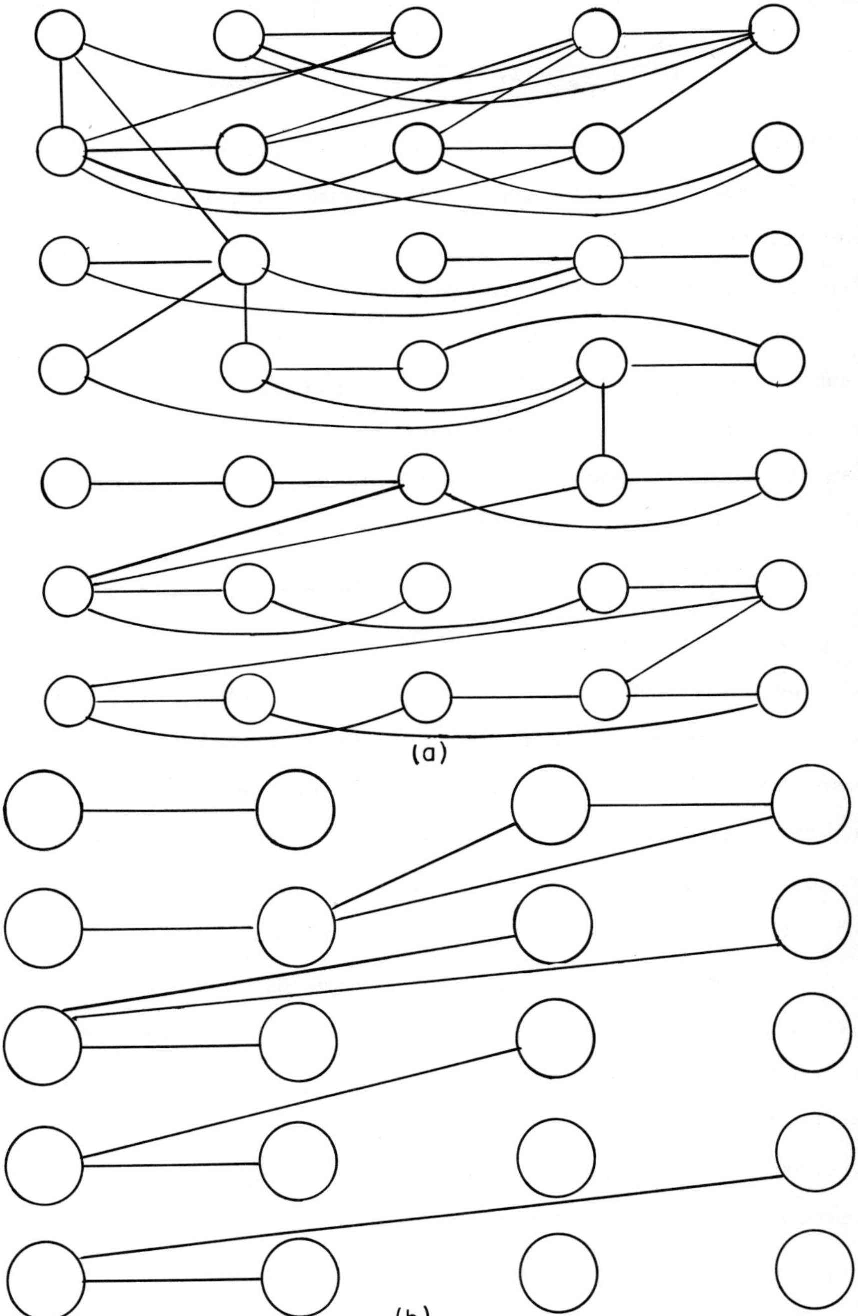

(a)

(b)

Figure 2.3

connections from one node on the average, then we can expect

$$\alpha(I^{\mathscr{D}}) \simeq r\alpha(I) \tag{2.37}$$

so that $I^{\mathscr{D}}$ will appear to be less connected on the average.

We can repeat this $\mathscr{D}$ with varying values of the sampling fraction r. In this way we again arrive at a semigroup of probabilistic deformation mechanisms.

Another sampling operation is also of interest. Select a sample of nodes by simple random sampling. From each node obtained go out along the segments of I, say by taking at most k segments each time. $I^{\mathscr{D}}$ will consist of all the nodes met in this way together with the segments that may connect them. This is called <u>snowball sampling</u>.

3. <u>Changes in Contrast</u>.

We now begin to consider some heteromorphic deformations that can be expressed in terms of contrast patterns, namely those where the change is in the contrast. It will be convenient to include point, line, and set patterns in this discussion, as pointed out in Chapter 3, and we shall even deal with a case where the generators are measures.

Consider a plane crystal image (see (3.5.2) and expose it to a jittering deformation as in (2.2). The generators here consist of points in the plane and $\mathscr{R}$ tells us how they must be located to form a regular lattice. It is clear that in general $\mathscr{D}$ will destroy $\mathscr{R}$ so that the mechanism is heteromorphic. This has already been discussed; see Case 2.3.

Two deformation mechanisms for line patterns will be introduced here.

Case 3.1 (change in curvature). Consider an image as in (5.16) given by the intrinsic equation

$$d\phi/ds = \rho(s) . \tag{3.1}$$

The deformed image will be defined by

$$\mathscr{D}': \rho(s) \to \rho^{\mathscr{D}}(s) = \rho(s)+n(s) \tag{3.2}$$

or

$$\mathscr{D}'': R(\phi) \to R^{\mathscr{D}}(\phi) = R(\phi)+n(\phi) \tag{3.3}$$

where we have written $R = 1/\rho$ for the radius of curvature. In complete specification of a probabilistic $\mathscr{D}$ the stochastic process n must be defined.

Note that Case 3.1 is constructed by additive noise on the curvature, $\mathscr{D}'$, or on the radius of curvature, $\mathscr{D}''$. The latter may be used when I is convex so that to each angle is associated a unique point (if flats are excluded, $R < \infty$).

If we ask that the images in $\mathscr{I}^{\mathscr{D}}$ should have certain global properties we can apply the reasoning in Section 2.10. To illustrate how this can be done, say that $\mathscr{R}$ specifies that the curves be closed and that $\mathscr{D}$ should not destroy this property.

The x-displacement along I is

$$\int_0^{2\pi} \cos\phi\, ds = \int_0^{2\pi} R(\phi) \cos\phi\, d\phi \tag{3.4}$$

so that $\mathscr{D}''$ should satisfy the constraint

$$\int_0^{2\pi} n(\phi) \cos\phi\, d\phi = 0 \tag{3.5}$$

and similarly

$$\int_0^{2\pi} n(\phi) \sin \phi \, d\phi = 0 \tag{3.6}$$

so that these two constraints are the conditions that will restrict the n-process.

To see when $\mathscr{D}$ is covariant in probability remember that with the similarity group of the present image algebra it is natural to assume that $n(\phi)$ be rotationally invariant in probability. If we write γ for the angle of a rotation s then for a pure image $I = I(\phi)$ we get

$$\begin{aligned} s^{-1}ds\, I(\phi) &= g^{-1}d[I(\phi+\gamma)] \\ &= s^{-1}[I(\phi+\gamma)+n(\phi)] = I(\phi)+n(\phi-\gamma) \end{aligned} \tag{3.7}$$

and a requirement alone on rotational invariance means that $\mathscr{D}$ is covariant in probability, something that we strive for in the general construction of deformation mechanisms (see Section 6.1).

Then we can write

$$n(\phi) = \sum_{\nu=0}^{\infty} \alpha_\nu \cos \nu\phi + \sum_{\nu=1}^{\infty} \beta_\nu \sin \nu\phi \tag{3.8}$$

with convergence in the mean if all the α_ν and β_ν are uncorrelated and satisfy

$$\begin{aligned} &\sum_0^{\infty} \sigma_\nu^2 < \infty \\ &\mathrm{Var}(\alpha_\nu) = \mathrm{Var}(\beta_\nu) = \sigma_\nu^2 \, . \end{aligned} \tag{3.9}$$

The constraints (3.7) - (3.8) should now be taken into account and the reader is reminded of the general discussion in Section 2.10. The probability distributions must now be completely specified, second-order properties are not enough.

Doing this by requiring normal distribution, it follows that all the Fourier coefficients α_ν, β_ν are independent so that constraints on α_1 and β_1 as in (3.7) - (3.8) do not influence the others. This implies that $\sigma_1 = 0$ but no change otherwise is needed, so that

$$n(\phi) = \alpha_0 + \sum_{\nu=2}^{\infty} [\alpha_\nu \cos \nu\phi + \beta_\nu \sin \nu\phi]. \tag{3.10}$$

To get some idea of how $\mathscr{D}''$ together with $\mathscr{R}$ deform a closed curve, we have simulated the case when I = circle and plotted the result in Figure 3.1 where the strength of the deformation is lowest in (a) and highest in (c). It will be

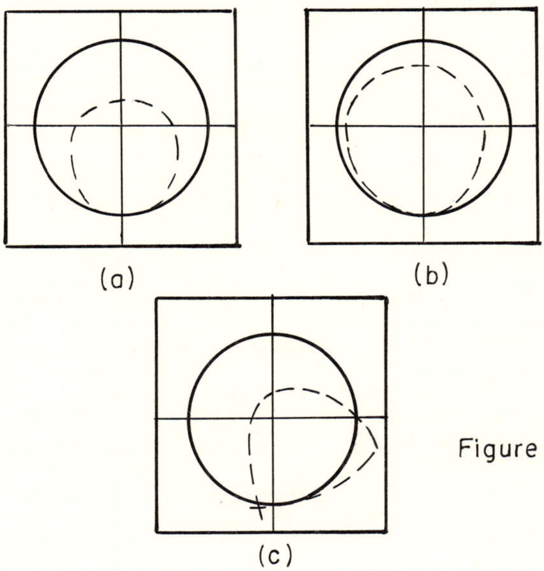

Figure 3.1

seen in (a) and (b) that the shape remains close to that of I, while in (c) more drastic changes are produced. The reason for this is that in (3.12) the value of $n(\phi)$ can very well become negative if the σ's are considerable. The deformed arcs then moved backwards, resulting in the appearance of cusps as in the figure.

We now pass to set image algebras and introduce three $\mathcal{D}$'s. In the first two the generators are points in the discrete plane $x = (\nu,\mu) \in X$ so that an image can be represented through its indicator function $I(x)$ and the contrast values are only 0 and 1.

<u>Case 3.2</u> (individual point changes). For each $x \in X$ the deformed image is defined through the probability of an error of the first kind

$$P(x \notin I^{\mathcal{D}} | x \in I) = e_1(x,I) \tag{3.11}$$

when a point is deleted from I, and through the probability of an error of the second kind

$$P(x \in I^{\mathcal{D}} | x \notin I) = e_2(x,I). \tag{3.12}$$

All these errors should be stochastically independent.

Note that here each point in the background space X is treated separately and is not influenced by what happens elsewhere. This is likely to result in a rather spotty appearance for $I^{\mathcal{D}}$ although I may have been quite regular. Also note that e_1 and e_2 may depend upon the image; for example, they may be <u>boundary-sensitive</u>: e_1 and e_2 take their largest values when x is close to ∂I, where the boundary has been defined as in 5.5.

The case when e_1 and e_2 do not depend upon I is particularly simple and we then speak of an <u>image-independent</u> $\mathcal{D}$. If $\mathcal{D}$ is image-independent it certainly does not mean that $I^{\mathcal{D}}$ is not influenced by I; it is d that is not dependent upon I.

Less chaotic deformations are produced by

Case 3.3 (clustering point changes). A function $\lambda = \lambda(x)$ is given on X and generates a cluster set c such that

$$P(x \in c) = p(x) \tag{3.13}$$

with stochastic independence between points $x \in X$. Given the cluster set the error probabilities are given as

$$e_1 = e_1(x;I,c), \quad e_2 = e_2(x;I,c) \tag{3.14}$$

with stochastic independence between what errors occur at different points conditional upon the cluster set c. Typically e_1 and e_2 would take their largest values close to c and decrease as one moves away from the cluster.

In Figure 3.2 we have illustrated the effect of these models. In (a) the pure image is shown and $I^{\mathcal{D}}$ from Case 3.2 appears in (b). It can be seen that the errors are spread over the entire area in an irregular manner. Case 3.3 is illustrated in (c) where the errors occur in a more regular and cohesive manner.

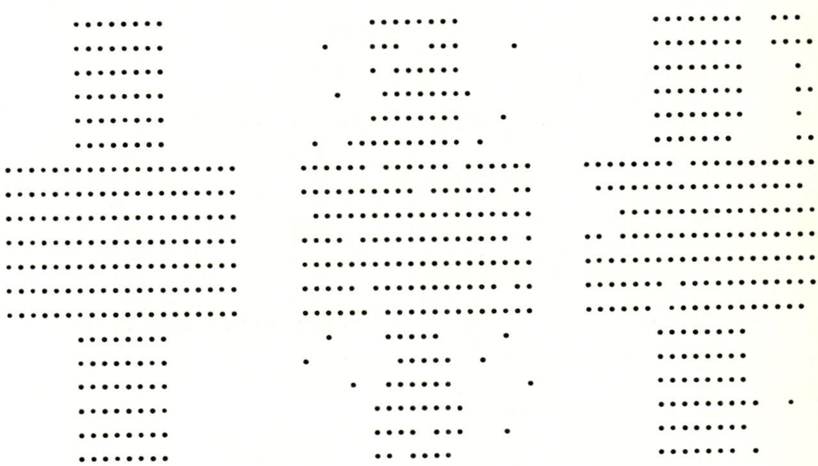

(a) (b) (c)

Figure 3.2

Case 3.4 (disturbed inequalities). Consider an image algebra of sets in the discrete plane of the type

$$I = \{(x,y) \mid f(x,y) \geq t\} \tag{3.15}$$

where the constant t is the threshold and f is the level function. The deformed image is defined through

$$I^{\mathscr{D}} = \{(x,y) \mid f(x,y) \geq t+n(x,y)\} \tag{3.16}$$

where the n's are stochastically independent.

As the reader may have noticed, this is a special case of 3.2 and we have mentioned it separately only because the formulation in (3.15) - (3.16) can be a convenient tool for letting the deformations express dependence upon the boundary ∂I.

We will observe strong fluctuations in the point density of $I^{\mathscr{D}}$ both when a clustering mechanism is at work (as in Case 3.3) and when $\mathscr{D}$ is image-dependent (as in Case 3.2 and Case 3.4) but the reasons are quite different. In the first case it is because of stochastic dependence, in the second because of dependence upon the pure image.

We now come to contrast patterns in the strict sense.

When we discussed general additive noise in Case 2.10 we assumed that the pure image algebra was a semigroup and that the deformations were simply elements in the same semigroup acting upon the pure image . Heteromorphic additive noise assumes only that $\mathscr{D}$ is a semigroup with $\mathscr{T} \subset \mathscr{D}$; dI can very well be outside of $\mathscr{T}$.

Two cases deserve special mention, both of them originating in communication engineering.

<u>Case 3.5</u> (signal detection, fixed signal). Consider a real, separable Hilbert space $\mathscr{A}$ with a closed subspace $\mathscr{T}$ = pure image algebra generated as described in Section 2. We define $\mathscr{D}$ through

$$\mathscr{D}: I \to I+d \tag{3.17}$$

where d has a given probability measure P over $\mathscr{A}$.

<u>Case 3.6</u> (signal detection, random signal). Same as Case 3.5 except that we start with diffuse patterns characterized through an a priori measure π over $\mathscr{T}$.

An important variation on this theme is given by

<u>Case 3.7</u> (disturbed equation). Let L be a linear operator in $\mathscr{A}$ and define $\mathscr{T}$ as the null-space

$$\mathscr{T} = \{I \mid LI = 0\} \tag{3.18}$$

and

$$\mathscr{D}: I \to I^{\mathscr{D}} \text{ such that } LI^{\mathscr{D}} = d \tag{3.19}$$

where d is an element in $\mathscr{A}$ chosen with a probability measure P.

To make the definition in (3.19) unique, it will be necessary to impose additional constraints on $I^{\mathscr{D}}$. If, for example, the elements of $\mathscr{A}$ are temporal images and L is a differential operator, such constraints could be given as initial conditions so that the initial values and derivatives up to a given order coincide for I and $I^{\mathscr{D}}$.

The following deformation combines Case 2.5 with 3.7. Given one mechanism that affects the attributes of the

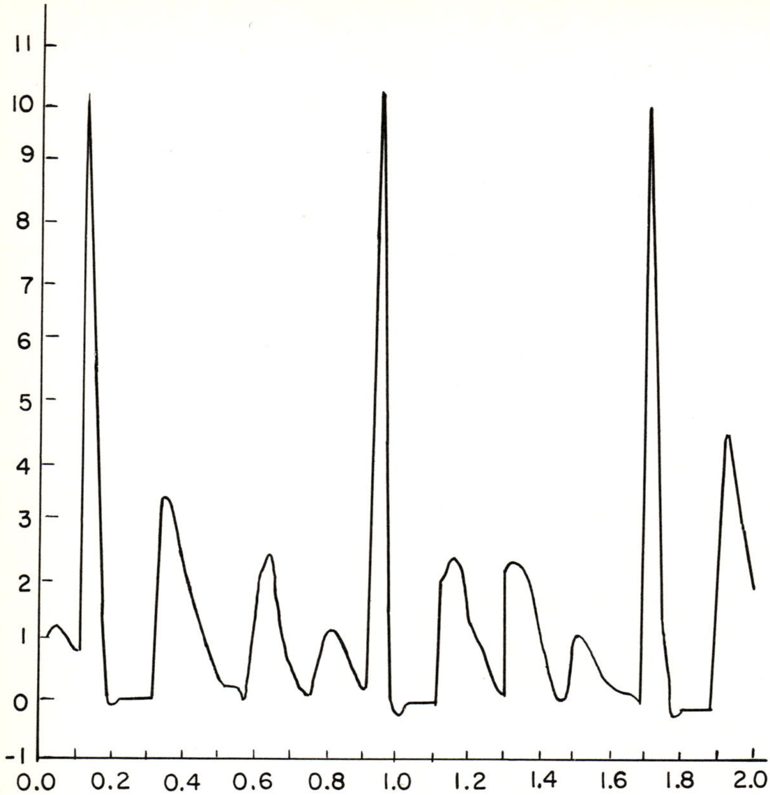

Figure 3.3

generator, and another like (3.19) in a regime, we get a regime process as the deformed temporal patterns. The length of the regime and the amplitude of the driving spike at the beginning of each regime are now random. A numerically synthesized deformed pattern of this type is given in Figure 3.3 which should be compared to Figure 3.4.7 a pure image.

Heteromorphic deformations destroy by definition the regularity of the image. This can be done in several ways, two of the most important ones being <u>breaking bond connections</u> and <u>disturbing bond relations</u>.

The first of these is illustrated by the following case where the connection type is violated by breaking up bonds and replacing them by others.

<u>Case 3.8</u> (breaking wave-form bonds). The pure image algebra is as in Case 2.4.1. A pure image I consisting of generators g_ν connected by continuity conditions is replaced by I consisting of generators g'_ν where the $L^{\alpha'}$ corresponding to g'_ν has been chosen by picking a value for α' from a probability distribution that may depend upon α: the generator index for g_ν. The new configuration does not satisfy Σ. Except for that the bond relations will be assumed to be the same.

This assumes that the generators can be determined from I (see the discussion in Section 3.4).

A similar case illustrates how bond relations can be disturbed.

<u>Case 3.9</u> (disturbing wave-form bond relations). Pure images as in Case 2.4.1. We now let the connection type be intact for $\mathcal{T}^{\mathcal{D}}$ with $\alpha \to \alpha+1$ structure, but let the initial value of a new waveform segment be replaced by a randomized version of the ending values of the previous (deformed) segment. Otherwise $\mathcal{T}^{\mathcal{D}}$ behaves as the pure image algebra.

The last case applies with small changes to configurations analyzed metrically in Theorem 2.10.3, if we no longer require that ρ = EQUALITY. Ask instead only that the two bonds x_1 and x_2 satisfy $|x_1-x_2| < \varepsilon$ where ε is a given positive number. The deformed images are no longer in $\mathcal{T}^{\mathcal{D}}$, but if ε is small the deformations are barely noticeable.

Cases 3.5 - 3.7 represent well-known models but in a more general setting. The next case is also familiar to the mathematician but its pattern analysis (see Chapter 17) has not been studied in depth in spite of repeated attempts in the literature.

<u>Case 3.10</u> (blurring of spectrum patterns). Starting from the image algebra in Case 2.9.5 with pure point spectra patterns, we define the family of deformations $\mathscr{D}(t)$ through the diffusion equation for $I^{\mathscr{D}} = I^{\mathscr{D}}(x,t)$

$$\frac{\partial I^{\mathscr{D}}}{\partial t} = \frac{\sigma^2}{2} \frac{\partial^2 I^{\mathscr{D}}}{\partial x^2}, \quad t > 0 \tag{3.20}$$

with $I^{\mathscr{D}} = I$ for $t = 0$.

If I has the spectrum with amplitudes A_ν at frequencies x_ν we get $I^{\mathscr{D}}$ with a continuous spectrum of density

$$\frac{1}{\sqrt{2\pi}\sigma} \sum_\nu \frac{A_\nu}{\sqrt{t}} \exp\left[-\frac{1}{2\sigma^2 t}(x-x_\nu)^2\right]. \tag{3.21}$$

We have illustrated this in Figure 3.4 where the pure image $t = 0$ has three spectral lines (see (a)) which become more blurred when the deformation strength parameter t increases. The ordinates have been scaled differently in the different figures.

The practical background that motivated Case 3.10 is mass spectroscopy (see, e.g., Ernst (1966) and Medgyessy (1961)) where one is led to this sort of superposition of intensity curves originating from more or less well concentrated elementary spectra. Variations of (3.20) may be of interest where

$$\partial I^{\mathscr{D}}/\partial t = LI^{\mathscr{D}}; \quad t > 0 \tag{3.22}$$

and L is some linear operator. When L is chosen as the infinitesimal operator of the Cauchy process (see Feller (1965), IX.6) one gets the <u>Lorentz spectra</u> in which case (3.21) will be replaced by

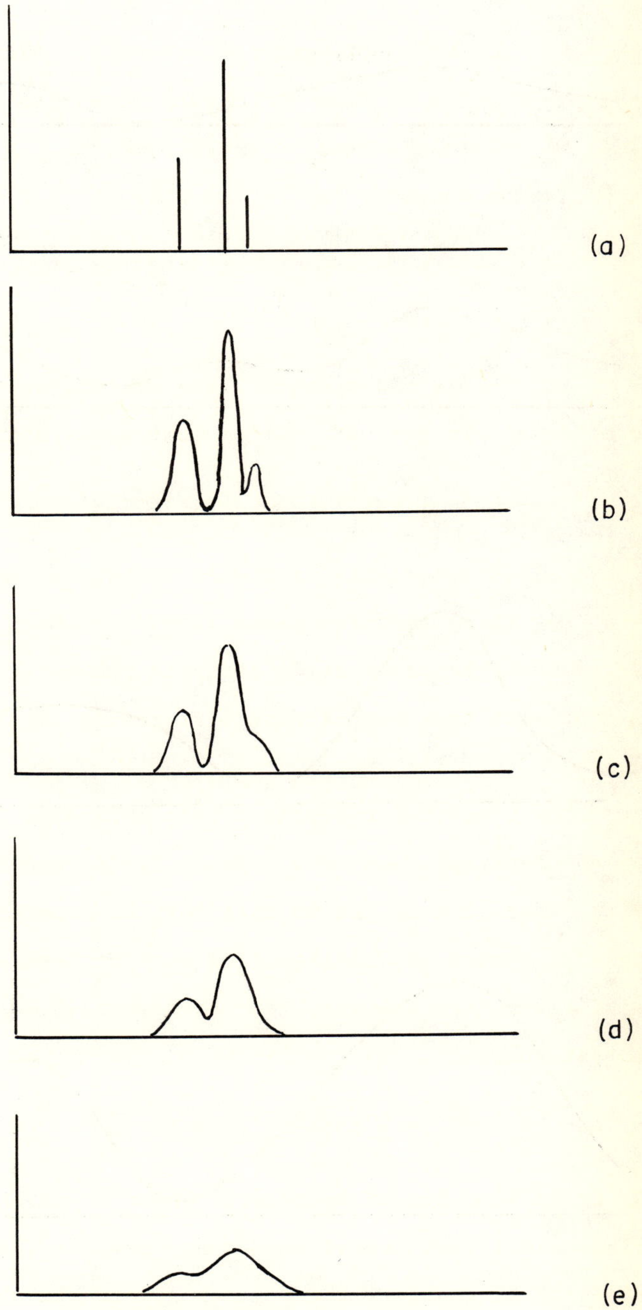

Figure 3.4

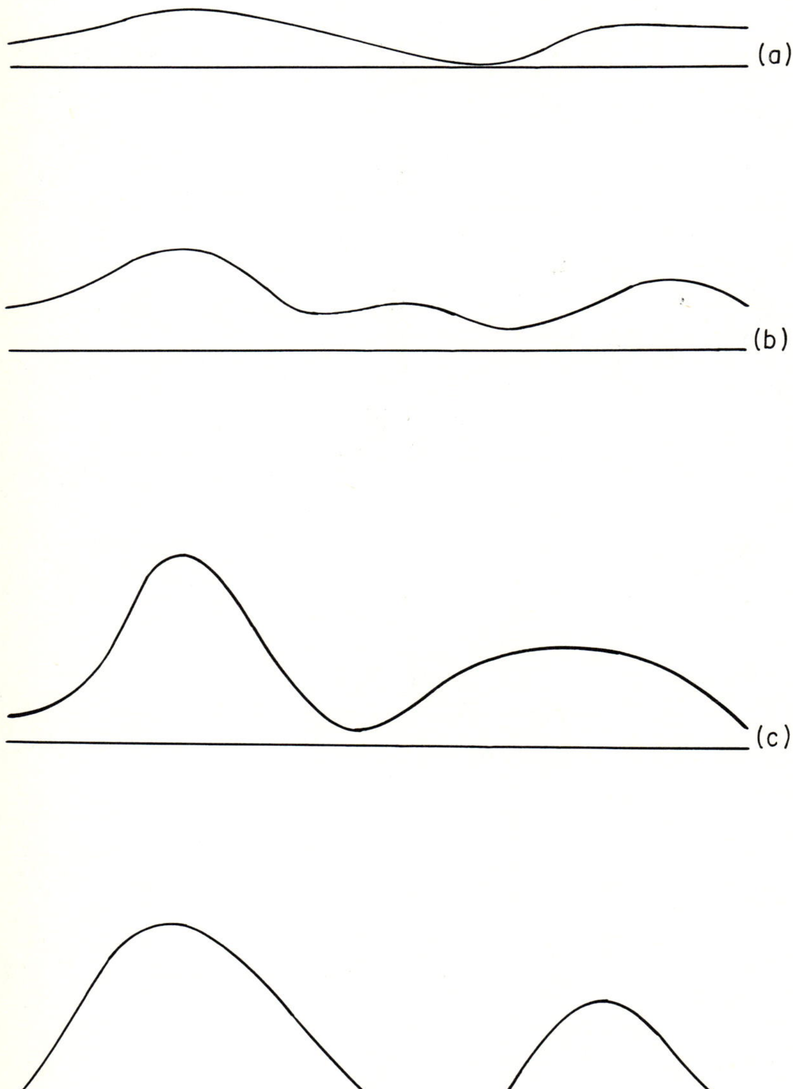

Figure 3.5

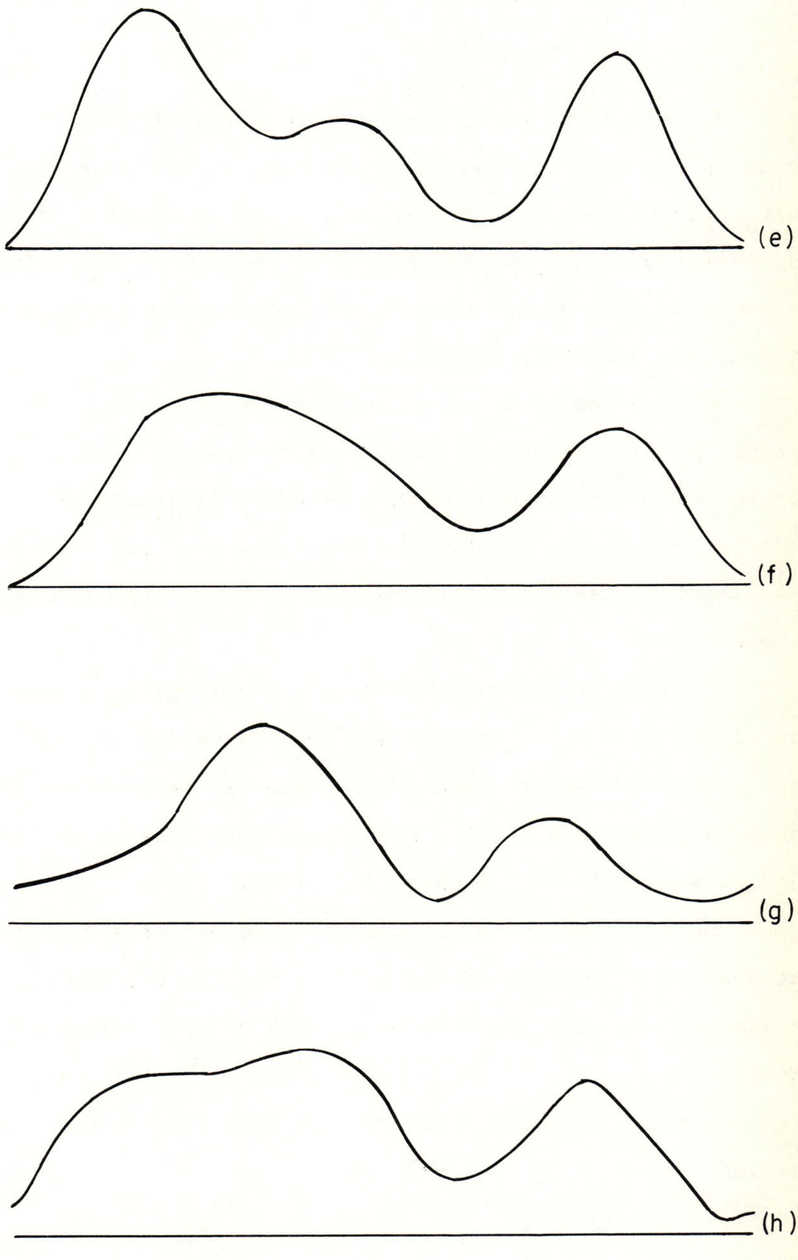

Figure 3.5

$$\frac{1}{\pi t} \sum \frac{A_\nu}{\sigma_\nu \left[1 + \left(\frac{x - x_\nu}{t\sigma_\nu}\right)^2\right]} \quad . \tag{3.23}$$

A similar but considerably more complicated deformation mechanism has recently been suggested for dynamic patterns in geomorphology (see Grenander (1975)). Let us consider diffuse patterns formed by tectonic uplift activity, occurring at some time points, and then deformed by an erosion mechanism that operated like Case 3.10.

We have two opposing mechanisms $\mathscr{E}$ and $\mathscr{S}$. The role of $\mathscr{E}$ is that of erosion, breaking down dominating features of the land form, with a tendency to levelling. Let us denote by z the height over some arbitrary zero level. We shall study the dynamic changes of z as a function of space and time.

To simplify the analysis we shall let space be one-dimensional with circular boundary conditions for z. In other words, the single space coordinate x denotes the location of a point on a circle, say of circumference 1, so that $x \in (0,1)$ where 0 and 1 are identified.

The restriction to one dimension is not essential for what follows but should of course be removed. One could, for example, let the location be a point on the surface of a sphere.

We shall postulate that $\mathscr{E}$ operates via the diffusion equation

$$\frac{\partial z}{\partial t} = \frac{\sigma^2}{2} \frac{\partial^2 z}{\partial x^2} \tag{3.24}$$

where σ^2 is a positive parameter that expresses the strength of the erosion deformation (see Sprunt (1972)).

If z is given as $z(x,0) = \zeta(x)$ at time $t = 0$, it is well known that we can write the solution of (3.24) as

$$z(x;t) = \frac{1}{\sigma\sqrt{2\pi t}} \sum_{\nu=-\infty}^{\infty} \int_0^1 \zeta(y) \exp - \frac{(x-y-\nu)^2}{2\sigma^2 t}\, dy$$

$$= \int_0^1 K(x,y;t)\zeta(y)dy \tag{3.25}$$

for $t > 0$. This will be abbreviated as

$$z(\cdot;t) = T^t\zeta \tag{3.26}$$

and the set of <u>erosion deformations</u> $\mathscr{E} = \{T^t, t>0\}$ is seen to form a one-parameter semigroup.

The generator of the diffuse seismic patterns $\mathscr{S}$, <u>the tectonic one</u>, is instantaneous and is applied only at discrete time points $\{t_n;\ n = \ldots -1,0,1,\ldots\}$. Later on we shall see how this can be relaxed without essentially changing the model. The effect of $\mathscr{S}$ applied at the time t_n is to add a term

$$\delta(t-t_n)g_n(x) \tag{3.27}$$

to the right-hand side of equation (3.24). This amounts to the same thing as to change z instantaneously at time t_n:

$$\phi:\ z \to z + g_n(x). \tag{3.28}$$

Note that the change is associated with one of the points $\{t_n\}$.

It will be assumed that these points form a Poisson process with some constant intensity λ.

For g_n we shall assume that they are all stochastically independent of each other and of the t_n's. Further, they shall be stationary and periodic processes on the circle which implies that any one of them can be written as

$$g(x) = \sum_{\nu=-\infty}^{\infty} g_\nu \, e^{2\pi i\nu x} \tag{3.29}$$

with uncorrelated Fourier coefficients g. Also g is assumed to have identically mean zero and we introduce the spectral power

$$\gamma_\nu = E[|g_\nu|^2] \tag{3.30}$$

which must have the sum $\Sigma\gamma_\nu$ convergent.

To any g we associate the total change, the stochastic variable

$$G = \int_0^1 g(x)dx. \tag{3.31}$$

We want the total mass of the landform, i.e. the integral of z, to remain constant (conservation of mass). It is obvious that the deformation $\mathscr{L}$ conserves mass, but this is true for $\mathscr{S}$ only if $G = 0$; see equation (3.28). Therefore we assume that $g = g_0 = 0$ which is the same thing as saying that $\gamma_0 = 0$.

Let us start with an arbitrary landform z_0 at time 0 and with such numbering of the t_n that

$$t_{-1} < 0 < t_1 < t_2 < \ldots < t_n < t < t_{n+1}. \tag{3.32}$$

It is no restriction to assume that the zero level has been chosen so that the integral of z_0 is zero. At time t_1 we have $z = T^{t_1} z_0$ before $\mathscr{S}$ is applied and $z = g_1 + T^{t_1} z_0$ afterwards. Between t_1 and t_2 we have

$$z = T^{s-t_1}[g_1+T^{t_1}z_0] = S^{s-t_1}g_1 + T^s z_0. \tag{3.33}$$

In the same way for s between t_2 and t_3 we get

$$z = T^{s-t_2}g_2 + T^{s-t_1}g_1 + T^s z_0 \tag{3.34}$$

and generally

$$z = T^{t-t_n}g_n + T^{t-t_{n-1}}g_{n-1} + \dots + T^{t-t_1}g_1 + T^t z_0. \tag{3.35}$$

Our first problem is to see what happens in (3.35) when time increases. The height z of our landform can be viewed as the solution of a stochastic partial differential equation driven by tectonic activity consisting of spikes (in time) and stationary (in space). Unfortunately, stochastic partial differential equations are not well understood at present so that their general theory does not help us much. It is still possible to deal directly with this simple stochastic partial differential equation, and we shall arrive at the result by a technique that seems capable of considerable generalization. If it should turn out that the solution does not tend to statistical equilibrium as t increases this would of course make the model less attractive, so we must first look into this question.

The expansion (3.33) involves the random elements (the t_n's and g_n's) intertwined with the semigroup of operators T. We shall see now that a direct approach will solve the problem.

Note that for fixed location x we can write

$$z = T^t z_0 + \int_0^t T^{t-s}Z(ds) \tag{3.36}$$

where the stochastic set function Z has the properties

$$\begin{aligned}&\text{(i) } Z(i_1) \text{ and } Z(i_2) \text{ are uncorrelated for disjoint intervals } i_1 \text{ and } i_2.\\&\text{(ii) } Z(i) = 0 \text{ with probability } 1+\lambda m(i)+o(m(i)).\\&\text{(iii) } Z(i) = g(\cdot) \text{ with probability } \lambda m(i)[1+o(1)].\end{aligned} \tag{3.37}$$

Note that our stochastic set function Z is <u>function-valued</u>.

The first term of (3.36) is deterministic and easy to deal with. It can be written as

$$\int_0^1 z_0(y)K(x,y;t)\,dy \to \int_0^1 z_0(y)\,dy = 0$$

using the fact that the theta series defining K

$$K(x,y;t) = \frac{1}{\sigma\sqrt{2\pi t}} \sum_{\nu=-\infty}^{\infty} \exp\left[-\frac{(x-y-\nu)^2}{2\sigma^2 t}\right] \tag{3.39}$$

is the Gaussian density on the circle. On the circle the central limit theorem means convergence to the uniform distribution (see Grenander (1963)) which implies (3.38).

To simplify the notation we now move the interval $(0,t)$ back t units of time to $(t,0)$. We shall prove that the stochastic integral

$$J_t = \int_{-t}^{0} T^{-s}Z(ds) = \int_{-t}^{0} K(x,y;-s)Z(ds) \tag{3.40}$$

converges in L_2-norm as $t \to \infty$. Because of Z being an orthogonal process this is the same as proving that the variance (the L_2-norm squared) remains bounded.

Direct calculation gives

$\mathrm{Var}(J_t) =$

$$= \lambda \int_{s=-t}^{0} \int_{y=0}^{0} \int_{y=-s}^{0} K(x,y;-s)K(x,y';-s)E[g(y)g(y')]dydy'ds. \quad (3.41)$$

But

$$E[g(y)\overline{g(y')}] = \sum_{\nu=-\infty}^{\infty} \gamma_\nu \, e^{2\pi i\nu(y-y')} \quad (3.42)$$

with absolute and uniform convergence. Hence, using (3.29), we get

$$\mathrm{Var}(J_t) = \lambda E\left(\int_{s=-t}^{0} \left|\int_{y=0}^{1} K(x,y;-s)g(y)dy\right|^2 ds\right) \quad (3.43)$$

and form

$$\int_{y=0}^{1} K(x,y;-s)g(y)dy = \frac{1}{\sigma\sqrt{2\pi(-s)}} \int_{-\infty}^{\infty} g(y)\exp\left(\frac{(x-y)^2}{\sigma^2 s}\right)dy. \quad (3.44)$$

This gives us

$$\sum_{\nu=-\infty}^{\infty} g_\nu \, \exp[2\pi i\nu x + \frac{\sigma^2 s}{2} 4\pi^2\nu^2]. \quad (3.45)$$

Evaluating the expected value and using the fact that the g_ν are uncorrelated, we find

$$\mathrm{Var}(J_t) = \lambda \int_{-t}^{0} \left[\sum_{\nu=-\infty}^{\infty} \gamma_\nu \, \exp \, 4\pi^2\sigma^2\nu^2 s\right] ds$$

$$= \lambda \sum_{\nu=-\infty}^{\infty} \frac{\gamma_\nu}{4\pi^2\sigma^2\nu^2} [1-\exp(-4\pi^2\sigma^2\nu^2 t)]. \quad (3.46)$$

As $t \to \infty$ this tends to

$$\lim_{t\to\infty} \mathrm{Var}(J_t) = \lambda \sum_{\nu=-\infty}^{\infty} \frac{\gamma_\nu}{4\pi^2\sigma^2\nu^2} \quad (3.47)$$

which is finite; remember that $\gamma_0 = 0$. This establishes convergence of the stochastic integral so that we have indeed convergence to statistical equilibrium.

To find the covariances of the stationary (in both time and space) field $z(x,t)$ representing the pattern of our landform we just repeat the same sort of operation but for the infinite integral that we now know exists:

$$z(x,t) = \int_{-}^{t} K(x,y;t-s)Z(ds). \tag{3.48}$$

In t and $t+h$, with $h > 0$, and for two locations x and x' we calculate

$$R = \mathrm{Cov}[z(x,t),z(x',t+h)]. \tag{3.49}$$

We get

$$R = \lambda \int_{-\infty}^{t} E\left[\int_{y=0}^{1} K(x,y;t-s)g(y)dy \int_{y'=0}^{1} K(x',y';t-s+h)g(y')dy'\right]$$

$$= \lambda \int_{-\infty}^{t} \left\{\sum_{\nu=-\infty}^{\infty} \gamma_\nu \, e^{2\pi i\nu(x-x')}\exp[-2\pi^2\sigma^2\nu^2(t-s+t-s+h)]\right\} ds$$

$$= \frac{\lambda}{4\pi^2\sigma^2} \sum_{\nu=-\infty}^{\infty} \frac{\gamma_\nu}{\nu^2} e^{2\pi i\nu(x-x')}\exp(-2\pi^2\sigma^2\nu^2 h) = R(x-x',y). \tag{3.50}$$

This proves the following result.

<u>Theorem 3.1</u>. <u>The stochastic partial differential equation that we have used to describe the geomorphological patterns defines a height field in statistical equilibrium with the covariance function in</u> (3.50). <u>In particular, the time covariance function</u> $(x = x')$ <u>is</u>

$$R_{time}(h) = \frac{\lambda}{4\pi^2\sigma^2} \sum_{\nu=-\infty}^{\infty} \frac{\gamma_\nu}{\nu^2} \exp(-2\pi^2\sigma^2\nu^2 h), \quad h > 0 \tag{3.51}$$

<u>and the space covariance is</u>

$$R_{space}(x-x') = \frac{\lambda}{4\pi^2\sigma^2} \sum_{\nu=-\infty}^{\infty} \frac{\gamma_\nu}{\nu^2} e^{2\pi i\nu(x-x')}. \tag{3.52}$$

It will be convenient to write

$$R(x-x',h) = \sum_{\nu=-\infty}^{\infty} \frac{\beta_\nu}{\nu^2} e^{2\pi i\nu(x-x')} a_\nu(h) \tag{3.53}$$

with

$$\beta_\nu = \frac{\lambda}{4\pi^2\sigma^2}\gamma_\nu , \qquad a_\nu(h) = \exp(-2\pi^2\sigma^2\nu^2 h). \tag{3.54}$$

This theorem enables us to formulate

Case 3.11 (dynamic geomorphological patterns). The diffuse patterns generated by $\mathcal{S}$ are deformed by $\mathcal{L}$ so that the deformed images satisfy the stochastic partial differential equation and converge to a steady state solution with properties given in (3.51-52).

The functioning of the stochastic partial differential equation is not so easy to visualize so that to gain some insight a simulation experiment was designed and carried out.

Several runs were obtained but we shall present only the results from one of them. In Figure 3.5 we represent successive solutions of the stochastic partial differential equation. The value of λ is low, .05, so that the seismic activity is not very frequent, but the γ's are fairly large. We therefore get rather drastic but seldom occurring changes, resulting in a hilly landscape.

If we make λ larger we get closer to a continuously ongoing seismic activity. We shall return to the analysis of this type of pattern in Part III.

4. Background Deformations.

Still assuming the pattern to consist of images of contrast type, we shall now study the important class of deformation mechanisms that change the background space.

Definition 4.1. Consider a semigroup $\mathcal{D}$ of mappings $d: X \to X$ and define the deformed images through

$$d: \quad I^{\mathscr{D}}(x) = I[d(x)], \quad d \in D. \tag{4.1}$$

<u>This is called a background deformation</u>.

In general we do not insist that the d's have inverses, although this will sometimes be the case. If a d-mapping is not onto X so that its range $R(d)$ is not the whole X, a part of the image I is lost, namely the set of values of $I(x)$ for which $x \notin R(d)$.

Note that (4.1) defines deformations independently of the pure image, in the sense that it is the background space that is affected without reference to I, and then the resulting change is carried along by I.

Here $I^{\mathscr{D}}$ has no bond structure which will be needed if deformed images shall be combined. This is not surprising, since deformations tend to destroy structure. However, in rare cases it may happen that a background deformation is automorphic. Such is the case if d is a generator deformation such that

i) Σ is monatomic and d preserves σ

ii) d maps G into G

iii) if g_1 and g_2 have a bond relation $\beta_1 \rho \beta_2$ then for $g_1' = dg_1$ and $g_2' = dg_2$ the relation $\beta_1' \rho \beta_2'$ holds. (4.2)

Under these conditions any regular configuration is deformed into another regular configuration so that $\mathscr{T}^{\mathscr{D}} \subseteq \mathscr{T}$ and the deformation mechanism is automorphic.

Let us now consider the case when X is the real line or a subset of it. In particular, for the time patterns $I = I(t)$, t = time, we get the notion of <u>subjective time</u> $\tau =$

$\tau(t)$ and the deformed image is $I(t) = I[\tau(t)]$. Here τ is some non-decreasing function which describes how the observer's time scale is changed. At the absolute time t the subjective time is off an amount $\tau(t)-t$ and the subjective time rate is changed by a factor $\tau'(t)$.

In order that a subjective time deformation be covariant, we need $sd = ds$ (in distribution) for any s, which here amounts to $s\tau(t) = \tau(st)$ in distribution. Note that this concerns all the finite-dimensional distributions, not just the one-dimensional marginals.

In the case of S = translation group we should have the same distribution for $h+\tau(t_1),h+\tau(t_2),\ldots,h+\tau(t_n)$ as for $\tau(t_1+h),\tau(t_2+h),\ldots,\tau(t_n+h)$. In other words, $\tau(t) = t +$ stationary stochastic process.

The further specification of τ will depend on <u>the clock that measures the time of the observer</u>. Only three examples will be mentioned here.

If the clock is a second-order oscillator, whose period is the absolute time unit, but with a nonlinearity that causes deviations from a purely harmonic oscillation, then

$$\tau(t) = \int_0^t p(s)ds \tag{4.3}$$

where p is a positive, continuous function with integral P over a period of length P. Hence $\tau(t)$ will fluctuate around $\tau = t$ without introducing any systematic deviations in the long run.

If the clock runs at constant rate except for sudden changes occurring at discrete time points $\{t_\nu\}$, then τ can still be written as in (4.3) but p is now piecewise constant.

More precisely, let $\{t_\nu\}$ form a Poisson process with intensity λ, let $p(s)$ be a positive stochastic variable of mean 1, and let the values of p in different t_ν-intervals be i.i.d. Then $E[\tau(t)] = t$ so that there are no systematic time changes in either direction. This is more irregular than the first example.

An even more drastic time deformation occurs when we assume that $\tau(t)$ forms a stochastic process with independent and non-negative increments satisfying $E[\tau(t_2)-\tau(t_1)] = t_2-t_1$ for $t_2 \geq t_1$. Then the realizations of τ will be step functions almost certainly so that the deformed image will appear discontinuous even when the pure one was not. Some of the information in I will then be lost. A special case of this is when $\tau(t+h)-\tau(h)$ has the type III frequency function

$$f(\tau,h) = \frac{1}{\Gamma(h)} \tau^{h-1} e^{-\tau}, \quad \tau > 0. \tag{4.4}$$

We combine these models into

<u>Case 4.1</u> (subjective time). The mappings $d: R \to R$ are nondecreasing and take a pure image $I = I(t)$ into $I^{\mathcal{D}} = I(\tau(t))$, where $\tau(t) = d(t)$. We assume that τ has been obtained in one of the following ways:

a) $\tau(t)$ is deterministic as in (4.8).

b) $\tau(t)$ is random, defined as in (4.8) but piecewise constant between Poisson points and with $E(p) = 1$.

c) $\tau(t)$ is random with independent increments satisfying $E[\tau(t_2)-\tau(t_1)] = t_2-t_1$, $t_2 > t_1$.

Keeping $X \subseteq R$, but not necessarily interpreting x as time, let us assume that for $x_1 < x_2 < \ldots < x_n$ the joint distribution of $y_1 = d(x_1)$, $y_2 = d(x_2), \ldots, y_n = d(x_n)$ is given through a frequency function $f(y_1,y_2,\ldots,y_n;\ x_1,x_2,\ldots,x_n) = f(y;x)$ and that $d(x)$ is almost certainly non-decreasing as a function of x.

Given $I = I(x)$ we can calculate, at least in principle, the distribution of $I^{\mathscr{D}}(x_1,x_2,\ldots,x_n)$. To actually do this we will have to take into account the global form of I. Say, for example, that I is increasing with a positive and continuous derivative over X. We could then get directly the frequency function $g(y_1,y_2,\ldots,y_n;\ x_1,x_2,\ldots,x_n)$ of $I^{\mathscr{D}}$ at the points $x_1,x_2,\ldots,x_n$ as

$$g(y;x) = f(J(y_1),\ldots,J(y_n);\ x_1,x_2,\ldots,x_n)\prod_{\nu=1}^{n} J'(y_\nu) \qquad (4.5)$$

where J is the inverse function of I.

On the other hand, if I is not monotonic we will get more terms on the right-hand side of (4.5) but considerable simplification occurs due to the order-preserving nature of the d's. In the case of a Tschebychev pattern (see Section 5.5) of structural complexity α we can have at most α branches of the multivalued function J. Indexing these branches in increasing order as $J_i(y)$, we get the relation

$$g(y;x) = \sum f[J_{i_1}(y_1),\ldots,J_{i_n}(y_n);x_1,x_2,\ldots,x_n]$$
$$\prod_{\nu=1}^{n} |J'_{i_\nu}(y_\nu)| \qquad (4.6)$$

summed over all indices $(i_1,i_2,\ldots,i_n)$ such that $i = 1,2,\ldots,\alpha$ and $i_1 \leq i_2 \leq \ldots \leq i_n$. Counting terms in (4.6), we get at most

$$\frac{(n+1)(n+2)\ldots(n+\alpha-1)}{(\alpha-1)!} \tag{4.7}$$

terms by a direct combinatorial argument. The number of terms can be smaller than (4.7) depending upon the values of y. Note that at exceptional points (measure zero) the derivatives $J'_{i_\nu}(y_\nu)$ need not exist.

The above is certainly not limited to Tschebychev patterns: for a given I the relation (4.6) still holds under appropriate regularity conditions. Depending upon the global properties of the contrast images, however, the corresponding relation may involve sums of cardinality that need not be bounded over $\mathcal{T}$.

Allowing for the moment X to be an arbitrary background space with a measure m on it, write $s(I)$ for the support of the pure image, i.e. the set where $I(\cdot)$ is defined. For a fixed deformation d the observer will see the values of $I(y)$ for which $y \in ds(I)$, $y = d(x)$. Calling the measure of y induced by the deformation the <u>exposure measure</u> m^d, it is clear that m^d describes how intensively the values of $I(y)$ are observed. For example, if a set $A \subset s(I)$ has m^d-measure zero this part of the pure image will not be available to the observer and we have lost information. On the other hand, it can happen that a single y-value has positive m^d-measure but m-measure zero: then the deformation emphasizes the information carried by the pure image at this point. If the Radon-Nikodym derivative dm^d/dm exists it will be denoted by ε and called the <u>exposure density of the deformation function</u>.

If $ds(I)$ contains points not belonging to $s(I)$ the deformed image will not be defined over the whole support of I.

We now return to the real line and assume that d maps $[0,1]$ into itself. The background deformation is partially described through the mean value and variance functions

$$\begin{aligned} m(x) &= E[d(x)] \\ V(x) &= \mathrm{Var}[d(x)]. \end{aligned} \tag{4.8}$$

If $m(x) < x$, $0 < x < 1$, $\mathscr{D}$ is called <u>left-systematic</u> and similarly <u>right-systematic</u> if $m(x) > x$, $0 < x < 1$. A particularly important case is when $\mathscr{D}$ is <u>fair</u>, meaning that $m(x) \equiv x$.

Introduce the kernel K of the deformation as $K(t,x) = P(d(x) \le T)$. We shall assume that

$$\begin{aligned} &d \text{ is non-decreasing and } 0 \le d(x) \le 1 \text{ a.c.} \\ &d(0) = 0,\ d(1) = 1 \text{ a.c.} \\ &d(x) \text{ is continuous in probability.} \\ &d(x) \text{ has a continuous distribution if } x \ne 0,1. \end{aligned} \tag{4.9}$$

<u>Lemma 4.1</u>. In order that there be a deformation mechanism with the properties (4.9) it is necessary and sufficient that

a) K be a d.f. in t for each x and non-increasing in x.

b) $K(t,0) = 1$ for $t > 0$ and $K(t,1) = 0$ for $t < 1$.

c) K be continuous in x for each t and in t for each positive x.

<u>Proof</u>: If $\mathscr{D}$ is as in (4.9) we have $d(x_1) \le d(x_2)$ a.c. if $x_1 < x_2$, which implies that $K(t,x_1) \ge K(t,x_2)$. The condition (b) follows because of the boundary conditions on $d(x)$, and

(c) from the two continuity conditions in (4.9).

Conversely, if K satisfies the stated conditions, define $d_\xi(x)$ through the equation

$$K(d_\xi(x),x) = \xi \tag{4.10}$$

where ξ is $R(0,1)$. If $K(t,x)$ has a t-interval where it is constant the value of $d_\xi(x)$ is not determined uniquely, but this happens only with probability zero. Of course $d_\xi(x)$ will have the kernel K and it satisfies the conditions in (4.9). Only continuity in probability may not be obvious. However, for arbitrary x and an $h > 0$ we have

$$0 = \lim_{h\downarrow 0} \int_0^1 t[K(dt,x+h) - K(dt,x)] =$$

$$= \lim_{h\downarrow 0} E[d_\xi(x+h)-d_\xi(x)] \tag{4.11}$$

and since the second integrand is non-negative it follows that $d_\xi(x)$ is indeed continuous in probability.

For a given K many $\mathscr{D}$'s may be formed. The one given by (4.10) is of limited interest: it is too deterministic for practical use since knowing $d(x)$ for a single x-value will determine the entire d.

In order to arrive at more representative $\mathscr{D}$'s we present two other cases. In the first one we shall have

$$K(t,x) = \exp[-\,xG(t)] \tag{4.12}$$

where G is non-negative, non-increasing and absolutely continuous for each $t < 1$ with $G(0) = +\infty$, $G(1) = 0$.

Then the conditions of Lemma 4.1 are satisfied and the construction there would give a $\mathscr{D}$ with K as the kernel.

A more flexible deformation mechanism can be obtained as follows.

Case 4.2 (successive maxima). Consider the points $0 = x_0 < x_1 < x_2 < \dots < x_n < x_{n+1} = 1$ and define z_i as independent stochastic variables with the d.f.

$$P(z_i \le t) = \exp[-(x_i - x_{i-1})G(t)];\, i=1,2,\dots,n. \tag{4.13}$$

Put

$$d(x_i) = \max(z_1, z_2, \dots, z_i),\ d(0) = 0,\ d(1) = 1, \tag{4.14}$$

so that

$$P(d(x_i) \le t) = \exp -x_i G(t) = K(t, x_i). \tag{4.15}$$

This defines all the finite-dimensional probability distributions of d in a consistent manner as can be seen by observing that $\mathcal{D}$ is Markovian with the transition probability given $d(x) = y$ to $d(x') = y'$ for $x' > x$,

$$\delta(y'-y)\exp[-(x'-x)G(y)] + (x'-x)g(y')\exp[-(x'-x)G(y')] \tag{4.16}$$

where $g = -G'$. The realizations of d will be step functions and the length of the interval in which $d(x)$ remains equal to $y = d(x)$ is exponential (the first term in (4.16)) with mean $1/G(y)$. In other words, $G(y)$ represents the conditional jump density.

The effect of this deformation mechanism is to evaluate the contrast $I(x)$ at certain points in background space and present these ordinates as a stepfunction to the observer. If the jump density is high enough the global form of I will be preserved, as in Figure 4.1 where (a) gives the pure

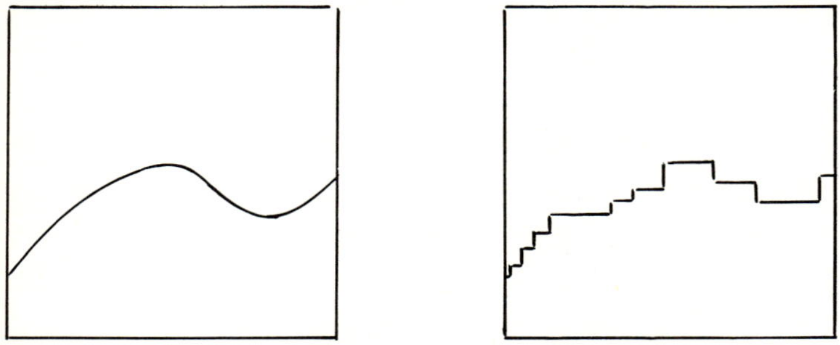

(a) (b)

Figure 4.1

image as a continuous function with two relative maxima and two relative minima and (b) shows $I^{\mathscr{D}}$.

A quite differently behaved $\mathscr{D}$ arises when $X = [0,1]$ is treated as an <u>elastic medium</u> with fixed endpoints and is subjected to a random force field in the longitudinal direction.

<u>Case 4.3</u> (elastic deformations). Define $d(x) = x+u(x)$ with the displacement u satisfying the differential equation

$$d^2u/dx^2 = F(x), \quad u(0) = u(1) = 0 \tag{4.17}$$

and $|F| \leq 1$ a.c.

Since the Green's function of (4.17) is

$$G(x,y) = \begin{cases} x(y-1), & x < y \\ -y(1-x), & x > y \end{cases} \tag{4.18}$$

we get the deformation

$$d(x) = x + \int_0^1 G(x,y)F(y)dy =$$

$$= x-(1-x)\int_0^x yF(y)dy + x\int_x^1 (y-1)F(y)dy. \tag{4.19}$$

If the force field F is continuous in the mean we get the mean square derivative

$$d'(x) = 1 + \int_0^1 yF(y)dy - \int_x^1 F(y)dy \geq 0. \tag{4.20}$$

A simulation of this case in a two-dimensional version is shown in Figure 4.2 where I is given in (a), and (b), (c) show deformed images. Here the F-field has been chosen as white noise so that the inequality $|F| \leq 1$ is not satisfied. A consequence of this is visible in (c) where the image has been folded over itself.

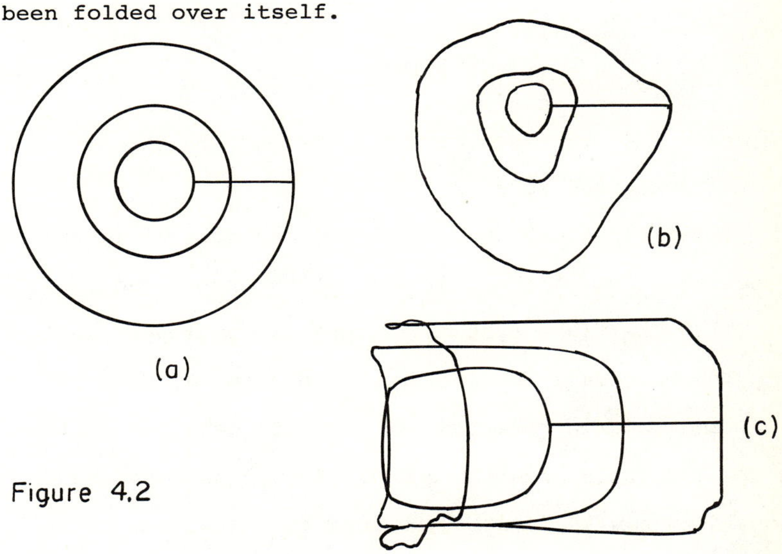

Figure 4.2

If $E[F] \equiv 0$ and $\mathrm{Cov}[F(y_1),F(y_2)] = R(y_1,y_2)$ it follows from (4.19) that $\mathscr{D}$ is a fair deformation with

$$V(x) = \int_0^1 \int_0^1 G(x,y_1)G(x,y_2)R(y_1,y_2)dy_1dy_2. \tag{4.21}$$

It is important to know what happens to the image when successive deformations (see Section 1) are applied, in particular what limit if any can be expected for

$$I_n^{\mathscr{D}} = \mathscr{D}(n)I, \quad n \to \infty. \tag{4.22}$$

For systematic deformations of [0,1] the answer is given as follows.

<u>Theorem 4.1</u>. <u>If</u> I <u>is a continuous function on</u> [0,1] <u>and</u> $\mathcal{D}$ <u>is left-systematic then for each</u> x <u>the limit of</u> I(x) <u>exists and is equal to</u> 0 <u>or</u> 1.

<u>Proof</u>: First it is clear that for a fixed x the sequences $y_n = d_n \ldots d_2 d_1(x)$ form a sub-martingale and converge a.c. to a limit y. If the d.f. of y is denoted by F we must have

$$0 = \int_0^1 [x-m(x)]F(dx) \tag{4.23}$$

which means that F must have all its mass in the set {0,1}, where the integrand is zero.

This means that successive deformations will tend to pick out only two of the ordinates of the pure image, I(0) and I(1). The right-systematic case is completely analogous.

The fair case is similar but requires a different method, and to gain some insight into its behavior we first simulate a fair deformation and display the successive deformations for n = 10,20,30,40,50 in Figure 4.3.

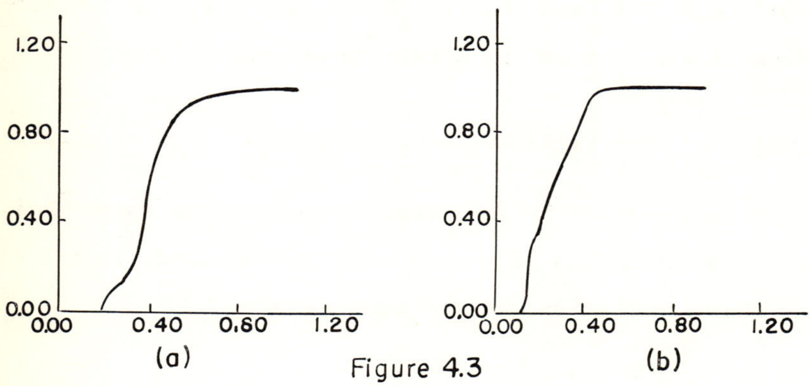

Figure 4.3

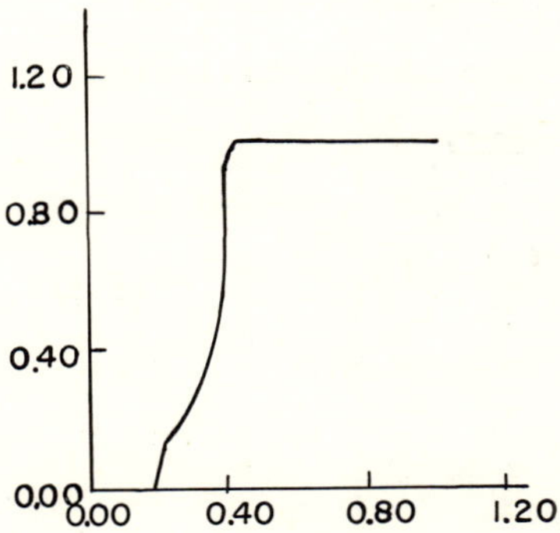

(c)

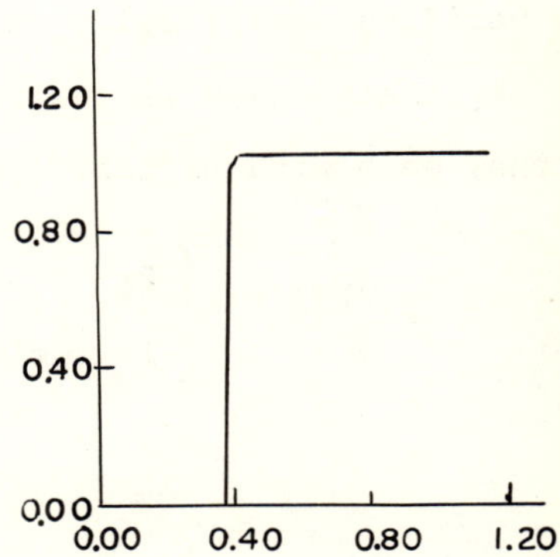

(d)

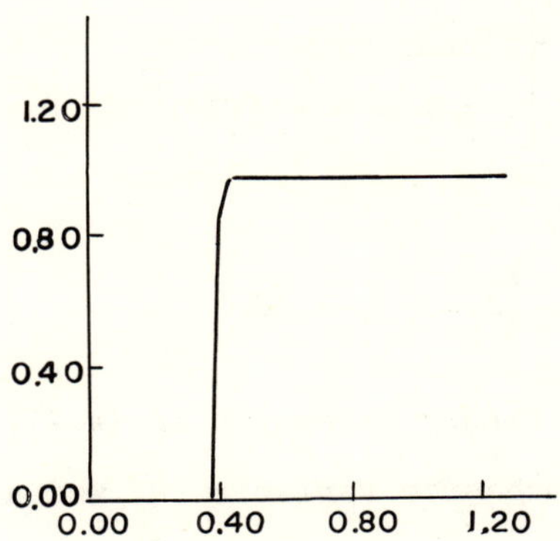

(e)

Figure 4.3

A series of mathematical experiments involving different fair deformations led to the same qualitative result: there seemed to exist a randomly given threshold value below which the motion was toward 0 and above to 1. This can be verified analytically in the following special case.

Say that $d(x)$ is realized as one of the following functions, each with probability one-half:

$$\phi^{(0)}(x) = \begin{cases} 2x & \text{for } 0 \leq x < 1/2 \\ 1 & \text{else} \end{cases} \tag{4.24}$$

and

$$\phi^{(1)}(x) = \begin{cases} 0 & \text{for } 0 \leq x \leq 1/2 \\ 2x-1 & \text{else} \end{cases} \tag{4.25}$$

If we associate a $z_i = 0$ for the choice $\phi^{(0)}$ and $z_i = 1$ for choosing $\phi^{(1)}$ in the ith successive deformation, we can write

$$\mathscr{D}(n): x \to \begin{cases} 0 \text{ for } 0 \leq x < \xi_n \\ \text{linear between} \\ 1 \text{ for } x \leq \xi_n + 2^{-n} \end{cases} \tag{4.26}$$

where

$$\xi_n = \frac{z_1}{2} + \frac{z_2}{4} + \dots + \frac{z_n}{2^n} \tag{4.27}$$

Then, continuing the binary expansion in (4.27) indefinitely, we get a limiting stochastic variable ξ which is $R(0,1)$ and the same qualitative behavior as observed empirically in several cases.

This is generalized as follows.

<u>Theorem 4.2</u>. <u>If</u> I <u>is continuous and if</u> $\mathscr{D}$ <u>is a fair</u>

<u>deformation with a continuous</u> $V(x)$ <u>vanishing only at</u> $x = 0$ <u>or 1, there exists a stochastic variable</u> $\xi = R(0,1)$ <u>such that</u>

$$\lim I_n^{\mathscr{D}}(x) = \begin{cases} I(0) & \text{if } x < \xi \\ I(1) & \text{if } x > \xi \end{cases} \tag{4.28}$$

<u>in probability</u>.

<u>Proof</u>: Consider the second moments for $\mathscr{D}(n)\colon x \to \psi_n(x)$

$$\alpha_n = E[\psi_n(x)]^2 \tag{4.29}$$

and note that since $\mathscr{D}$ is a fair deformation

$$\begin{aligned} 1 \geq \alpha_{n+1} &= EE[\mathscr{D}\psi_n(x)]^2 \\ &= E[V[\psi_n(x)] + \psi_n^2(x)] \geq \alpha_n \end{aligned} \tag{4.30}$$

so that $\alpha_n \uparrow \alpha$. This last inequality is strict unless $V[\psi_n(x)] = 0$ a.c. which only happens if $\psi_n(x) = 0$ or 1 a.c. For any convergent subsequence of distributions of $\psi_n(x)$ we must have with some $p(x)$

$$\begin{aligned} P(y=1) &= p(x) \\ P(y=0) &= 1 - p(x) \end{aligned} \tag{4.31}$$

with $p(0) = 0$, $p(1) = 1$ and p non-decreasing. But fairness implies that $p(x) \equiv x$ so that the limit is unique. The non-decreasing character of ψ_n implies that there is a variable ξ with the properties stated.

Under the conditions of Theorem 4.2, except that $V(x)$ now vanishes at points x_i, $0 = x_0 < x_1 < \ldots < x_r = 1$, it is clear that $\mathscr{D}$ takes $x_i \to x_i$, and because of the non-decreasing character each interval $(x_{i-1},x_i) \to (x_{i-1},x_i)$. Hence we can apply the above reasoning to each of the intervals and get the existence of stochastic variables $\xi_1,\xi_2,\ldots,\xi_r$ and $\xi_i = R(x_{i-1},x_i)$ with

$$\lim_{n\to\infty} I_n^{\mathscr{D}}(x) = \begin{cases} I(x_{i-1}) & \text{for } x_{i-1} \leq x < \xi_i \\ I(x_i) & \text{for } \xi_i < x < x_i \end{cases} \tag{4.32}$$

Hence several (= r+1) ordinates of the pure image will appear in the limit for $I^{\mathscr{D}}$.

To obtain a continuous family of deformations $\mathscr{D}(t)$, t positive real, let us approximate it by a discrete one.

<u>Case 4.4</u> (infinitesimal deformations). With $X = [0,1]$, define

$$d\colon x \to x \pm \varepsilon_n \phi(x,\xi) \tag{4.33}$$

where $\phi(\cdot,\xi)$ vanishes at $x = 0$ and 1, is continuous in x and uniformly bounded with some stochastic variable ξ. With equal probabilities one selects one of the two signs in (4.33).

Put $\mathscr{D}(t)$ as the limit in probability of $d_1,d_2,\ldots,d_{[nt]}$, where $\varepsilon_n = \sqrt{1/n}$. Then the infinitesimal speed (see Feller (1966), p. 323) is

$$\lim_{h\downarrow 0} \frac{1}{h} E[\mathscr{D}(h)x-x] = 0 \tag{4.34}$$

and the infinitesimal variance

$$b(x) = \lim_{h\downarrow 0} \frac{1}{h} E[\mathscr{D}(h)x-x]^2 = \int \phi^2(x,\xi)F(d\xi) \tag{4.35}$$

where F is the d.f. of ξ. Hence the variable $\mathscr{D}(t)x$ has a frequency function $p = p(x;t)$ satisfying

$$\frac{\partial p}{\partial t} = \frac{1}{2} \frac{\partial^2}{\partial x^2} (pb) \tag{4.36}$$

which should be started with $p = \delta$-function at x for $t = 0$.

Note, however, that this is <u>not</u> just a (one-dimensional) diffusion process. Instead, we have to deal with all values of x simultaneously so that we have a random <u>field</u> on $[0,1]$ developing in time according to the continuous semigroup of deformations $\{\mathscr{D}(t)\}$.

A special case of some interest is when ϕ is linear in x except for a break at $x = \xi$,

$$\phi(x,\xi) = \begin{cases} \frac{x}{\xi}, & 0 \leq x \leq \xi \\ \frac{1-x}{1-\xi}, & \xi \leq x \leq 1 \end{cases} \tag{4.37}$$

Direct calculation then gives

$$b(x) = 2x(1-x) \tag{4.38}$$

an equation known from population genetics, whose solution can be written in series form involving Gegenbauer polynomials (see Bharucha-Reid (1960)).

The higher-dimensional analogues of Theorem 4.2 and of the deformation (4.33) have not been investigated. The case X = unit square looks promising. It seems that the interior points would move toward the boundary and an appeal to potential theory is motivated. It should be remarked, however, that this is not enough here, since the deformation mechanism involves all the points of X simultaneously, not just their marginal (two-dimensional) distribution.

For a random deformation mechanism with kernel K and with m = Lebesgue measure, the <u>expected exposure</u>, i.e.

the expected value of the exposure measure, is simply given by (4.39), $F \subset X$,

$$E[m^d(F)] = \int_0^1 K(F,x)dx \tag{4.39}$$

so that if K has a continuous partial derivative with respect to its first argument we get the expected exposure density

$$E[\varepsilon(y)] = \int_0^1 \frac{\partial K(y,x)}{\partial y}\, dx. \tag{4.40}$$

Let us say that $X = R^3$, the case of R^2 being analogous. If the deformation of X is through a <u>flow</u> given by a velocity field $v(x) = (v_1(x), v_2(x), v_3(x))$, then a point x is carried into $x' = x+v(x)dt$ during a time interval of length dt. A volume element $dv = (dx_1, dx_2, dx_3)$ situated at x will not only be subject to this <u>translation</u> but also to more essential changes, as is well known from the mechanics of deformable media.

Indeed, the volume will be subject to a <u>rotation</u> around the axis of the vector

$$\text{curl } v = \left(\frac{\partial v_3}{\partial x_2} - \frac{\partial v_2}{\partial x_3}, \frac{\partial v_1}{\partial x_3} - \frac{\partial v_3}{\partial x_1}, \frac{\partial v_2}{\partial x_1} - \frac{\partial v_1}{\partial x_2}\right) \tag{4.41}$$

and with an instantaneous angular velocity equal to $\frac{1}{2}||\text{curl } v||$. The remaining part of the change can then be written as

$$dx_i' = \sum_j \frac{\partial v_j}{\partial x_i}\, dx_j - \frac{1}{2}(\text{curl } v \times dx)_i \tag{4.42}$$

or, written out fully,

$$
\begin{aligned}
dx_1' &= \frac{\partial v_1}{\partial x_1}\,dx_1 + \frac{1}{2}\left(\frac{\partial v_1}{\partial x_2} + \frac{\partial v_2}{\partial x_1}\right)dx_2 + \frac{1}{2}\left(\frac{\partial v_1}{\partial x_3} + \frac{\partial v_3}{\partial x_1}\right)dx_3 \\
dx_2' &= \frac{1}{2}\left(\frac{\partial v_1}{\partial x_2} + \frac{\partial v_2}{\partial x_1}\right)dx_1 + \frac{\partial v_2}{\partial x_2}\,dx_2 + \frac{1}{2}\left(\frac{\partial v_2}{\partial x_3} + \frac{\partial v_3}{\partial x_2}\right)dx_3 \qquad (4.43) \\
dx_3' &= \frac{1}{2}\left(\frac{\partial v_3}{\partial x_1} + \frac{\partial v_1}{\partial x_3}\right)dx_1 + \frac{1}{2}\left(\frac{\partial v_2}{\partial x_3} + \frac{\partial v_3}{\partial x_2}\right)dx_2 + \frac{\partial v_3}{\partial x_3}\,dx_3
\end{aligned}
$$

and the matrix of the right-hand side determines the *dilatation* of dv. Along the eigenvectors of the matrix the deformation consists of a pure dilatation along these axes, and the relative change in volume element is given by the divergence

$$\text{div } v = \sum_{i=1}^{3} \frac{\partial v_i}{\partial x_i}\,. \qquad (4.44)$$

If the flow preserves volume then this divergence must be zero.

In the phase space of a Hamiltonian mechanical system expressed in canonical coordinates the classical theorem of Liouville says that volume is preserved. As a simple example let us look at a two-dimensional phase space $x = (x_1, x_2)$, where the flow is along circles with the origin as center:

$$
\begin{aligned}
r &\to r \\
\phi &\to \phi + a(r)dt
\end{aligned}
\qquad (4.45)
$$

where $a(r)$ is a continuous function of the radius. This flow has divergence zero so that it is area-preserving and, for example, a circle will be transformed into some set of area equal to that of the circle. This has been simulated in Figure 4.4 where the circle (actually a polygon with many sides for the sake of the computing) has been changed into a

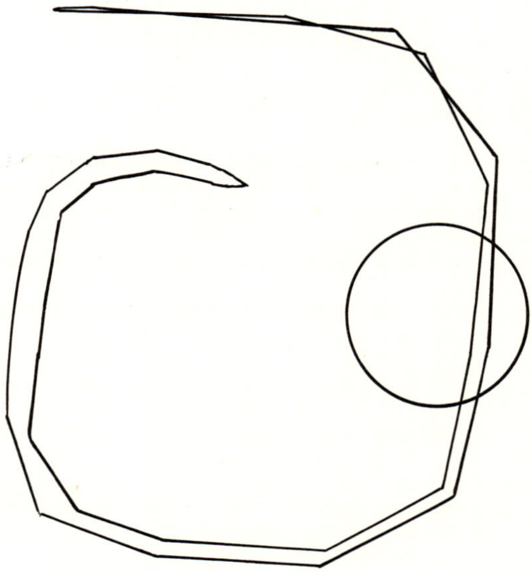

Figure 4.4

long and thin worm-like shape. A contrast pattern subjected to the deformation in (4.45) will become highly irregular when it increases in spite of the smooth appearance of the pure image.

In higher dimensions the situation is similar but more difficult to analyze. Let us consider a linear chain of coupled oscillators, unit mass, and with positions $x_k = x_k(t)$, $k = 0,1,2,\ldots,n$, satisfying the system of equations

$$\begin{aligned} \ddot{x}_k &= F(x_k - x_{k-1}) + F(x_k - x_{k+1}), \quad k = 1,2,\ldots,n-1 \\ x_0 &\equiv 0, \quad x_n \equiv 1 \end{aligned} \tag{4.46}$$

where F is a force function derived from a potential U, $F = -U'$, and the system is started at $t = 0$ with $x_k(0) =$

k/n and some given values for $\dot{x}_k(0)$.

With the pure image of contrast type on $X = [0,1]$, we define, for a fixed value of the deformation parameter t,

<u>Case 4.5</u> (linear lattice deformation)

$$\mathscr{D}: I \to \{I(x_k(t), k=0,1,\ldots,n\} \tag{4.47}$$

for those k for which the expression is defined so that the deformed image is a vector, possibly with some missing components.

Now some anomalies can occur that we did not encounter in the earlier deformations. First, it may happen that the ordering of the values $x_k(t)$ is not monotonically increasing. This will drastically <u>affect the global appearance of</u> $I^{\mathscr{D}}$. If $\mathscr{T}$ has some global property, say the images are monotonic or are built from generators with the Tschebychev property, then we may lose that property after $\mathscr{D}$ has been applied.

Second, it may happen that some $x_k(t)$ is less than 0 or greater than 1. Then the corresponding component of the $I^{\mathscr{D}}$-vector <u>is not defined</u>.

Whether this will happen or not depends upon the form of the potential U. If $U(x)$ is decreasing in $|x|$ (repulsive force) with $U(0) = +\infty$, since the total energy is constant, it follows that x_k cannot pass through any of the values $0, x_{k-1}, x_{k+1}, 1$. Therefore the anomalies mentioned cannot occur in this particular case whatever the initial velocities are. If, however, U is bounded they may occur for some initial velocities.

The latter case occurs for the harmonic oscillators

with $U = \frac{1}{2} sx^2$ so that we have attractive forces with the spring coefficient s. Then, changing variables to $y_k = x_k - k/n$, the system in (4.46) becomes

$$\ddot{y}_k + s(2y_k - y_{k-1} - y_{k+1}) = 0, \quad k-1,2,\ldots,n-1$$
$$y_0 \equiv y_n \equiv 0 \tag{4.48}$$

started with the initial condition $y_k(0) = 0$ for $k = 1,2,\ldots,n-1$, or in matrix notation

$$\ddot{y} + sMy = 0 \tag{4.49}$$

with $y = (y_1, y_2, \ldots, y_{n-1})$ and

$$M = \begin{Bmatrix} 2 & -1 & 0 & 0 & . \\ -1 & 2 & -1 & 0 & . \\ 0 & -1 & 2 & 0 & . \\ 0 & 0 & 0 & 0 & . \\ . & . & . & . & \end{Bmatrix} \tag{4.50}$$

The solution of (4.48) is then a linear combination of $\exp \pm it\sqrt{s\lambda_\nu}$ where λ_ν are eigenvalues of M,

$$\lambda_\nu = 2 \cos \frac{\nu\pi}{n} \quad , \quad \nu=1,2,\ldots,n-1. \tag{4.51}$$

Diagonalizing $M = O^T \Lambda O$ with the λ_ν's as the entries in the main diagonal of Λ, we can write the solution in terms of the orthogonal matrix O

$$y(t) = O^T D(t) O \dot{y}(0) \tag{4.52}$$

where $D(t) = \operatorname{diag}[(s\lambda_\nu)^{-1/2} \sin (s\lambda_\nu)^{1/2} t]$.

A stochastic version of this $\mathcal{D}$ will have an equation as in (4.48) but driven by white noise, say. The system is no longer conservative and if we want it to settle down to statistical equilibrium we must include a friction term that will dampen the oscillations

$$\begin{aligned} &\ddot{y}_k + f\dot{y}_k + s(2y_k - y_{k-1} - y_{k+1}) = \dot{w}_k \\ &y_0 \equiv y_n \equiv 0 \end{aligned} \tag{4.53}$$

with the positive friction coefficient f.

An equilibrium solution of (4.53) can be written in its spectral representation

$$y(t) = \int_{-\infty}^{\infty} e^{it\lambda} dZ(\lambda) \tag{4.54}$$

with an orthogonal vector-valued process Z. Solving for Z we get from (4.53)

$$dZ(\lambda) = (-\lambda^2 I + fi\lambda I + sM)^{-1} dZ_w(\lambda) = N(\lambda) dZ_w(\lambda) \tag{4.55}$$

with dZ_w corresponding to the driving term $\dot{w}$. This gives us the covariance matrix $R(h) = \mathrm{Cov}\{y_k(t), y_\ell(t+h)\}$ as

$$R(h) = \sigma^2 \int_{-\infty}^{\infty} e^{ih\lambda} N(\lambda) N^T(\lambda) d\lambda \tag{4.56}$$

and we can calculate the probability measure of the deformed image $I^{\mathcal{D}}$ using (4.56) and (4.6). This will be useful for the pattern analysis in Part III.

Background deformations occur in other contexts, one of which is when X is finite. If the deformations are 1-1 we simply permute the 2^n points of X, so that $\mathcal{D}$ operates on the symmetric group over n objects. Let's say, more generally, that we have a compact group Γ with elements γ and that the deformation mechanism is specified by

<u>Case 4.6</u> (compact group deformations). For $h > 0$ we define

$$\mathscr{D}(h): \gamma \to \begin{cases} \gamma & \text{with probability } 1-\lambda h \\ \gamma_1\gamma & \text{with probability } \lambda h(1+o(1))\Pi(d\gamma_1) \end{cases} \tag{4.57}$$

where Π is a probability measure over Γ. Introduce the set $\{M^r(\gamma);\ r=0,1,\ldots\}$ of irreducible and non-equivalent unitary representations of Γ, $M^{(0)} \equiv$ identity operator. Denoting the probability distribution over Γ due to $\mathscr{D}(t)$ by $P(d\gamma,t)$ we have, introducing their Fourier transforms

$$\hat{P}_r(t) = E[M^{(r)}(\gamma)] = \int_\Gamma M^{(r)}(\gamma)P(d\gamma,t), \quad r=0,1,2,\ldots \tag{4.58}$$

the relation

$$\hat{P}_r(t+h) = (1-\lambda h)\hat{P}_r(t) + \lambda h(1+o(1))\hat{\Pi}_{(r)}\hat{P}_r(t) \tag{4.59}$$

where $\hat{\Pi}_r$ is the Fourier transform of Π; see Grenander (1963), Chapter 3, for more details. Hence we get the system of differential equations

$$\frac{d\hat{P}_r(t)}{dt} = -\lambda\ \hat{P}_r(t) + \lambda\ \hat{\Pi}_r\hat{P}_r(t) \tag{4.60}$$

with the solution

$$\hat{P}_r(t) = \hat{P}_r(0)\ \exp[\lambda(I-\hat{\Pi}_r)] \tag{4.61}$$

so that if we start at the unit element $e \in \Gamma$ then

$$\hat{P}_r(t) = \exp[\lambda(I-\hat{\Pi}_r)] \tag{4.62}$$

from which $P(t)$ can be obtained using the orthogonality relations between the elements of the representation matrices $M^{(r)}$ (see, e.g., Pontjagin (1946), section 28).

Returning to the special case of a Boolean pattern,

the above deformation mechanism means that two or several questions have been exchanged: what is $I(x)$, answered by the value of $I(d(x))$. On the other hand, it can happen that the deformations are not one-one but rather change x into $d(x)$ for each x separately and independently. Say that

$$\mathscr{D}: x \to \begin{cases} x \text{ with probability } q \\ y\ (\neq x) \text{ with probability } \frac{1-q}{n-1} \end{cases} \tag{4.63}$$

Then for a given $\xi \in X$ the exposure density can be different from one, but the expected exposure density is still equal to one.

Other deformations can affect the background space, not through a mapping but by changing other structures such as ordering or distance. Since these deformations behave quite differently from the ones studied in this section, they will be discussed separately in Section 7.

5. Incomplete Observations.

In the last two sections we were concerned with deformations in which either the contrast or the background were affected. Of course combinations of both cases can also occur. Anyway, the important thing is that in both instances the observer's measurements were incorrect to some extent, so that his values of the contrast and/or background position were subject to error.

We now turn to the situation where the observer can determine accurately what some of the background and contrast values are, but can do this only partly for the pure image. We then speak of incomplete observations.

Definition 5.1. *Given a family* W *of sets* $w \subset X$ *let* $I^{\mathcal{D}}$ *be given as the restriction of the contrast image* I *to* w,

$$I^{\mathcal{D}}(x) = I(x) \quad \text{for} \quad x \in w. \tag{5.1}$$

Then w *is called the window of the deformation and is identified with* d *and* $\mathcal{D}$ *is called a mask deformation*.

The further specification of such a $\mathcal{D}$ will describe the properties of W. The family may consist of fixed or random w's, the windows may be regular or more chaotic, and the topology of w may vary in terms of connectedness and dimension. When the window is regular it will be natural to look at it as a pattern from some image algebra, and the result of $\mathcal{D}$ will depend upon how this image and the pure image under observation relate to each other.

Observing $I^{\mathcal{D}}$ will in general not determine I uniquely. An I that can lead to the given $I^{\mathcal{D}}$ for the given window w is said to be *consistent with the deformed image*. In particular, if a pattern $\mathcal{P} = \{I \mid I = sI_0\}$ is generated from a single prototype I_0 the subset L of S for which I is consistent with $I^{\mathcal{D}}$ is called the *leeway set* and denoted by $L(I^{\mathcal{D}};I_0)$. If the windows are large we will expect the leeway set to be small, a statement that will be made precise later on in this section in concrete cases.

Viewed through a window w the images will typically lose bond properties so that $\mathcal{T}^{\mathcal{D}}$ will not be as structured as $\mathcal{T}$. A case where this is not true occurs as follows.

Theorem 5.1. *Consider a single window* w *and assume that it is a bond-determining set. Then the mask deformation* (5.1) *is a homomorphism*.

Proof: Define the new image algebra $\mathcal{T}^{\mathcal{D}}$ considering its elements as formed by $I \to I^{\mathcal{D}}$ via (5.1), with the same similarity transformations, and by giving an element $I^{\mathcal{D}}$ the bond structure and bond values of any $I \to I^{\mathcal{D}}$. This is unique since if I_1 and I_2 lead to $I^{\mathcal{D}}$ they coincide on w and hence have the same bonds determined on w. $\mathcal{T}^{\mathcal{D}}$ is then determined by giving it the same connection type and bond relation as $\mathcal{T}$ has.

According to Theorem 3.1.2 it is then enough to show that we use an identification rule R' which is coarser than the one that led to $\mathcal{T}$. But then it is clear that R' is actually a legitimate identification rule (see the four conditions in Definition 3.1.1). First, notice that R' operates in two steps, first R and then (5.1) so that $c_1 R' c_2$ means that c_1 and c_2 have the same external bonds and represent the same function on the window set w, which implies that R' is an equivalence relation. Also c_1 and c_2 must then have the same external bonds and bond values. The two last conditions of Definition 3.1.1 are then satisfied so that R' is indeed an identification rule. Of course R is finer than R' so that the proof is complete.

As an example, let $\mathcal{T}$ be generated by waveform generators on intervals on the real line with integral end points, say with differential operators of order one. Let w consist of all the integers so that bonds are determined by w. Then the deformations obtained by just observing I(x) for those x-values in the support of I that are integers we get a deformation mechanism that maps $\mathcal{T}$ homomorphically into $\mathcal{T}$ and hence preserves image algebra structure.

Note that in this example we would not get a homomorphism if the generators were allowed to have arbitrary intervals as their supports. Then the bonds need not be uniquely determined and we could not combine deformed images directly with the same bond relation as in the pure image algebra.

As a second example, consider the configurations in Case 2.9.6 identified as functions when adding the components of the configurations pointwise. We then have full graph structure and ρ guarantees that we combine only different powers. Let the maximum allowed number of α be n and introduce $w = \{x_1, x_2, \ldots, x_{n+1}\} \subset [0,1]$ with all x_i different. Then observing $I(x)$ on w will determine all the powers that are used in its configurations so that w is a bond-determining set and the deformation (5.1) will form a homomorphism again. Actually no information is lost here when $\mathscr{D}$ is applied, in contrast to the case where we do not impose any constraint on the values of α allowed in the generation of the pure image algebra; this will be discussed below.

Now we shall no longer assume that w is a bond-determining set. We can look upon w as the parameter of the deformation and write $\mathscr{D} = \mathscr{D}(w)$. Successive deformations will satisfy $\mathscr{D}(w_1)\mathscr{D}(w_2) = \mathscr{D}(w_1 \cap w_2)$ and will lose more and more information. On the other hand, we can also <u>combine windows</u> which means that $\mathscr{D} = \mathscr{D}(w_1 \cup w_2)$, increasing the information obtained compared to when we view the pure image only through one of the windows w_1 or w_2.

When X is Euclidean space R^k let us look at what happens when w is one or several points, lines or planes.

We will start with a highly regular window.

Case 5.1 (invariant windows). Let S_0 be a subgroup of S such that w has S_0 as its group of symmetries (see Section 5.1).

This deformation does not depend explicitly upon the pure image and is usually superimposed on X by a random choice with probabilities invariant with respect to S. It is a matter of convenience whether we look at w or I, as random, but usually it is more natural to think of the image as fixed, although unknown, and the window as randomly located.

To randomize regular windows one proceeds as follows. Let w_0 be a fixed regular window invariant with respect to a normal subgroup S_0. A difficulty is that we cannot immediately apply the Haar measure on S since it will be infinite in many interesting cases. Instead let us deal with the factor group $F = S/S_0$ and assume that it is compact so that its Haar measure is finite and can be normalized to total measure one. Pick a random element f from F according to this measure and define the randomized window as $w = fw_0$. The element f represents a coset, and can be looked at as an arbitrary s-element belonging to this coset.

It then follows that $\mathcal{D}$ is covariant in probability, since sdI means the function $I(s^{-1}x)$ with x restricted to the set sw, and dsI means the function $I(s^{-1}x)$ restricted to the set w. But since $sw = sfw_0$, $w = fw_0$, and because sf has the same distribution over F as f the two random sets have the same probability distribution and the assertion follows.

Depending upon the symmetries of the situation it may be necessary to randomize the window still further. We then apply one more random transformation.

Say that R^k is the Euclidean plane with S = translation group and S_0 consists of all translations $(x_1,x_2) \to (x_1+h_1,x_2+h_2)$ with h_1,h_2 equal to integers. Let w be the set of all the corresponding lattice points superimposed at random as described and I some set image in the plane. To start let I be a circle with radius r. We can just as well say that the window is fixed and the circle has center at (c_1,c_2) where c_1 and c_2 are uniformly distributed over a unit square. Hence

$$I = \{(x_1,x_2)\,|\,(x_1-c_1)^2 + (x_2-c_2)^2 \leq r^2\}. \tag{5.2}$$

Introduce $N(x)$ = # points in I, $x = r^2$. The study of $N(x)$ is a classical problem in number theory. Of course $N(x)$ can be expected to behave as πx for large x, but the more detailed properties of the function are less obvious. We cite one result of relevance to the present discussion.

<u>Theorem 5.2</u> (<u>D.G. Kendall</u> (1948)). <u>The quantity</u> $N(x)$ <u>is a stochastic variable with mean</u> πx <u>and a variance</u> $V(x)$ <u>such that</u>

$$V(x) = O(r) \tag{5.3}$$

<u>and</u> $V(x)$ <u>is at least of this order for some arbitrarily large</u> r.

<u>Proof</u>: The value of $N(x)$ depends upon c_1,c_2 and is periodic in both with period one, and we use its formal Fourier

series over a fundamental cell Q

$$N(x) = N(x;c_1,c_2) \sim \sum a_{n_1n_2} \exp 2\pi i(n_1c_1+n_2c_2) \quad (5.4)$$

where

$$a_{n_1n_2} = \iint_Q N(x;c_1,c_2)\exp\{-2\pi i(n_1c_1+n_2c_2)\}dc_1dc_2$$

$$= \sum \int_{n_1-1}^{n_1} \int_{n_2-1}^{n_2} F(u_1,u_2)\exp\{2\pi i(n_1u_1+n_2u_2)\}du_1du_2 \quad (5.5)$$

where F is the indicator function of the circle centered at $(0,0)$ and with radius $r = \sqrt{x}$. Hence

$$a_{n_1n_2} = \iint_Q F(u_1,u_2)\cos(2\pi n_1u_1)\cos(2\pi n_2u_2)du_1du_2$$

$$= \frac{r}{\sqrt{n_1^2+n_2^2}} J_1\left[2\pi r \sqrt{n_1^2+n_2^2}\right] \quad (5.6)$$

if n_1 and $n_2 \neq 0$, and πx else.

Whatever the convergence properties of the formal expression in (5.4), we are at least in L_2, so that Parseval's relation gives

$$\iint_Q N^2(x;c_1,c_2)dc_1dc_2 = \pi^2x^2 + x\sum{}' \frac{J_1^2\left[2\pi r\sqrt{n_1^2+n_2^2}\right]}{n_1^2+n_2^2} \quad (5.7)$$

(leaving out the term $n_1 = n_2 = 0$)

$$= \pi^2x^2 + x \sum_{n=1}^{\infty} \frac{r(n)}{n^2} J_1^2(2\pi\sqrt{xn})$$

where $r(n)$ is the number of ways n can be written as the sum of two squares and the rearrangement of terms is made legitimate using the bound of the J_1-function, $|J_1(x)| < cx^{-1/2}$.

Hence the mean of N is, of course,

$$E(N) = \iint_Q N(x;c_1,c_2)dc_1dc_2 = \pi x \tag{5.8}$$

and the variance bounded by

$$\operatorname{Var}(N) = \iint_Q [N(x;c_1,c_2)-\pi x]^2 dc_1dc_2 = O(r) \tag{5.9}$$

On the other hand,

$$\operatorname{Var}(N) > xJ_1^2(2\pi\sqrt{x}) \tag{5.10}$$

so that using the asymptotic behavior of the J_1-function, the variance is at least of the order of magnitude r for some arbitrarily large values of r, which proves (5.3).

This result has been extended to smooth convex sets, and the interested reader should consult D.G. Kendall (1948).

In the above example we formed w_0 by applying all the elements of the subgroup $(x_1,x_2) \to (x_1+h_1,x_2+h_2)$, h_1 and h_2 integers, to the element $x = (0,0)$. Now let us do the same but with the subgroup of all transformations of the form $(x_1,x_2) \to (x_1+h_1,x_2+h_2)$ where h_1 is an integer and h_2 a real number. Then the window w_0 will consist of parallel lines parallel to the x_2-axis and $w = fw_0$ will be a randomly displaced (in the x_1-direction) copy of it. In order to get rotational invariance we rotate w by an angle uniformly distributed over $(0,2\pi)$.

Let I be a line image, a piecewise smooth curve, and of finite length ℓ. The deformed image will then consist of the points in the plane where the window of parallel lines intersects the curve. Denoting the number of intersections by N, which is a stochastic variable, we recall the classical

<u>Theorem 5.3</u> (<u>Buffon's needle problem</u>). <u>The stochastic variable</u> N <u>has mean</u> $2\ell/\pi$.

The proof is well known from most elementary textbooks on probability and we only sketch it. Say that I consists of a straight line segment of length $\ell < 1$. To compute the mean of N in this case we can assume that the segment is located with one endpoint at $(0,y)$, with y uniformly distributed between $(0,1)$ and making an angle ϕ with the positive x-direction, with ϕ uniform on $(0,\pi)$. To get an intersection we must have $y + \ell \sin \phi > 1$ so that

$$E(N) = P(\text{one intersection}) = \frac{1}{\pi} \iint_F d\phi dy \qquad (5.11)$$

where F is the (y,ϕ)-set with $y + \ell \sin \phi > 1$. Hence $E(N) = 2\ell/\pi$. For an arbitrary curve, we approximate it by a polygon with many sides, all smaller than 1. Since means add this gives us the celebrated result stated above.

Now let $X = R^3$ and S the translation group with S_0 consisting of the transformations $(x_1,x_2,x_3) \to (x_1+h_1, x_2+h_2,x_3+h_3)$ where h_1 is an integer and h_2 and h_3 are real numbers. The same construction as before will give us a window consisting of equidistant and parallel planes displaced randomly in the x_1-direction.

If the pattern is built from non-overlapping spheres with the frequency function $p(D)$ for the diameter D (see Case 2.9.2), the deformed image will consist of circles resulting from the intersection of the planes with the spheres. One can observe the distribution of the diameters of these circles but it is certainly not equivalent to p, although this fact has not always been recognized. The relation is given by an integral equation.

<u>Theorem 5.4</u> (<u>Wicksell</u> (1925-6)). <u>Denote the frequency function of the diameters of the circles by</u> q. <u>Then if</u> p <u>has a finite mean</u> D_0 <u>we have</u>

$$p(D) = -\frac{2DD_0}{\pi}\int_D^\infty \frac{1}{\sqrt{x^2-D^2}}\frac{d}{dx}\left(\frac{1}{x}q(x)\right)dx. \tag{5.12}$$

<u>Proof</u>: Consider a sphere of diameter D intersecting the plane and denote the distance between the center and the plane by y so that $0 \le y \le D/2$. The density of y is then $2/D$ and the diameter of the resulting intersection circle is then $x = (D^2-4y^2)^{1/2}$. But the density of the diameter of a sphere intersecting the plane must be proportional to $Dp(D)$ so that this density is

$$Dp(D)/D_{mean} = Dp(D)/\int_0^\infty Dp(D)dD. \tag{5.13}$$

We then get for the joint density of y and D

$$\frac{2}{D}\frac{Dp}{D_0} = \frac{2p(D)}{D_0} \tag{5.14}$$

and changing variables to x and D, including the Jacobian of the substitution, we get the density

$$\frac{x}{D_0\sqrt{D^2-x^2}}\,p(D). \tag{5.15}$$

Since $D \ge x$ we get for the q-density

$$q(x) = \frac{x}{D_0}\int_x^\infty \frac{p(D)}{\sqrt{D^2-x^2}}\,dD. \tag{5.16}$$

This integral equation is of Abel's type and can be solved directly (see, e.g., Kanwal (1971), pp. 167-170), leading to the solution (5.1).

If the volumes are ellipsoids instead of spheres a similar result can be obtained (see Wicksell (1926)).

Consider now the opposite case where the windows are irregular sets. An extremely irregular situation occurs in the next case.

Case 5.2 (Poisson windows). For a given contrast image $I = I(x)$ in X consider the set W of realizations w of a Poisson process over X with intensity $\lambda = \lambda(x;I)$ with respect to an S-invariant measure dx.

To fix ideas, suppose that $X = R^2$ with S = translation group and dx = Lebesgue area. If $\mathcal{T}$ consists of convex sets in the plane (see Case 3.5.9), the deformed image will consist of points in the plane, and the density of these p points will be governed by the function λ.

An important special case is when

$$\lambda(x;I) = \begin{cases} \lambda & \text{for } x \in I \\ 0 & \text{else} \end{cases} \tag{5.17}$$

so that the entire deformed image is contained in I. This $\mathcal{D}$ will be studied in detail in Chapter 12 (see Appendix).

As λ increases so will the number n of points in $I^{\mathcal{D}}$ (in probability) and it is well known that conditional upon n the distribution over I is uniform. For this case we shall see how the leeway set behaves.

Theorem 5.5. Consider a pattern $\mathcal{P} = \{I | I=sI_0\}$ formed from a prototype I_0 consisting of a convex set with positive and continuous curvature. With s consisting of translations

<u>the expected area of the leeway set</u> L <u>for</u> n <u>points in</u> $I^{\mathcal{D}}$, <u>deformed from</u> I_0, <u>satisfies</u>

$$\lim_{n\to\infty} n^2 E[m(L_n)] = 2A_0^2 \int_0^{\pi} \frac{d\phi}{D_\phi^2} \tag{5.18}$$

<u>where</u> $A_0 = m(I_0)$ <u>and</u> D_ϕ <u>is the diameter in the direction</u> ϕ. <u>Furthermore, the linear dimension of</u> L, ℓ_ϕ <u>in the direction</u> ϕ,

$$\ell_\phi = \max_t \{t \mid (t\cos\phi,\ t\sin\phi) \in L\} \tag{5.19}$$

<u>has a d.f.</u> F_n <u>with</u>

$$\lim_{n\to\infty} F_n(nx) = 1 - \exp\left\{-\frac{tD_\phi}{A_0}\right\}. \tag{5.20}$$

<u>Proof</u>: Write $f = f(x,y)$ for the indicator function of I_0. But the leeway set can be expressed in terms of the points (x_ν, y_ν) in the deformed image

$$\begin{aligned} L &= \{(a,b) \mid (x_\nu - a, y_\nu - b) \in I_0 \text{ for } \nu = 1,2,\ldots,n \\ &= \bigcap_{\nu=1}^{n} \{(a,b) \mid (a,b) \in (x_\nu,y_\nu) - I_0\} \end{aligned} \tag{5.21}$$

and it follows that L is also convex.

Now the indicator function ψ of L is seen to be

$$\psi(x,y) = \prod_{\nu=1}^{n} f(x_\nu - x, y_\nu - y) \tag{5.22}$$

so that

$$m(L) = \iint \psi(x,y)\,dxdy = \iint \prod_{\nu=1}^{n} f(x_\nu - x, y_\nu - y)\,dxdy. \tag{5.23}$$

But the points (x_ν, y_ν) are stochastically independent and with the density $(1/A_0)f$ so that

$$m(L) = \iint \frac{1}{A_0} \left[\iint f(u-x,v-y)f(u,v)\,du\,dv\right]^n dx\,dy$$
$$= \iint K^n(x,y)\,dx\,dy \tag{5.24}$$

where the kernel K is

$$K(x,y) = \frac{1}{A_0} m(I_0 \cap sI_0), \quad s = (x,y). \tag{5.25}$$

Hence it is non-negative and bounded by 1 which value is attained for $s = e = (0,0)$.

If the vector $s = (x,y)$ is small and points in the direction ϕ we have

$$K(x,y) = 1 - \sqrt{x^2+y^2}\;\frac{D_\phi}{A_0}\,(1+o(1)) \tag{5.26}$$

and a change of variable followed by a standard limit argument gives

$$\lim_{n\to\infty} n^2 E[m(L)] = \iint \exp\left[-\frac{D_\phi}{A_0}\sqrt{x^2+y^2}\right] dx\,dy \tag{5.27}$$

or in polar coordinates

$$\lim_{n\to\infty} n^2 E[m(L)] = \int_{\phi=0}^{2\pi} \int_{\rho=0}^{\infty} \exp\left[-\frac{D_\phi}{A_0}\rho\right] \rho\,d\rho\,d\phi \tag{5.28}$$

which proves the first part of the theorem. The proof of (5.19) is quite similar and will not be given here.

The second part implies that, asymptotically, the standardized linear extent $n\ell_\phi$ of the leeway set is exponentially distributed with mean A_0/D_ϕ. But this means that we will not get convergence to a fixed limit when n grows. Instead, it would be of interest to see if the correlation coefficient between ℓ_ϕ and ℓ_ψ converges; this remains to be done.

The same problem, but with other similarity groups,

has been studied (see Grenander (1973)).

A related $\mathcal{D}$ appears in the study of patterns of hypotheses and inductive reasoning (see Section 5.9). Then X is the Boolean space with elements $x = (x_1, x_2, \ldots, x_\ell)$ and we take dx as counting measure. A hypothesis, represented by a contrast image $I = I(x)$, has a truth set, which can be identified with I, where the Boolean function $I(x)$ takes the value TRUE.

As λ increases so will $I^{\mathcal{D}}$. Actually if we introduce the size of the difference set

$$N = \#(I - I^{\mathcal{D}}) \tag{5.29}$$

it is clear that N is a binomial stochastic variable $N = B(n,p)$, with $n = \#(I)$ and $p = e^{-\lambda}$. Hence $E(N) = ne^{-\lambda}$ and $N \to 0$ in probability for $\lambda \to \infty$.

A variation of (5.17) is to let λ take a value λ_1 inside I and another λ_2 outside, with $\lambda_1 > \lambda_2$. This deformation mechanism was originally introduced to model the pictures obtained in radiography of the brain. It was assumed that tumors would correspond to large values of λ, i.e. in the set I, and in the rest λ_2 would represent the remaining noise.

We mention in passing similarly constructed deformation mechanisms, but where the windows consist of Poisson lines, Poisson planes, or Poisson flats in higher dimensions.

In these deformations we sample from a random mechanism depending on the pure image. A more general situation is when we start directly from measure generators (see Case 1.3.21), and build the pure image algebra from them. We are then led to

<u>Case 5.3</u> (sampling deformations). For I = a probability measure over X, let $I^{\mathscr{D}}$ consist of the points in X obtained by i.i.d. sampling n times from I.

Such a deformed image can be written as $I = (x_1, x_2, \ldots, x_n)$. Much of classical statistics deals with how $I^{\mathscr{D}}$ approximates I as n becomes large in various senses, and it would be superfluous to go into that here. Let us just mention one special result.

Introduce a diffuse pattern of measure images as follows. Let α be a finite measure over X and define the measure over $\mathscr{T}$ by demanding that for any partition $B_1, B_2, \ldots B_k$ of X the stochastic vector $z = (z_1, z_2, \ldots, z_k)$ with $z_i = P(B_i)$ have a <u>Dirichlet distribution</u> with parameters $\alpha(B_1), \alpha(B_2), \ldots, \alpha(B_n)$. We remind the reader that this means that if $\alpha(B_i) > 0$, all i, the vector $(z_1, z_2, \ldots, z_{k-1})$ has a frequency function

$$f(z_1, z_2, \ldots, z_{k-1}; \alpha_1, \alpha_2, \ldots, \alpha_k)$$

$$= \frac{\Gamma(\alpha_1+\alpha_2+\ldots+\alpha_{k-1})}{\Gamma(\alpha_1)\Gamma(\alpha_2)\ldots\Gamma(\alpha_{k-1})} \prod_{i=1}^{k-1} z_j^{\alpha_j - 1}\left(1 - \sum_{i=1}^{k-1} z_j\right)^{\alpha_k} \tag{5.30}$$

in the simplex $\left\{z_1, z_2, \ldots, z_{k-1} \mid z_i \geq 0, \sum_j^{k-1} z_i \leq 1\right\}$ and zero outside. If one of the α's is zero the distribution of the corresponding z_i degenerates and has its mass at zero.

These marginal distributions satisfy Kolmogorov's consistency theorem and determine a probability measure over the probability distributions over X. For a proof of this and a discussion of the measure-theoretic aspect, see Ferguson (1973). A crucial result is the following description of how the conditional distribution of the probability measure behaves after observing $I = \{z_1, z_2, \ldots, z_n\}$.

Theorem 5.6 (Ferguson (1973)). *Under the stated conditions the conditional distribution of the probability measures is Dirichlet with the* α *measure modified to the measure* $\alpha + \sum_{i=1}^{n} \delta_{z_i}$, *where* δ_z *means the point mass* 1 *situated at the point* z.

The proof is a direct consequence of the classical result that the a posteriori distribution of a multinomial distribution, when the a priori distribution of its parameter is Dirichlet, is again Dirichlet but with the α-parameters changed by adding one for each realized value in the multinomial vector.

Another type of deformation occurs for contrast images in the strict sense. Say that the contrast space $Y = R'$ and that the deformed images are defined throughout X with values as in (5.31).

Case 5.4 (filtering deformations). For a contrast image $I = I(x)$ we define

$$I^{\mathscr{D}}(x) = \int K(x,z)I(z)dz = KI \tag{5.31}$$

with integration over X with respect to an S-invariant measure dx, and K is a given real-valued *kernel*.

Of course, the usual sort of precautions are needed, to insure that the integral in (5.31) is meaningful, but this will be done in each separate case in Part III.

As usual, our main interest is in deformations that destroy information, in particular when the linear operator K in (5.31) does not possess an inverse.

With the additional invariance condition that $K(s^{-1}x,z) = K(x,sz)$, we see that sd corresponds to the image

$$sI^{\mathcal{D}}(x) = \int K(s^{-1}x,z)I(z)dz \tag{5.32}$$

while ds means the image with values

$$\int K(x,z)I(s^{-1}z)dz = \int K(x,sz)I(z)dz \tag{5.33}$$

using the invariance of the measure dz. Hence this deformation mechanism is covariant.

We mention three examples of filtering deformations familiar from communication engineering.

For $X = R'$, dx = Lebesgue measure, say that the pure images are in $L_2(-\infty,\infty)$. Let $K(x,z)$ be a convolution kernel, $K(x,z) = L(x-z)$ with L in L_2. Then, taking Fourier transforms, we have a.e.

$$\hat{I}^{\mathcal{D}}(\lambda) = \hat{L}(\lambda)\hat{I}(\lambda). \tag{5.34}$$

If $\hat{L}$ is bounded and zero for large $|\lambda|$ we have a <u>low-pass filter</u>. If it is zero for small $|\lambda|$ we have a <u>high-pass filter</u>, and if it is non-zero only in a middle range of $|\lambda|$ we have a <u>band-pass filter</u>. In each case information is lost so that the pure image cannot be completely recovered from an observed $I^{\mathcal{D}}$.

Instead let $z_1,z_2,\ldots,z_n$ be a set of disjoint real numbers and define

$$K(x,z) = \sum_{i=1}^{n} \delta(z-z_i)\frac{(x-z_1)\ldots(x-z_{i-1})(x-z_{i+1})\ldots(x-z_n)}{(z_i-z_1)\ldots(z_i-z_{i-1})(z_i-z_{i+1})\ldots(z_i-z_n)} \tag{5.35}$$

where $\delta(z)$ is the Dirac function centered at zero. Then the deformed image is a polynomial of degree $n-1$ passing through the values taken by I at $x = z_j$, $j = 1,2,\ldots,n$; $I^{\mathscr{D}}(z_j) = I(z_j)$. The formula (5.35) is <u>Lagrange's interpolation method</u> and again we will lose information unless the pure images are uniquely determined by their values on the z_j's.

In the third example let $X = [0,1]$ and consider diffuse patterns of images being $L_2(0,1)$ real-valued functions. Assuming the means to be identically zero and introducing the covariance function $r(x,y) = \mathrm{Cov}(I(x),I(y))$, we assume r to be continuous in the unit square, and hence we can use the Karhunen-Loève expansion

$$I(x) = \sum_{k=1}^{\infty} \sqrt{\lambda_k}\, z_k\, \phi_k(x) \tag{5.36}$$

where λ_k and ϕ_k are the eigenelements of the symmetric integral equation

$$\lambda\phi(x) = \int_0^1 r(x,y)\phi(y)dy, \quad ||\phi|| = 1. \tag{5.37}$$

It will be convenient to index the λ's so that they become non-increasing. The z's are uncorrelated stochastic variables of variance one, and the expansion converges in the mean for any $x \in [0,1]$.

Define the kernel K by

$$K(x,z) = \sum_{k=1}^{n} \phi_k(x)\phi_k(z). \tag{5.38}$$

That implies, using (5.36), that the deformed image is given as

$$I^{\mathscr{D}}(x) = \int_0^1 K(x,z)I(z)dz = \sum_{k=1}^{n} \sqrt{\lambda_k}\, z_k\, \phi_k(x) \tag{5.39}$$

in other words as a truncated Karhunen-Loève series. It is of interest to look at this from the point of view of <u>linear feature selection</u> which means that we try to approximate $I(x)$ by a finite linear expression

$$I^a(x) = \sum_{k=1}^{n} \zeta_k f_k(x) \tag{5.40}$$

in such a way that the expected mean square error is minimized (for fixed n)

$$E\left\{\int_0^1 [I^a(x)-I(x)]^2 dx\right\} = \min. \tag{5.41}$$

<u>Theorem 5.7</u>. <u>Among all linear approximations as in</u> (5.40), <u>the one in</u> (5.38) <u>realizes the minimum</u>.

<u>Proof</u>: It is no restriction to assume that the ζ_k are orthonormal since this could always be achieved by a linear transformation, possibly leading to a smaller number of ζ_k's. If they are, it is obvious that the best f_k's would be its "Fourier coefficients"

$$f_k(x) = E[\zeta_k I(x)] \tag{5.42}$$

in which case

$$Q = E\left\{\int_0^1 I^2(x)dx\right\} - \sum_{k=1}^{n} \int_0^1 f_k^2(x)dx. \tag{5.43}$$

We then have to choose the ζ_k so that the sum in the second term on the right-hand side is maximized. But

$$f_k(x) = \sum_{\ell=1}^{\infty} \sqrt{\lambda_\ell}\ (\zeta_k, z_\ell)\phi_\ell(x) = \sum_{\ell=1}^{\infty} \sqrt{\lambda_\ell}\ c_{k\ell}\phi_\ell(x) \tag{5.44}$$

and

$$\sum_{k=1}^{n} \int_0^1 f_k^2(x)dx = \sum_{\ell=1}^{\infty} \sum_{k=1}^{n} \lambda_\ell c_{k\ell}^2 = \sum_{\ell=1}^{\infty} \lambda_\ell c_\ell^2. \tag{5.45}$$

However, $1 \geq c_\ell^2 \geq 0$, and Bessel's inequality gives us

$$\sum_{\ell=1}^{\infty} c_\ell^2 \leq n.$$

Hence the extremum is obtained if we make $c_{k\ell} = \delta_{k\ell}$ for $k = 1,2,\ldots,n$ and 0 else (remember that the λ's are non-increasing), which implies that $\zeta_k = z_k$ for $k = 1,2,\ldots,n$ and then $f_k(x) = \phi_k(x)$.

This result is actually old, and goes back to the early work on symmetric integral equations, but its relevance to feature selection had to await Karhunen's and Loève's representation.

If X in Case 5.4 is a finite set $\{1,2,\ldots,n\}$ with dx = counting measure, then the images are vectors and K can be represented by a matrix. In particular, if K does not have an inverse, so that R^n is mapped into a proper subspace $\mathscr{L} \subset R^n$, information is lost when $\mathscr{D}$ is applied. This type of deformation is typical for systems theory.

6. Indirect Observations.

We now direct our attention to situations when the observer uses some exploring device to obtain the image. Usually this will lead to less than perfect determination of the image and the resulting deformation may be quite severe if the device is not chosen so that it fits the structure of the image algebra.

The class of deformations caused by indirect observations overlaps to some extent with those treated in the last section. For example, if the device consisted of a mask put over I and giving precise information inside the window, then we would again be confronted by Definition 5.1. Usually, however, the device has some typical property that makes it natural to treat these deformations separately.

We shall consider only two types of devices here, but many others occur in practice. One type uses a device made of some rigid material, the other one exposes the image to waves of some form.

Case 6.1 (material probes). Let $\mathscr{T}$ consist of set images in a background space X. A particular fixed subset $P \subset X$ called a probe is transformed by similarity transformations in all ways such that sP does not intersect I. Formally, the deformed image is given as

$$I^{\mathscr{D}} = \{s \mid sP \cap I = \phi\} \subset S. \tag{6.1}$$

Hence the deformed image is of quite different nature from the pure one. While I is a subset of X the deformed one is a subset of S. Of course to a given $I^{\mathscr{D}}$ we can associate the pure images consistent with $I^{\mathscr{D}}$ (see Section 4.5). In the special case when the pure image is known to be similar to a fixed prototype, let Γ be the leeway set when we start from a prototype I_0. Then for an arbitrary $s_1 \in S$ we get

$$\begin{aligned} ds_1 I_0 &= \{s \mid (sP) \cap (s_1 I_0) = \phi\} = \{s \mid (s_1^{-1} sP) \cap I_0 = \phi\} \\ &= \{s_1 t \mid (tP) \cap I_0 = \phi\} = s_1 dI_0 \end{aligned} \tag{6.2}$$

so that probe deformations are covariant. Also, if Γ_0, Γ_1

are the leeway sets of dI_0 and ds_1I_0 respectively then it follows, using the covariant property, that

$$\begin{aligned}\Gamma_1 &= \{s \mid ds_1I_0 \text{ consistent with } sI_0\}\\ &= \{s \mid dI_0 \text{ consistent with } s_1^{-1}sI_0\}\\ &= \{s_1t \mid dI_0 \text{ consistent with } tI_0\} = s_1\Gamma_0. \qquad (6.3)\end{aligned}$$

In the case of $X = R^2$ and $S = EG(2)$, the form of the probe in practical applications to <u>measurement machines</u> can be a half plane, it can be ball-shaped at one end, or it can have some special purposes such as locating an edge or corner. Let us consider the important case when the probe is a long thin rod or, idealized, a ray extending from the origin in the direction of the x-axis. Say that the images are connected, simply connected sets bounded by a piecewise smooth boundary ∂I. Then we shall say that I is <u>explorable by linear probes</u> if for any point on the boundary there is a ray completely outside of I and with its endpoint arbitrarily close to the given point on ∂I (see Figure 6.1).

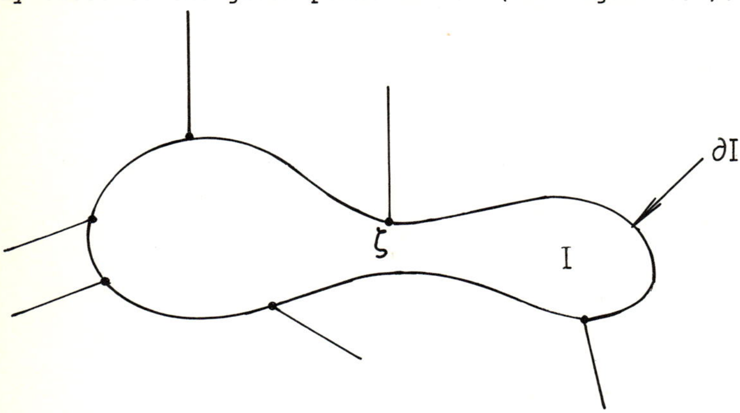

Figure 6.1

<u>Theorem 6.1. Let I be as described. Then it is explorable by linear probes if and only if it is 2-feature closed in the feature logic of half planes (see 3.5).</u>

<u>Proof</u>: Assume that I is 2-feature closed so that

$$I = \bigcap_{\alpha_1,\alpha_2 \in A} (H_{\alpha_1} \cup H_{\alpha_2}) \qquad (6.4)$$

with half planes H_α, and with the intersection extended over some generator index set A. Let ζ be an arbitrary point on ∂I. Then we can find another point ζ', arbitrarily close to ζ, such that ζ' belongs to the complement of some 2-feature $H_{\alpha_1} \cup H_{\alpha_2}$. But each point of this complement can be reached by a linear probe. This means that probes can reach points arbitrarily close to ∂I so that I is explorable by linear probes.

Conversely, assume that I is explorable by linear probes. Surround the outside of ∂I by a set $\{P\}$ of points, such that to any point on ∂I there are points P arbitrarily close to the given one. To each P we can associate a ray originating in P and extending outwards, entirely in I^c. Let us extend the ray with the same endpoint into the largest open angle that is contained in I^c. This means that I^c can be written as the intersection of the complements of these angles. But such a complement can be written either as $H_{\alpha_1} \cup H_{\alpha_2}$ if the angle is less than π or $H_{\alpha_1} \cap H_{\alpha_2}$ else. Hence I has the form (6.4) which proves the theorem.

The deformation caused by the observer being limited to a particular probe assumes that exact values of s are recorded for which $sP \cap I = \phi$. In practical cases only a finite set of s-values are measured and each with some measurement error. In the analysis of patterns probed by measuring machines (see Chapter 12) we shall allow for this further deformation.

In the remainder of this section we shall study some of the deformations that can occur when the observer explores the image using some sort of radiation satisfying the wave equation

$$\nabla^2 u = \frac{1}{c^2} \frac{\partial u}{\partial t} \tag{6.5}$$

where c is the wave velocity and u is the amplitude of the wave. The physical interpretation of u depends upon the nature of the waves, but here we shall deal with the most important case, illumination with light, and treat the light field as a scalar one. For more detailed information on this subject the reader is referred to Goodman (1968) and O'Neill (1963).

The linearity of (6.5) allows for superposition of waves - a fact of major importance for the study of optical processes. In R^3 let r denote the distance from the origin to a point (x,y,z) and put a point source of monochromatic light of wave length λ at the origin. In a homogeneous medium, constant c, this results in a spherical wave with

$$u = \frac{A}{r} \exp i(kr-\omega t) \tag{6.6}$$

where k is the wave number $2\pi/\lambda$ and ω is the angular frequency kc. This monochromatic type of wave can be used to build up more complicated wave fronts.

The relation between I and $I^{\mathcal{D}}$ will depend upon the properties of the medium situated between the object I and the deformed image $I^{\mathcal{D}}$. Also the size of the wave length λ is important, since a large value of λ will prohibit the observer from seeing details in the pure image that are much smaller than λ.

The background space will now be the (x_1,y_1)-plane at $z = 0$ and $I^{\mathcal{D}}$ will be situated in the (x,y)-plane at z

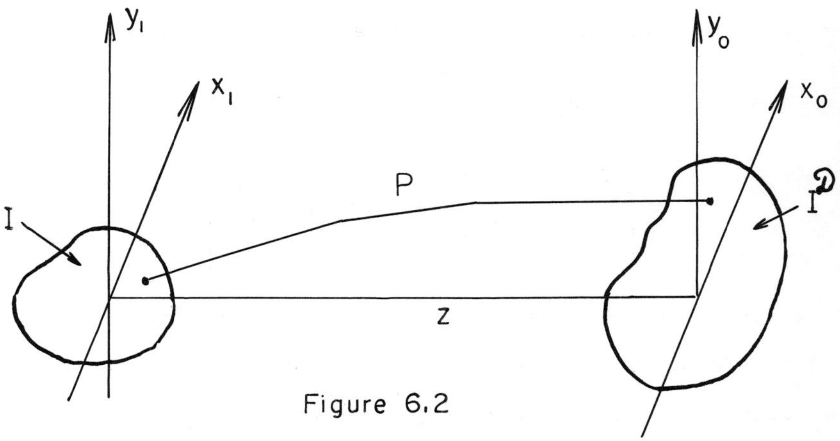

Figure 6.2

(see Figure 6.2, where we have indicated one of the light paths P starting at (x_1,y_1) in the object plane and ending at (x_0,y_0)). Denote by $\Phi(x_0,y_0;x_1,y_1)$ the phase shift caused by following P. If we only consider such P as lie near the optical axis (<u>the paraxial approximation</u>), the z-axis, we need not consider the variation in the denominator r in (6.6), only the phase shift. Using superposition, we get for the complex amplitude in the (x_0,y_0)-plane with some constant C

$$I^{\mathcal{D}}(x_0,y_0) = C \iint \exp\{i\Phi(x_0,y_0;x_1,y_1)\}I(x_1,y_1)dx_1dy_1 \quad (6.7)$$

and this is

Case 6.2 (monochromatic illumination). A plane contrast image represented by the field amplitude I in the (x_1,y_1)-plane results in the deformed image represented by the field amplitude in (6.7).

The further specification of the function Φ will depend upon the geometry and optical properties of the observational setup. Let us recall a few classical special cases.

First, say that I consists of an aperture in an opaque infinite screen and that the intervening medium is homogeneous. Then the phase shift will be proportional to

$$\begin{aligned}\text{distance from } (x_1,y_1,0) \text{ to } (x_0,y_0,z) \\ &= ((x_0-x_1)^2 + (y_0-y_1)^2 + z^2)^{1/2} \\ &= z\left[1 + \frac{1}{2z}\left(\frac{x_0-x_1}{2}\right)^2 + \frac{1}{2z}\left(\frac{y_0-y_1}{2}\right)^2\right] \qquad (6.8)\end{aligned}$$

if z is large compared to the dimensions along the x and y directions. The proportionality factor is k (see (6.6)). The kernel in (6.7) is then proportional to

$$\exp\{\frac{ik}{2z}[(x_0-x_1)^2 + (y_0-y_1)^2]\} \qquad (6.9)$$

known as the Fresnel approximation. This implies that $\mathscr{D}$ is given approximately as a convolution

$$I^{\mathscr{D}}(x_0,y_0) \simeq C\iint I(x_1,y_1)\exp\{\frac{ik}{2z}[(x_0-x_1)^2+(y_0-y_1)^2]\}dx_1dy_1 \qquad (6.10)$$

integrated over the aperture in the (x_1,y_1)-plane.

This convolution corresponds to a multiplication (in the spatial frequency domain) by a function proportional to

$$\exp\{-[i\pi\lambda z(f_x^2+f_y^2)]\}. \qquad (6.11)$$

Expanding the square in (6.6), we get $I^{\mathscr{D}}$ proportional to

$$\exp\{\frac{ik}{2z}(x_0^2+y_0^2)\}\iint I(x_1,y_1)\exp\{\frac{ik}{2z}(x_1^2+y_2^2)\}$$

$$\cdot \exp\{-[\frac{2\pi i}{\lambda z}(x_0x_1+y_0y_1)]\}dx_1dy_1. \tag{6.12}$$

With an even stronger assumption, z is large compared to $k(x_1^2+y_1^2)/2$ in the aperture, we get the <u>Frauenhofer approximation</u>

$$I^{\mathscr{D}}(x_0,y_0) \simeq C \exp \frac{ik}{2z}(x_0^2+y_0^2)\iint I(x_1,y_1)$$

$$\cdot \exp\{-\frac{2\pi i}{\lambda z}[x_0x_1+y_0y_1]\}dx_1dy_1. \tag{6.13}$$

where we have used $k = 2\pi/\lambda$. This means that <u>this optical system carries out a two-dimensional Fourier transform</u> at the spatial frequencies $x_0/\lambda z$, $y_0/\lambda z$.

A celebrated example is the Frauenhofer diffraction pattern from a circular aperture of diameter ℓ, uniformly illuminated by monochromatic coherent light. The resulting field in the (x_0,y_0) plane is then obtained from (6.13) and we have to evaluate the double integral

$$\iint \exp\{-\frac{2\pi}{\lambda z} r_0r_1 \cos\phi\} r_1dr_1d\phi \tag{6.14}$$

where $r_0 = (x_0^2+y_0^2)^{1/2}$ and r_1 varies from 0 to $\ell/2$ and ϕ varies uniformly over $(0,2\pi)$. Using the definition of the Bessel functions of the first kind, we find that $I^{\mathscr{D}}$ is proportional to

$$\exp(\frac{ikr_0^2}{2z}) \frac{k\ell^2}{i4z} J_1(\frac{k\ell r_0}{2z}) \frac{2z}{k\ell r_0}. \tag{6.15}$$

Second, let us consider a thin lens of thickness

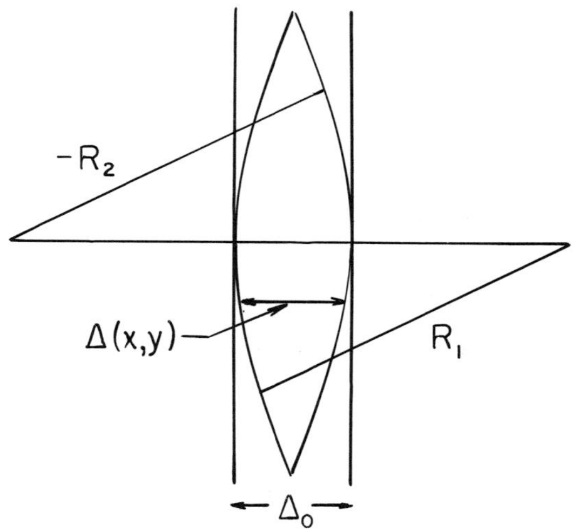

Figure 6.3

$\Delta(x,y)$ at the point (x,y) in its plane and let $\Delta_0 = \max \Delta(x,y)$, and let n be the index of refraction of the lens. The lens then causes the phase shift $kn\Delta(x,y)$ and the region between the lens and the two tangential planes causes another phase shift $k[\Delta_0-\Delta(x,y)]$ (see Figure 6.3). Hence the total phase shift will be

$$\Phi(x,y) = kn\Delta(x,y) + k[\Delta_0-\Delta(x,y)]. \tag{6.16}$$

The lens in the figure is assumed to be bounded by two spherical surfaces, where the sign of the radius is positive if a ray left-to-right meets a convex surface, negative otherwise. The total thickness function is then split into two, say $\Delta_1(x,y)$ and $\Delta_2(x,y)$ with maxima Δ_{01} and Δ_{02} respectively. Then $\Delta = \Delta_1+\Delta_2$ with

$$\begin{aligned} \Delta_1(x,y) &= \Delta_{01} - [R_1 - (R_1^2-x^2-y^2)^{1/2}] \\ \Delta_2(x,y) &= \Delta_{02} - [-R_2 - (R_2^2-x^2-y^2)^{1/2}] \end{aligned} \tag{6.17}$$

so that if we add Δ_1 and Δ_2 and use the paraxial approximation again, we get

$$\Delta(x,y) = \Delta_0 - \frac{x^2+y^2}{2}\left(\frac{1}{R_1} - \frac{1}{R_2}\right). \tag{6.18}$$

With the help of (6.7) and (6.16) this leads to a kernel proportional to

$$\exp\left[-\left\{ik(n-1)\frac{x^2+y^2}{2}\left(\frac{1}{R_1} - \frac{1}{R_2}\right)\right\}\right] \tag{6.19}$$

or, introducing the focal length f

$$\frac{1}{f} = (n-1)\left(\frac{1}{R_1} - \frac{1}{R_2}\right) \tag{6.20}$$

the kernel is proportional to

$$\exp[-\{i \frac{k}{2f}(x^2+y^2)\}] \tag{6.21}$$

which will now be used to study the situation when the object with the complex field $I(x,y)$ is placed in front of the lens at a distance d and we observe the field $I^{\mathscr{D}}$ in the focal plane at the distance f behind the lens.

Denote the two-dimensional Fourier transform of I by $\hat{I}$; then the Fresnel approximation (6.11) tells us that the Fourier transform of the complex field at the frontal focal plane of the lens is

$$\hat{I}_1(f_x,f_y) = \hat{I}(f_x,f_y)\ \exp\{-\ [i\pi\lambda d(f_x^2+f_y^2)]\}. \tag{6.22}$$

The action of the lens is to multiply the complex field at the frontal focal plane by (6.21). Hence (see (6.12)),

$$I^{\mathscr{D}}(x_0,y_0) = \exp[\frac{ik}{2f}(x_0^2+y_0^2)]\iint I_1(x,y)\exp[-\frac{ik}{2f}(x^2+y^2)]$$

$$\cdot \exp[\frac{ik}{2f}(x^2+y^2)]\exp[-\frac{2\pi i}{f}(xx_0+yy_0)]\,dxdy$$

$$= \exp[\frac{ik}{2f}(x_0^2+y_0^2)]\iint I_1(x,y)\exp[-\frac{2\pi i}{\lambda f}(xx_0+yy_0)]dxdy \quad (6.23)$$

integrated over the aperture of the lens. If we do not take the finite extent of it into account we get

$$I^{\mathscr{D}}(x_0,y_0) = \exp[\frac{ik}{2f}(x_0^2+y_0^2)]\hat{I}_1(\frac{x_0}{\lambda f}, \frac{y_0}{\lambda f}). \quad (6.24)$$

Using (6.21), this leads us to

$$I^{\mathscr{D}}(x_0,y_0) = \exp[\frac{ik}{2f}(1 - \frac{d}{f})(x_0^2+y_0^2)]\ \hat{I}(\frac{x_0}{\lambda f}, \frac{y_0}{\lambda f}). \quad (6.25)$$

In the important case when the object is placed in the frontal focal plane, $d = f$, the above reduces to the important relation

$$I^{\mathscr{D}}(x_0,y_0) = \hat{I}(\frac{x_0}{\lambda f}, \frac{y_0}{\lambda f}). \quad (6.26)$$

In other words, we have proved

Theorem 6.2. An object placed in the frontal plane of a thin lens illuminated by monochromatic light is mapped (approximately as described) into the spatial Fourier transform by the relation (6.26).

Optical systems built from thin lenses and transparencies with linear transmission can hence be analyzed in terms of successive Fourier transforms and scalar multiplications either in the spatial or in the frequency domain. This will be further used in Chapter 13 for the filtering of optical images.

Under somewhat idealized conditions we have seen how a pure contrast image represented as a complex field is deformed by the mechanism described in Case 6.2. It is a linear operator on the complex field. In most cases this deformation is followed by another which depends upon the fact that most recording devices are sensitive only to the average power of the field, not its phase. This power is proportional to $|I|^2$, the <u>intensity</u>, so that only the absolute value will be observable.

<u>Case 6.3</u> (loosing phase). If $I = I(x,y)$ is the complex amplitude field we define $I^{\mathcal{D}}(x,y) = |I(x,y)|$ = the absolute value of the complex field.

It now depends upon the phase differences of the field what happens with the resulting intensity. If the phases vary together so that relative phases remain fixed the illumination is called <u>spatially coherent</u>. Point sources of light would typically be coherent and of course the laser should be mentioned. On the other hand, if at different locations the phases vary in a stochastically independent manner one speaks of completely <u>spatially coherent illumination</u>.

Say that we use spatially coherent illumination with a complex field I_1. Passing it through an optical system as discussed above, we will get another field I_0 linearly related to the first:

$$I_0(x_0,y_0) = \iint K(x_0,y_0;x_1,y_1)I_1(x_1,y_1)dx_1dy_1 \tag{6.27}$$

with some kernel K. The corresponding intensity is then

$$J_0(x_0,y_0) = |I_0(x_0,y_0)|^2 =$$

$$\iiiint K(x_0,y_0;x_1,y_1)\overline{K}(x_0,y_0;x_2,y_2)I(x_1,y_1)\overline{I}(x_2,y_2) dx_1dx_2dy_1dy_2 . \tag{6.28}$$

On the other hand, if the illumination is completely incoherent with the intensity J_1 then

$$J_1(x_0,y_0) = \iint |K(x_0,y_0;x_1,y_1)|^2 J_0(x_1,y_1)dx_1dy_1 \tag{6.29}$$

so that in this case it is the quadratic quantity intensity that transforms linearly.

Once the resulting intensity has been obtained, it is possible to discuss the effect on a photographic detector, e.g. photographic film. It consists of a transparent base made of glass or acetate and a layer of photographic emulsion (see Figure 6.4). The emulsion has a large number of photo-

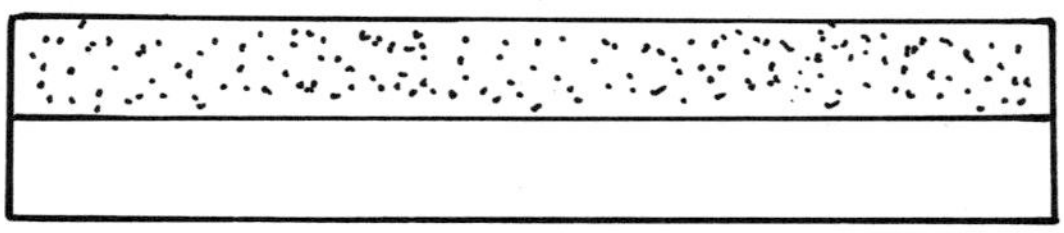

Figure 6.4

sensitive particles suspended in gelatin. When exposed to light some of the particles are reduced to silver. It is the density of those reduced particles that will determine the <u>optical transmittance</u> T. The transmittance is the ratio between input and output intensities of light. The <u>photographic density</u> is defined as $D = -\log T$.

The granular structure of the emulsion and the statistical fluctuations in the reduction to silver introduce <u>photographic deformations</u> and the following model gives some information on their effect (see Yu (1974), Chapter 8).

Assume that the emulsion is uniform in the sense that grain size is constant and that the grains are regularly distributed throughout the emulsion. Consider a small volume element containing N grains and expose it for t time units. Consider the process with intensity $q(t)$ of being developed per time unit and grain. With

$$P_k(t) = P(k \text{ grains undeveloped at time } t) \tag{6.30}$$

starting of course at $t = 0$ with all the N grains undeveloped, we get

$$\begin{aligned} \dot{P}_k(t) &= -kq(t)P_k(t)+(k+1)q(t)P_{k+1}(t), \quad k = 0,1,\ldots,N-1 \\ \dot{P}_N(t) &= -Nq(t)P_N(t). \end{aligned} \tag{6.31}$$

We recognize this as the same process as was studied in Case 2.11, and can therefore write down the solution directly:

$$\begin{aligned} P_k(t) &= \binom{N}{k}[1-e^{-p(t)}]^{N-k}e^{-kp(t)} \\ p(t) &= \int_0^t q(s)ds \end{aligned} \tag{6.32}$$

which has the mean $m(t) = Ne^{-p(t)}$ and variance $v(t) = Ne^{-p(t)}(1-e^{-p(t)})$.

If one assumes that the photographic density $D = N-m(t)$ we get

$$dD(t)/dt = -\dot{m}(t) = Nq(t)e^{-p(t)}. \tag{6.33}$$

It has been suggested that there is an approximate relation

$$dD(t)/dt \simeq N \frac{t}{\alpha^2} \exp\left(-\frac{1}{2}\left(\frac{t}{\alpha}\right)^2\right) \tag{6.34}$$

with α a constant = time when the above rate reaches its maximum. Then we should have

$$N \frac{t}{\alpha^2} \exp\left(-\frac{1}{2}\left(\frac{t}{\alpha}\right)^2\right) \simeq Nq(t)\, e^{-p(t)} \tag{6.35}$$

leading to

$$q(t) = t/\alpha^2, \quad p(t) = \frac{1}{2}\left(\frac{t}{\alpha}\right)^2 \tag{6.36}$$

which can be used to calculate the approximate mean, variance, and signal-to-noise ratio associated with photographic noise.

We formalize the above as

Case 6.4 (photographic deformation). For $X = R^2$, consider a contrast pattern with intensity $I = I(x,y)$. In X is given a regular arrangement of points $\{z\}$. For each z a death process with density $q(t)$ per time unit is allowed to operate t time units, all points being treated independently of each other. The resulting set F of developed points defines a coarse-grained stochastic density $D(x,y)$ and

$$I^{\mathscr{D}}(x,y) = I(x,y) \cdot 10^{-D(x,y)}. \tag{6.37}$$

This means that we perform an averaging operation over the set F and use the result as the coarse-grained density. The spatial extent of the averaging procedure should be a larger order of magnitude than the horizontal distances between the points of F.

Up till now we have treated the pure image as an object represented in a plane in terms of its complex light field or its illuminated intensity. This implies that we have neglected its depth as a three-dimensional object. If the depth is considerable we must take into account the mappings $R^3 \to R^2$ if the observer lacks depth perception.

Consider the situation depicted in Figure 6.5 with a central point C located at a distance ℓ behind the (x,y)-plane with $z = 0$. An image point $P = (x,y,z)$ is trans-

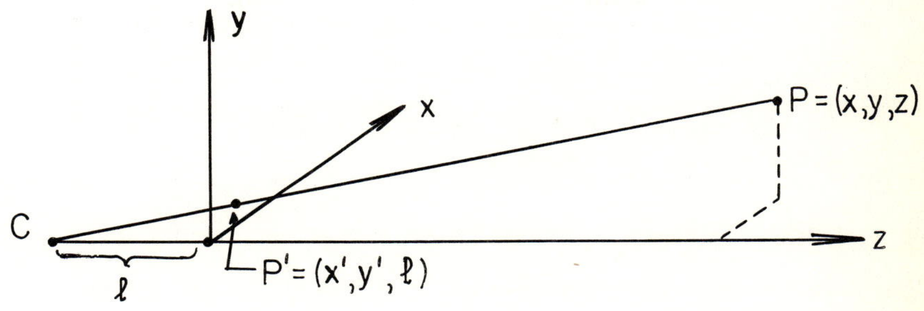

Figure 6.5

formed into an observed point P' where the ray $\overrightarrow{CP}$ intersects the (x,y)-plane.

It is clear that we then have

$$x'/\ell = \frac{x}{\ell+z}$$
$$y'/\ell = \frac{y}{\ell+z} \tag{6.37}$$

from which we can solve x',y' in terms of x,y,z. Conversely, given x',y' we can solve for the set of possible image points lying on the ray with the equation

$$x = \frac{\ell+z}{\ell} x'$$
$$y = \frac{\ell+z}{\ell} y' \tag{6.38}$$

and z plays the role of a parameter with $z > 0$.

Case 6.5 (perspective deformations). Let $\mathcal{T}$ consist of set images in the background space R^3 and define the deformed image as

$$I^{\mathcal{D}} = \{(\frac{\ell}{\ell+z} x, \frac{\ell}{\ell+z} y) \mid (x,y,z) \in I\}. \tag{6.39}$$

Notice that this mapping $R^3 \to R^2$ is different from those involving (orthogonal) projections or plane sections. However, if ℓ is large, it can be approximated by an orthogonal projection deformation, since the ratio $\ell/(\ell+z) \simeq 1$.

If only certain distinguished points of I can be observed via (6.39) the result of the deformation will depend upon whether the image represents an opaque or transparent object. In the former case difficulties will arise since some of the distinguished points may be hidden by that part of I that is situated between the points and the observer.

We end this section by mentioning that if the images are discrete structures, rather than representable by continuous irradiation functions, then the Fourier integrals above will have to be replaced by the analogous Fourier sums. The most thoroughly studied situation when this occurs is in crystallography using X-ray diffraction.

We then get deformations of the type

$$I^{\mathscr{D}}(x,y,z) = \sum_{h,k,\ell} F(h,k,\ell)\exp\{2\pi i(hx+ky+\ell z)\} \qquad (6.40)$$

where $I^{\mathscr{D}}$ is as before the amplitude of the deformed image and h,k,ℓ are the indices of the reciprocal lattice (see Sections 3.5 and 3.6) and the F-coefficients are the scattering factors. We shall return to some of the analysis problems resulting from (6.40) in Chapter 10 (see Appendix).

7. Other Deformations.

The types of deformations introduced in Sections 2-6 represent most of the situations that have been encountered in pattern synthesis so far. As the theory is gradually extended to other applications, however, it is likely that other sorts of deformation mechanisms will suggest themselves naturally. At present we shall only mention two $\mathcal{D}$'s that do not quite fit in with the ones discussed before.

The underlying idea is the same for both, that the background space loses some of its structure. This is not meant in the sense of changes of background as in Section 4, but as a more drastic destruction of some basic structural properties. We have two such properties in mind here: <u>order</u> and <u>metric</u>.

Consider an image algebra $\mathcal{T} = \langle G,S,\Sigma,\rho,R\rangle$ and introduce the monotonic extension $\mathcal{M}(\Sigma)$ of Σ (see Section 2.1). We shall let $\mathcal{D}$ take $\mathcal{T}$ into the related image algebra $\mathcal{T}^{\mathcal{D}} = \langle G,S,\mathcal{M}(\Sigma),\rho,\overline{R}\rangle$ where $\overline{R}$ is an identification rule over $\mathcal{C}^{\overline{\mathcal{R}}}(\overline{\mathcal{R}} = \langle\mathcal{M}(\Sigma),\rho\rangle)$ that reduces to R on $\mathcal{C}^{\mathcal{R}}$, $\mathcal{R} = \langle\Sigma,\rho\rangle$.

Consider an arbitrary pure image I and assume that it can be written as a combination of the pure images $J_1=s_1I_1$, $J_2=s_2I_2,\ldots,J_n=s_nI_n$. This means that for the corresponding configurations, one representing I will consist of others representing $J_1,J_2,\ldots,J_n$ respectively, joined by bonds that are external to the J's but internal to I. Let us denote the set of these bonds by B.

If for a $c \in I$ we delete all the bonds in B, the resulting configuration c' need not be regular so that we cannot claim that $c' \in \mathcal{C}^{\mathcal{R}}$. On the other hand, it is clear that its connection type belongs to $\mathcal{M}(\Sigma)$ and all the remaining bonds remain true for the bond relation ρ. Hence $c' \in \mathcal{C}^{\overline{\mathcal{R}}}$, where $\overline{\mathcal{R}} = \langle\mathcal{M}(\Sigma),\rho\rangle$ and the corresponding image $I^{\mathcal{D}}$ (mod $\overline{R}$) is in $\mathcal{T}^{\mathcal{D}}$.

Now we can do the same thing for an arbitrary subset $I_{k_1}, I_{k_2}, \ldots, I_{k_m}$, $m \leq n$. Since the bonds between generators in, say, I_{k_1} and I_{k_2} have been opened we need not satisfy the ρ-relation for them and the subset can be represented by configurations in $\mathscr{C}^{\bar{\mathscr{R}}}$ as before. The subset $I_{k_1}, I_{k_2}, \ldots, I_{k_m}$ of sub-images that are not mutually connected gives rise to a deformed image $I^{\mathscr{D}}$ in $\mathscr{T}^{\mathscr{D}}$. Note that we lose not only sub-images but also the s-values in the pure image.

<u>Case 7.1</u> (breaking images). Consider the set $\mathscr{D}$ of all deformations $I \to I^{\mathscr{D}}$ as above, so that

$$\mathscr{D}: \quad \langle G,S,\Sigma,\rho,R\rangle \to \langle G,S,\mathscr{M}(\Sigma),\rho,\bar{R}\rangle. \tag{7.1}$$

It should be remarked that if the connection type is monotonic, so that $\mathscr{M}(\Sigma) = \Sigma$, then $\mathscr{C}^{\mathscr{R}} = \mathscr{C}^{\bar{\mathscr{R}}}$ and since $\bar{R}$ should equal R on $\mathscr{C}^{\mathscr{R}}$ it follows that $\mathscr{T}^{\mathscr{D}} = \mathscr{T}$. In other words, breaking images on monotonic connection types leads to automorphic deformations.

Now look at a few examples of this $\mathscr{D}$, starting with linear connection type (see Section 2.4), and where ρ is an order relation between real-valued bonds. A pure image represented by a configuration as in Figure 2.4.1 will then be broken up into sub-images, each of which has linearly connected generators. Let S consist of translations of the time parameter. Some of the sub-images may get lost when $\mathscr{D}$ operates and for each sub-image we do not know what similarity transformation should be applied to it in order to recover the original sub-image.

This situation arises in the dating (and ordering in time) of archaeological objects. The bond values, the dates, are normally not directly observable so that we then have an additional deformation mechanism in addition to the one that destroys the order. It may be possible to group some of the objects in correct order using other information. This leads to the sub-images mentioned. The time relation between these groups has been lost, however.

When the bonds are not directly observable we will have to exploit other attributes of the objects to recover the ordering. In Chapter 17 we shall return to this question from the point of view of pattern analysis.

Quite similarly, the connection type could be TREE and (7.1) would then lead to collections of sub-trees. As an example we mention the study of <u>handwritten manuscripts</u> that have been copied from each other. The order of copying gives rise to the bond relations in the configuration with tree connection type. The deformation mechanism in (7.1) corresponds to the lack of complete knowledge concerning who copied one manuscript from another and when.

On the other hand, if we deal with background images, say in the plane with $S = EG(2)$, with the connection type FULL and with the bond relations as in Case 2.9.3 (joining contrast generators), then (7.1) means that we <u>partition the pure image into subsets</u>, perhaps throw away some of the pieces, and keep the rest. Of course, we do not know the original positions they had in I, the s-values. This leads to the pattern analysis problem of how to piece together the broken fragments in such a way that the original image is restored as faithfully as possible (see Figure 7.1, where (a) is the pure image and (b) is deformed). Artifacts found

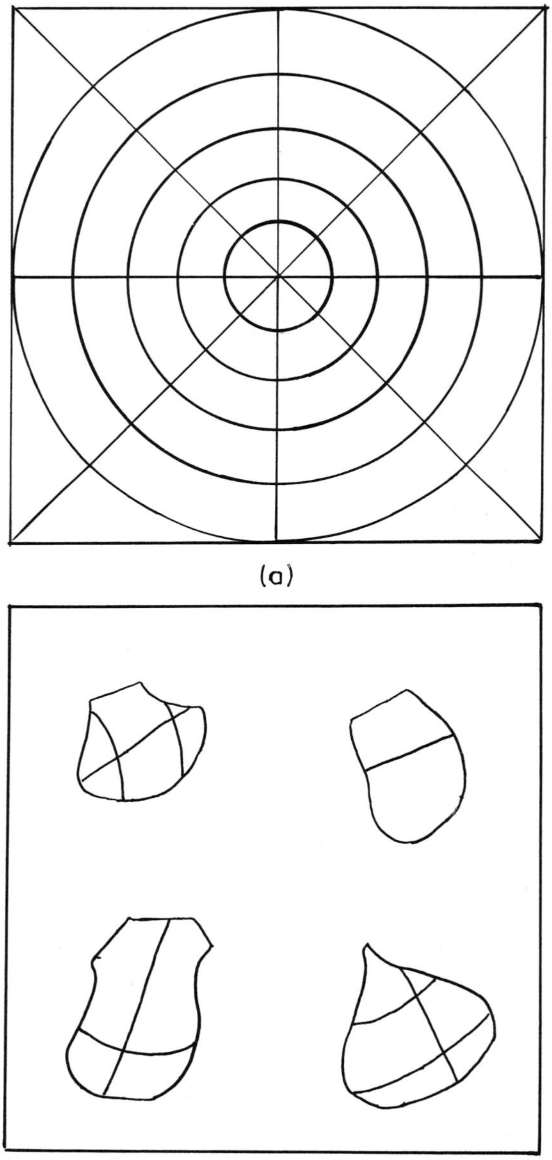

(a)

(b)

Figure 7.1

in archaeological excavations represent one instance of this problem.

Finally, a quite different example illustrating Case 7.1. Consider again Σ = FULL and the configurations of Case 2.9.7 (orthogonal representations).

A configuration is represented as a combination of sub-configurations and is then broken up into a set of sub-configurations which are all fully connected internally but separated from each other. Let $\bar{R}$ identify two such subsets if a) they have the same number of connected components c_i and c_i', $i = 1,2,\ldots,m$, and b) if $c_i R c_i'$ where R is the usual identification by equating the sums of generators involved. The effect of $\mathcal{D}$ is then to drop some of the orthogonal components of the configuration, but also to lose the "amplitudes": we are only given a set of orthogonal vectors with given directions but unknown lengths. This should be contrasted with the case in (5.38) (projections) where a milder deformation is applied, consisting only of a projection down to a <u>given</u> subspace. In the present case we do not necessarily know which orthogonal components will get lost and we may also lose the "amplitudes".

For a more complete discussion of image-breaking deformations we would have to examine the probabilistic structure of $\mathcal{D}$: how do the properties of the pure image influence the likelihood of the different breaking patterns? Very little is known about this at present and we will not pursue this topic.

Instead we shall examine briefly the other type of deformation to be studied in this section: <u>changes in metric</u>.

In the background space $X = R^d$ we consider an image algebra $\mathscr{T}$ of contrast images. For a given pure image I we have a set of $n(I)$ distinguished points $\{P_\nu(I);\ \nu = 1,2,\ldots,n(I)\}$. $\mathscr{T}$ has been generated from prototypes I_γ, $\gamma \in \Gamma$, so that we have the partitioning into disjoint prototype classes (see Section 5.1)

$$\mathscr{T} = \bigcup_\gamma \mathscr{T}_\gamma \quad , \quad \mathscr{T}_\gamma = \{sI_\gamma; s \in S\}. \tag{7.2}$$

The identification rule shall have been chosen so that in order that two pure images I and I' belong to the same prototype class it is necessary and sufficient that $n(I) = n(I')$ and that the respective distinguished points P_ν and P'_ν can be enumerated such that $d(P_\nu, P_\mu) = d(P'_\nu, P'_\mu)$. Here d is an S-invariant Riemannian metric based on the length element

$$(ds)^2 = \sum_{i,j=1}^{d} g_{ij}(x)\,dx_i dx_j. \tag{7.3}$$

<u>Case 7.2</u> (changes in metric). For any $I \in \mathscr{T}$ the deformed version is given as the set of mutual distances between its distinguished points

$$I^{\mathscr{D}} = \{d^{\mathscr{D}}(P_\nu, P_\mu);\ \nu,\mu = 1,2,\ldots,n(I)\} \tag{7.4}$$

and where $d^{\mathscr{D}}$ means a Riemannian metric with a tensor $g^{\mathscr{D}}$ instead of g in (7.3).

Suppose that the original metric is based on the Euclidean distance so that g_{ij} is simply the identity tensor δ_{ij}. If we measure the distance between two points P_1 and P_2 by recording the time it takes for a wave starting at the point source P_1 to travel to P_2, it is clear that the result depends upon the speed c of propagation. Inhomogeneities in the medium of propagation will lead to variations in $c = c(x)$ and the tensor of $d^{\mathscr{D}}$ will be

$$g_{ij}^{\mathscr{D}}(x) = \frac{1}{c^2(x)} \delta_{ij}. \tag{7.5}$$

The total length will be the integral of $d^{\mathscr{D}}$ along the geodesic from P_1 to P_2. Random inhomogeneities will hence lead to a randomly distributed metric.

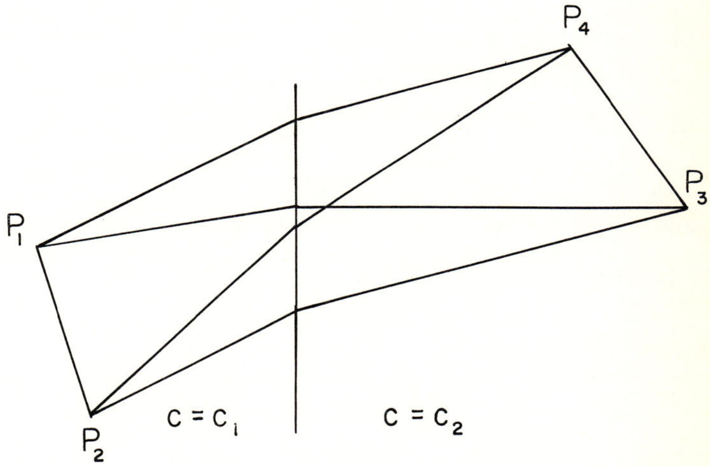

Figure 7.2

Consider the situation in Figure 7.2 where the propagation velocity is c_1 to the left of the dividing line and $c = c_2$ to the right of it. The distinguished points form a quadrilateral $P_1P_2P_3P_4$. The light rays will be linear except for breaks at the dividing lines where the angles will satisfy <u>Snell's law of refraction</u>. The light rays will of course be the geodesics: <u>Fermat's principle</u>.

This means that if we measure the six mutual distances $d^{\mathscr{D}}(P_\nu, P_\mu)$ we will typically not be able to reconcile the result with the original, Euclidean metric $c(x) \equiv$ constant. The pattern analysis will then have to include the estimation of the unknown tensor g.

Of course Figure 7.2 represents a very special case but it illustrates the basic problem encountered when the metric is changed. We shall return to the analysis of the deformed images in Chapter 17.

This concludes the present study of deformation mechanisms and of pattern synthesis. We know a good deal about the way patterns can be generated and deformed, but we have barely touched upon the inverse problem: how can we restore the pure image structure when we can only observe its deformed versions? We now have the formal means to express such problems in a precise mathematical form and their solution will be the main subject to be dealt with in the next volume, which will deal with pattern analysis.

APPENDIX

This work has been carried out according to a plan that has also been used to organize the later, more definitive volume on pattern theory. This volume will contain the material described by the following <u>Table of Contents</u>.

PART II. PATTERN SYNTHESIS

PART III. PATTERN ANALYSIS

Chapter 8. Analysis of Abstract Patterns

8.1. Numerical sequences
8.2. Paradigmatic and syntactic deformations
8.3. Style and flexibility in formal language
8.4. Abduction and grammatical inference

Chapter 9. Temporal and Other One-dimensional Patterns

9.1. Neural firing patterns
9.2. Image algebra for electrophysiology
9.3. Generalized splines and segmentation
9.4. Speech patterns

Chapter 10. Point Patterns

10.1. Lattice analysis
10.2. Crystal patterns and X-ray techniques
10.3. Clustering phenomena
10.4. Taxonomic methods
10.5. Distinguished points

Chapter 11. Line and Surface Patterns

11.1. Deformed arcs and their analysis, especially in physics
11.2. Textile and paper structure
11.3. Poisson flats
11.4. Root configuration
11.5. Ocean wave patterns

Chapter 12. Set Patterns

12.1. Statistical geometry and structure preservation
12.2. Deformed convex sets
12.3. Probing boundaries
12.4. Some stereology
12.5. Biological shapes

Chapter 13. Contrast Patterns

13.1. Geomorphological and topographic images
13.2. Craters
13.3. Analysis of elastically generated images
13.4. Feature selection for contrast patterns
13.5. Analysis of optical images

Chapter 14. Patterns of Behavior

14.1. Time-motion studies
14.2. Learning in strategy space
14.3. Planned behavior

Chapter 15. Patterns of Growth

15.1. General ideas on growth patterns
15.2. Stylized growth
15.3. Organizers in growth patterns
15.4. Convex growth

NOTES

1.1. The terminology and notation in this chapter differ somewhat from the ones used in earlier publications on pattern theory, Grenander (1967), (1969). The generators, G, were earlier called signs and denoted by S. The similarity transformations that had been denoted by G are now denoted by S. This does not affect the conceptual set up which remains unchanged.

1.2. The generators presented here serve only as examples. More complicated cases will be met in the following chapters.

1.4. The stochastic sources mentioned will suffice in more cases than one may think. The reason for this is that the more complicated dependencies that will be needed will be introduced indirectly, in terms of the configuration structure $\mathcal{R}$; see especially Section 2.10.

2.2. In earlier publications by this author the rules leading to the configuration spaces and to deformed images were called grammars. This terminology has been avoided here since it seems to have misled readers to think in terms of the usual formal languages, finite state, context free, and so on. It is a truism that if the grammar is powerful enough, it would generate any structure needed. Experience has convinced the author, however, that the combinatory formalism employed here is often more natural, and can easily be tailored to the particular application. We shall only speak of grammars in the usual, more restricted sense in connection with languages, natural or formal.

Attempts have been made to use grammars to define web structures etc., and the reader is referred to e.g., Narasimhan (1964), Shaw (1969), Rosenfeld (1971).

The algebraic approach used in Chapters 2 and 3 can be related to universal algebras and partial universal algebras, see Cohn (1965). A more detailed study of the algebraic properties of configurations and images seems motivated.

We use the term regularity to mean that the topology of the configuration is correct in terms of the given connection type, and that connected bonds agree via the bond relation. It is this combinatory approach that makes it possible to introduce the algebraic structures called image algebras, see Chapter 3.

2.5. Growth patterns are defined in space-time, so that if we fix the value of the time parameter we get, of course, a spatial pattern. This will be called a _snapshot_ of the growth pattern.

2.6. When dealing with network configurations, we have the choice of defining the generators as given nodes or as segments between pairs of nodes. Both possibilities will be met later and give rise to dual patterns.

2.9. Function patterns forming Tchebycheff systems will be met with later. A presentation of Tchebycheff systems will be found in Karlin-Studden (1966).

2.10. As pointed out after (2.10.48) the derivation is not rigorous. One should try to fill in this gap by proving convergence starting from finite configurations as was done in the linear case.

The later part of the section on the probabilistic aspects of finite state and context free languages is from Grenander (1966) and (1968). See, also Brodda (1967), Booth and Thompson (1973).

The approach in 2.10 opens up wide possibilities for studying metric properties of configurations. It is tempting to try to continue with other connection types than those discussed in this section.

Rather than imposing a measure on configuration space one could try to use the notion of "fuzziness", see Zadeh (1965).

The proof of Theorem 10.4 assumes $r > 0$. It is strange that this should be required; does it mean that negative dependence is physically impossible for these configurations?

3.1. The basic pattern theoretic notions used in this section were introduced in Grenander (1967) and extended in Grenander (1969), (1970a). We only give some results that will be needed later on, and a thorough examination of the abstract theory of image algebras remains to be carried out.

In connection with derived patterns it deserves to be mentioned that we often have the following situation. A device, perhaps a machine or algorith, produces images according to certain rules. The device itself is made up of given elements, so that it can be considered itself as <u>one</u> image from some other image algebra. This point of view will be fruitful if the device is complicated enough. We will then study relations between the image of the device and the images that it produces.

The notion of symmetry is much more restricted than that of regularity. All the configurations that we will study will be regular, but the symmetry properties of the resulting images will only be considered in certain cases.

3.2. The results in the second half of 3.2 are from Grenander (1974b). It is not known if the same technique can be applied to the syntax-controlled probability model in the general context free case.

3.3. For more information on the Nurngin system, see Lévi-Strauss (1969), in particular the Appendix to Part One by A. Weil.

3.4. Random topologies will be encountered in Part IV, Grenander-Silverstein (1975b), Silverstein (1975). Temporal patterns of regime have been treated recently in Ang (1974), Freiberger, Grenander, Sampson (1975) for compumetric studies.

3.5. Theorem 3.5.3 has been discovered repeatedly under different smoothness assumptions. Vitale's result seems to be the first covering the general case.

The construction in (3.5.112) has a long history. The first rigorous treatment of this model with random phase seems to be due to P. Lévy.

A proof of the second half of Theorem 3.5.4 can be found in Blaschke (1956), p. 115.

A clear presentation of crystallographic groups can be found in Burckhardt (1947).

A good reference to Moiré patterns can be found in Theocaris (1969). The case in (114) and Figure 3.5.22 are due to A. Grenander (1975).

The approximation of convex patterns is treated in McClure, Vitale (1974) which will be used in Part III.

The images in Cases 3.5.13 were introduced, and studied in detail in Grenander, Lavin (1972). We shall return to their pattern analysis in Chapter 12 (see Appendix).

3.6. Many more spatial patterns could have been included in this section. In particular symmetry properties of three dimensional objects have a long and distinguished history. The interested reader is referred to Weyl (1952) for an elegant presentation, and for further references.

3.7. More information on the description of hand motions etc., can be found in Larkin (1967). See, also Barnes (1949).

The form of weaving patterns is discussed qualitatively in e.g., Blumenau (1964), and Potter, Corbman (1954).

Figure 3.7.11 is one of many computed by D. Barer by solving equation (3.7.7) for different choices of generators forming the repertoire of motions.

A formalization of behavior in insurance activities was given in Bourke (1974). We shall return to this in Part IV (see Appendix) when discussing learning and memory, Grenander (1972a) and (1974c).

3.8. For a critical examination of Turing's model see Bard,

Laader (1974).

To see that (19) is true say that γ is an eigenvalue with the eigenvector $\{z_k\}$ so that

$$\sum_{\ell} c_{k-\ell} z_{\ell} = \gamma z_k, \quad \text{all} \quad k.$$

Multiply this by w^k and add, with $w = e^{2\pi i \frac{\nu}{n}}$. Then

$$\sum_{k,\ell} c_{k-\ell} w^{k-\ell} z_{\ell} w^{\ell} = \gamma \sum_{k} z_k w^k$$

or

$$\gamma(\lambda) \sum_{\ell} z_{\ell} w^{\ell} = \gamma \sum_{\ell} z_{\ell} w^{\ell}$$

with λ a multiple of $\frac{2\nu}{n}$. If $\{z_k\}$ is non-trivial, there is some ν such that the sum on the right is not zero which implies $\gamma = \gamma(\lambda)$.

The relation (28) can be derived as follows. Say that z is an eigenvector of the matrix in curly brackets in (27) with the eigenvalue λ (say even n)

$$\begin{aligned} \lambda z_1 &= r_2 z_2 + r_2 z_n \\ \lambda z_2 &= r_1 z_1 + r_1 z_3 \\ \lambda z_3 &= r_2 z_2 + r_2 z_4 \\ &\cdots\cdots\cdots \end{aligned}$$

If we write $z = (u_1, v_1, u_2, v_2, \ldots)$ we get

$$\begin{aligned} \lambda u &= r_2 M v \\ \lambda v &= r_1 M^T u \end{aligned}$$

with

$$M = \begin{Bmatrix} 1 & 0 & 0 & \ldots & 0 & 1 \\ 1 & 1 & 0 & \ldots & 0 & 0 \\ 0 & 1 & 1 & \ldots & 0 & 0 \\ \ldots & \ldots & \ldots & \ldots & \ldots & \ldots \\ 0 & 0 & 0 & \ldots & 1 & 1 \end{Bmatrix}.$$

Hence

$$\lambda^2 v = r_1 r_2 M^T M v$$

where

$$M^T M = \begin{Bmatrix} 2 & 1 & 0 & \ldots & 0 & 1 \\ 1 & 2 & 1 & \ldots & 0 & 0 \\ 0 & 1 & 2 & \ldots & 0 & 0 \\ \ldots & \ldots & \ldots & \ldots & \ldots & \ldots \\ 0 & 0 & 0 & \ldots & 1 & 2 \end{Bmatrix}$$

Then $\lambda^2 \leq 4r_1 r_2$ and the result follows. This gives the bound $2\sqrt{r_1 r_2}$.

To see that (44) holds let $\chi_i(g) = 1$ if g is i and zero else. Then

$$\sum_j n_{ij} = \sum_{\nu=1}^{\ell-1} \chi_i(g_\nu)$$

$$\sum_j n_{ji} = \sum_{\nu=1}^{\ell-1} \chi_i(g_{\nu+1})$$

so that

$$\sum_j n_{ij} = \sum_j n_{ji} + \chi_i(g_1) - \chi_i(g_\ell).$$

Most of the material in Section 3.8 is from Grenander (1973b). We start from a configuration space with a fairly loose bond structure, and then look for configurations that realize the extremum at least asymptotically. This leads to configurations with tighter bond structure. Hence, the extremum principle could be said to tighten the notion of regularity.

Theorem 3.8.4 can be extended in several directions. The generators can be assumed to have positive size, set generators, which will lead to other extremal arrangements. One can also deal with the practically important and highly interesting case when more than one generator class occurs in the configurations.

An up-to-date presentation of the methods in morphometrices can be found in Blackith, Reyment (1971).

3.9. Most of the material in this section comes from Grenander (1968a).

4.1. The basic ideas in this section were presented in Grenander (1967b), (1969), and (1970a). In particular, the last reference introduced explicitly the idea of a unified theory of patterns: one should try to consider the whole process of pattern synthesis from generator to deformation, and study the interdependence of the elements of this process. An example of this would be covariance in probability which relates similarity transformations to deformations. Another example is the relation between bond structure and identification rules. In Volume 2, we shall continue to explore this by relating the pattern analytic methods to the underlying image algebra.

4.2. More information about the law of large numbers for random compact sets can be found in Artstein, Vitale (1974).

Recently, much attention has been paid to random sets in general. The reader is referred to two important works: Matheron (1975), and Harding, Kendall (1973). In PART III we will have occasion to use some of the theory of random sets.

4.3. Figure 6.3.1 and the preceding discussion is from Freiberger, Grenander (1969).

The geomorphological patterns were introduced in Grenander (1975c).

4.5. The deformations in Case 4.5.2 have been treated analytically, especially for convex sets, see Rényi-Sulante (1964). Another deformation mechanism for convex sets based on supporting half planes was treated in Carlsson, Grenander (1967).

4.6. This section is based, in part, on Grenander (1970a).

4.7. Some deformations occurring in archaeological studies are discussed in Hodson, Kendall, Tantu (1971).

BIBLIOGRAPHY

The abbreviation R. P. A. stands for Reports on Pattern Analysis, Division of Applied Mathematics, Brown University, Providence, R. I. Some of the later reports in the series are still available and can be obtained by writing to the author.

1. Z. Artstein, R. A. Vitale (1975): A strong law of large numbers for random compact sets, Annals of Probability, 3, no. 5.

2. J. Bard, I. Lauder (1974): How well does Turing's theory of morphogenesis work? J. Theor. Biology, 45, pp. 501-531.

3. R. M. Barnes (1949): Motion and Time Study, John Wiley and Sons, Inc., New York.

4. A. T. Barucha-Reid (1960): Elements of the Theory of Stochastic Processes and Their Applications, McGraw-Hill, New York.

5. R. E. Blackith, R. A. Reyment (1971): Multivariate Morphometrics, Academic Press, New York.

6. W. Blaschke (1955): Vorlesungen über Integral-Geometrie, V. E. B. Deutscher Verlag der Wissenschaften, Berlin.

7. W. Blaschke (1956): Kreis und Kugel, Walter de Gruyter and Company, Berlin.

8. H. Blum (1967): A transformation for extracting new descriptors of shape, in Symp. Models for the Perception of Speech and Visual Form, pp. 362-280, M.I.T. Press.

9. L. Blumenau (1964): The Art and Craft of Hand Weaving, Crown Publishers, Inc., New York.

10. H. Bohman (1971): A mathematical model of insurance business, unpublished manuscript.

11. T. L. Booth, R. A. Thompson (1973): Applying probability measures to abstract languages, I. E. E. E. Trans. on Computers, C-22, pp. 442-450.

12. E. Bouchard (1967): Word-frequency distributions, unpublished manuscript.

13. P. Bourke (1974): The design and improvement of heuristics for an insurance management game, R. P. A. 27.

14. J. Boutkan (1968): ABC of the ECG, Philips Technical Library, Eindhoven.

15. B. Brodda (1967): The entropy of recursive Markov processes, Research Group for Quantitative Linguistics, KVAL PM 325, Stockholm.

16. J. J. Burckhardt (1947): Die Bewegungsgruppen der Kristallographie, Verlag Birkhäuser, Basel.

17. S. Carlsson, U. Grenander (1967): Statistical approximation of plane convex sets, Skand. Aktuarie Tidskrift, pp. 113-127.

18. P. M. Cohen (1965): Universal Algebras, Harper and Row, New York.

19. H. Cramér, H. Wold (1936): Some theorems on distribution functions, J. London Math. Soc., 11.

20. H. Cramér (1947): Mathematical Methods of Statistics, Princeton University Press, Princeton.

21. O. D. Creatzfeldt, S. Watanabe, H. D. Lux (1966): Relations between EEG phenomena and evoked potentials of single cortical cells, Electroencept. Clin. Neurophysiol. 20, 1-37.

22. R. T. Dehoff, F. N. Rhines (eds.)(1968): Quantitative Microscopy, McGraw-Hill Book Company, Inc., New York.

23. J. L. Doob (1953): Stochastic Processes, John Wiley and Sons, New York.

24. R. C. Elandt-Johnson (1971): Probability Models and Statistical Methods in Genetics, John Wiley and Sons, New York.

25. R. E. Ernst (1966): Sensitivity enhancement in magnetic resonance, in Advances in Magnetic Resonance (ed. J. S. Waugh), pp. 1-137, Academic Press, New York.

26. L. Fejes Tóth (1953): Lagerungen in der Ebene, auf der Kugel und im Raum, Springer-Verlag, Berlin.

27. L. Fejes Tóth (1964): Regular Figures, Pergamon Press, London.

28. W. Feller (1957): An Introduction to Probability Theory and Its Applications, Second Edition, John Wiley and Sons, New York.

29. T. S. Ferguson (1973): A Bayesian analysis of some non-parametric problems, Ann. Math. Stat., 1, pp. 209-230.

30. J. L. Flanagan (1972): Speech Analysis, Synthesis and Perception, Springer-Verlag, Berlin.

31. G. R. Fowles (1968): Introduction to Modern Optics, Holt, Rinehart and Winston, Inc., New York.

32. W. Freiberger, U. Grenander (1969): Computer generated image algebras, Information Processing 68, North-Holland Publishing Company, Amsterdam, pp. 1397-1404.

33. W. Freiberger, U. Grenander, P. D. Sampson (1975): Patterns in program references, I. B. M. J. of Research and Development, 19, pp. 230-243.

34. J. W. Goodman (1968): Introduction to Fourier Optics, McGraw-Hill, New York.

35. W. C. Graustein (1932): On the average number of sides of polygons of a net, Ann. Math. 32, pp. 149-153.

36. A. Grenander (1975): Moiré patterns: theory and applications, unpublished manuscript.

37. U. Grenander (1963): Probabilities on Algebraic Structures, John Wiley and Sons, New York.

38. U. Grenander (1966): Can we look inside an unreliable automaton? J. Neyman Festschrift, John Wiley and Sons, New York.

39. U. Grenander (1967a): Syntax-controlled probabilities, R. P. A. 2.

40. U. Grenander (1967b): Toward a theory of patterns, Symp. on Probability Methods in Analysis, Loutraki, Greece, Springer-Verlag, Berlin.

41. U. Grenander (1968a): A feature logic for clusters, R. P. A. 1.

42. U. Grenander (1969): Foundations of pattern analysis, Quarterly of Applied Mathematics, 27, pp. 1-55.

43. U. Grenander (1970a): A unified approach to pattern analysis, in Advances in Computers, Vol. 10 (W. Freiberger, ed.), Academic Press, New York, pp. 175-216.

44. U. Grenander (1970b): Looking at convolutions on certain groups with the help of a computer, in Les Probabilités sur les Structures Algébriques, Clermont-Ferrand, 1969.

45. U. Grenander (1972a): Reflections on Pavlovian memory, R. P. A. 6.

46. U. Grenander (1972b): On the analysis of Boolean patterns, R. P. A. 20.

47. U. Grenander (1973b): Extremum principles leading to regular patterns, R. P. A. 9.

48. U. Grenander (1973c): Statistical geometry: a tool for pattern analysis, Bull. Am. Math. Soc., 79, pp. 829-856.

49. U. Grenander (1974a): Probing a boundary, R. P. A. 11.

50. U. Grenander (1974b): Style and flexibility in certain language patterns, R. P. A. 24.

51. U. Grenander (1974c): Abduction machines that learn syntactic patterns, R. P. A. 25.

52. U. Grenander (1974d): Bond configurations and operator images, R. P. A. 32.

53. U. Grenander (1975a): Restoration of deformed lattice patterns, M. S. Bartlett Festschrift.

54. U. Grenander (1975c): Dynamic models in geomorphological patterns, Int. J. Math. Geology, 7, pp. 267-277.

55. U. Grenander, J. M. Porres (1972c): Clustering for bacteriological screening, R. P. A. 8.

56. U. Grenander, P. Lavin (1973a): Structure-preserving image restoration, R. P. A. 7.

57. U. Grenander, J. Silverstein (1975b): Spectral analysis of networks with random topologies, R. P. A. 33.

58. J. S. Griffith (1971): Mathematical Neurobiology, Academic Press, New York.

59. J. K. Hale (1969): Ordinary Differential Equations, Wiley-Interscience, New York.

60. G. H. Hardy, J. E. Littlewood, G. Polya (1924): Inequalities, Cambridge University Press, Cambridge.

61. T. E. Harris (1963): The Theory of Branching Processes, Springer-Verlag, Berlin.

62. F. R. Hodson, D. G. Kendall, P. Tautu, eds. (1971): Mathematics in the Archaeological and Historical Sciences, Edinburgh University Press, Edinburgh.

63. J. E. Hopcroft, J. D. Ullman (1969): Formal Languages and Their Relation to Automata, Addison-Wesley Publishing Company.

64. R. E. Horton (1945): Erosional development of streams and their drainage basins: hydrophysical approach to quantitative morphology, Geol. Soc. Am. Bull., 56, 275-370.

65. R. P. Kanwal (1971): Linear Integral Equations, Academic Press, New York.

66. S. J. Karlin, W. J. Studden (1966): Tchebycheff Systems: with Applications in Analysis and Statistics, Interscience Publishers, New York.

67. S. Kawabak (1972): Theoretical considerations in open-end spinning, Textile Inst. J., 63, pp. 150-170.

68. O. Kempthorne (1957): An Introduction to Genetic Statistics, John Wiley and Company, New York.

69. D. G. Kendall (1948): On the number of lattice points inside a random oval, Quart. J. of Mathematics, 19, pp. 1-26.

70. R. E. Kowalczyk (1972): Stochastic elastic deformation models, R. P. A. 4.

71. A. J. Larkin (1969): Work Study, McGraw-Hill Book Company, New York.

72. C. Lévi-Strauss (1969): The Elementary Structures of Kinship, Beacon Press, Boston.

73. A. M. Lieberman-F.S., Cooper-D.P. Shankweiler and M. Studdert-Kennedy (1967): Perception of the speech code, Psychological Rev., 74, pp. 431-461.

74. D. E. McClure (1970): Feature selection for the analysis of line patterns, R. P. A. 16.

75. D. E. McClure (1973): Nonlinear segmented function approximation and analysis of line patterns, Quarterly Appl. Math. 33, pp. 1-37.

76. D. E. McClure (1974): Problems and methods of nonlinear feature generation in pattern analysis, Proc. 8th Annual Princeton Conf. on Information Sciences and Systems, pp. 244-247.

77. D. E. McClure, R. A. Vitale (1974): Polygonal approximation of plane convex bodies, J. Math. Anal. Applic., 51, 2, pp. 326-358.

78. G. Matheron (1975): Random Sets and Integral Geometry, John Wiley and Sons, New York.

79. G. Malécot (1948): Les mathématiques de l'hérédité, Masson et Cie., Paris.

80. J. G. Martindale (1945): A new method of observing the irregularity of yarns with some observations on the origin of irregularities in worsted silvers and yarns, J. Textile Inst., 36, p.

81. B. Matérn (1960): Spatial Variation. Stochastic Models and Their Applications to Some Problems in Forest Surveys and Other Sampling Investigations, Statens Skogsforskningsinstitut, Stockholm.

82. J. C. Maxwell (1870): On hills and dales, Phil. Mag., pp. 421-427.

83. P. Medgyessy (1961): Decomposition of Superpositions of Distribution Functions, Hungarian Acad. Sci., Budapest, 1961.

84. R. E. Miles (1972)· The random division of space, Suppl. Adv. Appl. Prob., pp. 243-266.

85. R. Narasimhan (1964): Labelling schemata and syntactic description of pictures, Inf. and Control 7, pp. 151-179.

86. J. Neyman (1939): On a new class of contagious distributions, applicable in entomology and bacteriology, Ann. Math. Stat. 10, pp. 35-57.

87. E. L. O'Neill (1963): Introduction to Statistical Optics, Addison-Wesley Publishing Company, Reading, Mass.

88. G. Oster, M. Wasserman, C. Zwerling (1964): Theoretical interpretation of Moiré patterns, J. Optical Soc. Am., 54, pp. 169-175.

89. H. G. Othmer, L. E. Scriven (1974): Non-linear aspects of dynamic patterns in cellular networks, J. Theor. Biology, 43, pp. 83-112.

90. W. J. Pierson (1955): Wind generated gravity waves, Adv. in Geophysics, 2, Academic Press, New York.

91. L. Pontrjagin (1946): Topological Groups, Princeton University Press, Princeton, N. J.

92. M. D. Potter, B. P. Corbmaner (1954): Fiber to Fabric, Gregg Publishing Division, New York.

93. J. Sudarsana Rao (1961): A mathematical model for the ideal sliver and its application to the theory of roller drafting, Textile Institute J., 52, pp. 571-601.

94. J. Sudarsana Rao (1966): An application of stationary point processes to queuing theory and textile research, J. Appl. Prob., 3, pp. 231-246.

95. A. Renyi, R. Sulanke (1964): Über die konvexe Hülle von n zufällig gewählten Punkte, Z. Wahrsch. und verwandte Gebieten, Vol. 3, pp. 138-147.

96. R. Rosen (1972): Morphogenesis, <u>in Foundations of Mathematical Biology</u> (ed. R. Rosen), Vol. II, pp. 1-77.

97. R. M. Rosenberg, C. H. Chao, and J. Abbott (1972): A new mathematical model of electrical cardiac activity, Math. Biosciences, 14, pp. 367-394.

98. A. Rosenfeld (1971): Isotonic grammars, parallel grammars and picture grammars, Machine Intelligence, 6, pp. 281-294.

99. C. Shannon (1949): <u>The mathematical theory of communication</u>, University of Illinois Press, Urbana, Ill.

100. A. C. Shaw (1969): A formal picture description scheme as a basis for picture processing systems, Inf. and Control 14, pp. 9-52.

101. R. L. Shreve (1966): Statistical law of stream numbers, J. Geology, 74, pp. 17-37.

102. R. L. Shreve (1967): Infinite topologically random channel networks, J. Geology, 75, pp. 178-186.

103. R. L. Shreve (1969): Stream lengths and basin areas in topologically random channel networks, J. Geology, 77, pp. 397-414.

104. J. Silverstein (1975): Asymptotics applied to a neural network, R. P. A. 35.

105. J. S. Smart (1972): Channel networks, <u>Adv. in Hydroscience</u>, 8, pp. 305-346.

106. C. S. Smith (1952): Grain shapes and other metallurgical applications of topology, Metal Interfaces, Am. Soc. for Metals, pp. 65-113.

107. J. H. Steele (1972): Application of topological concepts in stereology, in _Stereology and Quantitative Metallography_, Am. Soc. for Testing and Materials, pp. 39-58.

108. J. C. Thacker (1974): Time series models in pattern analysis, R. P. A. 19.

109. P. S. Theocaris (1969): _Moiré Fringes in Strain Analysis_, Pergamon Press, London.

110. R. Thom (1973): _Stabilité Structurelle et Morphogénèse_, Benjamin, Addison-Wesley, Reading, Mass.

111. A. M. Turing (1954): On the chemical basis of morphogenesis, Phil. Trans. Royal Society of London, B, 237, pp. 37-72.

112. E. E. Underwood (1972): The mathematical foundations of quantitative stereology, _Stereology and Quantitative Metallography_, Am. Soc. for Testing and Materials, pp. 3-38.

113. R. A. Vitale (1974): Representation of convex, compact subsets of the plane, R. P. A. 12.

114. R. A. Vitale (1974): Linear transformations of convex sets, R. P. A. 22.

115. R. A. Vitale (1974): Convolutions in a group arising in pattern analysis, R. P. A. 29.

116. C. W. Wardlaw (1968): _Morphogenesis in Plants_, Methuen and Company, Ltd., London.

117. E. R. Weibel, H. Elias (eds., 1967): _Quantitative Methods in Morphology_, Springer-Verlag, Berlin.

118. A. Werritty (1972): The topology of stream networks, in _Spatial Analysis in Geomorphology_, ed., R. J. Chorley, Methuen and Company, London, pp. 167-196.

119. H. Weyl (1952): _Symmetry_, Princeton University Press, Princeton, N. J.

120. S. D. Wicksell (1925-26): The corpuscle problem, Biometrika 17, pp. 84-99 and 18, pp. 151-172.

121. M. J. Woldenberg (1972): The average hexagon in spatial hierarchies, in R. J. Chorley (1972), pp. 323-352.

122. L. Wolpert (1970): Positional information and pattern formation, in _Towards a Theoretical Biology_, Vol. 3 (ed. C. H. Waddington), pp. 198-230.

123. S. Wright (1921): Systems of mating, Genetics, 6, pp. 111-178.

124. U. Yule (1944): _The Statistical Study of Literary Vocabulary_, Cambridge, England, The University Press.

125. F. T. S. Yu (1973): _Introduction to Diffraction, Information Processing, and Holography_, M.I.T. Press, Cambridge, Mass.

126. L. A. Zadeh (1965): Fuzzy sets, Information and Control, 8, pp. 338-353.

INDEX